INTRODUCTION TO Statistics IN HUMAN PERFORMANCE

An understanding and working knowledge of the basic principles of statistics are of central importance in understanding the sport and health sciences. *Introduction to Statistics in Human Performance: Using SPSS and R* provides students facing statistical problems for the first time with an accessible and informal introduction to the key concepts and procedures of statistical analysis.

Now in its second edition, the book covers processes involved in using both SPSS and R, and includes chapters on:

- research methods
- descriptive statistics
- the normal curve and standard scores
- correlation and regression
- inferential statistics introduction
- issues in inferential statistics
- t-tests
- anova, factorial anova and manova
- advanced statistics, and
- nonparametric statistics

Including examples relevant to the field, review questions, practice computer problems and activities throughout, and online materials, the book offers students all the tools they need to understand statistical concepts in sport and exercise. This is a vital resource for any students of sport and exercise science, kinesiology, physical therapy, athletic training, and fitness and health taking classes in statistics.

Dale P. Mood is a Professor Emeritus at the University of Colorado, USA, where he taught for 47 years in the Department of Integrative Physiology. He is a former Department Chair, Associate Dean for Student Academic Affairs, coordinator of the Responsible Conduct of Research program on the Boulder campus and interim Director of the Post Baccalaureate Pre-Medical program. He has authored five textbooks, three book chapters, nearly 50 articles, and he has presented both nationally and internationally in the areas of measurement and research.

James R. Morrow, Jr. is a Regents Professor Emeritus in the Department of Kinesiology, Health Promotion, and Recreation at the University of North Texas, USA. He has authored more than 200 peer-reviewed manuscripts and five books, and he has made nearly 300 professional presentations. He received the 2011 President's Council on Fitness, Sports & Nutrition Honor Award. His teaching and research focus on measurement issues related to physical fitness and activity assessment.

Matthew B. McQueen is the founding Director of the Public Health Certificate Program and teaches introduction to epidemiology for undergraduate students as well as advanced biostatistics for graduate students at the University of Colorado Boulder. Dr. McQueen's research objectives are focused on the development and application of epidemiological and biostatistical methods to advance our understanding of human disease from genes to populations. His research program is highly interdisciplinary involving studies of genetics and the microbiome, cardiometabolic disease, substance abuse as well as neurological studies of Alzheimer's disease and mild traumatic brain injury.

INTRODUCTION TO Statistics IN HUMAN PERFORMANCE

Using SPSS and R

Second edition

Dale P. Mood,
James R. Morrow, Jr. and
Matthew B. McQueen

Routledge
Taylor & Francis Group
NEW YORK AND LONDON

Second edition published 2020
by Routledge
52 Vanderbilt Avenue, New York, NY 10017

and by Routledge
2 Park Square, Milton Park, Abingdon, Oxon, OX14 4RN

Routledge is an imprint of the Taylor & Francis Group, an informa business

First edition published by Holcomb Hathaway, Publishers, Inc. 2015
Second edition published by Routledge 2020

Library of Congress Cataloging-in-Publication Data
A catalog record for this book has been requested

ISBN: 978-0-8153-8119-8 (hbk)
ISBN: 978-0-8153-8120-4 (pbk)
ISBN: 978-1-351-21106-2 (ebk)

Typeset in Times
by Apex CoVantage, LLC

Visit the eResources: www.routledge.com/9780815381204

Dedication

All other goods by fortune's hand are given:
A wife is the particular gift of Heav'n.

ALEXANDER POPE

We dedicate this work to our grandchildren, our children, but mostly to our wives. It is not the number of classes taught, papers published, grants received, or meetings attended that are important in life. It is that your children (and ultimately, grandchildren) are doing well. We have each been blessed with wives who accomplished so much, including that all-important goal of preparing our children for life. We are deeply appreciative. Thank you, heaven, for Maureen, Melba, and Mary.

Contents

Figures

Tables

Symbols

ROMAN SYMBOLS

N The number of scores in a distribution
$\bar{X}$ The mean
X_i A score in a distribution
H The high score of a distribution
L The low score of a distribution
R The range of a distribution of scores
x The deviation score X from the mean
s^2 The sample variance
s The sample standard deviation
e Mathematical constant of 2.7183
z A standard score with a mean of 0 and a standard deviation of 1
T-score A standard score with a mean of 50 and a standard deviation of 10
r The Pearson Product Moment correlation coefficient
R The multiple correlation coefficient
r^2 The coefficient of determination
$\hat{Y}$ The predicted value of Y
a The Y intercept
b The slope of the line or weight given to a predictor variable
$s_{\bar{x}}$ The standard error of the mean for a sample
ES Effect size
PI Percent improvement
C Number of paired comparisons possible
K Number of means available
P Probability of an error with multiple t-tests
$\bar{\bar{X}}$ The Grand Mean
ss_T Sum of squares total
ss_B Sum of squares between
ss_W Sum of squares within
J Number of levels of the Independent Variable
MS Mean Square
F-ratio The ratio of two variances (e.g., MS_B/MS_W)
S Scheffe's Multiple Comparison Test
T Tukey Multiple Comparison Test
q Critical value from Studentized range table
SNK Student-Newman-Keuls Multiple Comparison Test
R^2 Effect size in ANOVA
SS_C Sum of squares for columns
SS_R Sum of squares for rows

MS_C	Mean Square for columns
MS_R	Mean square for rows
MS_E	Mean square for error
X^2	Test statistic used with phi coefficient
P	Percent agreement
W	Kendall coefficient of concordance
U	Mann–Whitney U test
H	Kruskal–Wallis test for n independent samples
Q	Cochran's Q
T	Friedman 2-way ANOVA by ranks

GREEK SYMBOLS

Σ	Read as "the sum of"
σ^2	The population variance
μ	The population mean
σ	The population standard deviation
π	Mathematical constant of 3.1416
$\sigma_{\bar{x}}$	Standard error of the mean for a population
Δ	Delta—Difference between the means that a researcher thinks is important when estimating power
ω^2	Omega squared
μ^2	Eta-Square
ε	Epsilon
φ	Phi coefficient
χ^2	Chi-Square
K	Kappa
ρ	Spearman rank correlation coefficient

Equations

Preface

Learning to read, interpret, use, and conduct statistical analyses are often daunting tasks for undergraduate and master's students. With *Introduction to Statistics in Human Performance: Using SPSS and R (2nd)* we hope to help alleviate fear associated with those tasks. Our goal is to allow readers to see the big picture and grasp the underlying logic of descriptive and inferential statistics and hypothesis testing. We want to give students the knowledge and skill to use statistics effectively in their professional lives and feel comfortable doing so. We believe that statistics can be fun and interesting. For students, we foresee a course delivery supported by the textbook's online tools, through the use of videos, graphics, and interactive experiences that help them learn by using the variety of provided resources. Essentially, we view this textbook as the foundation to your course, working in conjunction with other tools to build students' practical experience.

As an area of academic study, the field of human performance is relatively new. Evolving from physical education departments over the past 50 to 60 years, the field has assumed many forms and many names, including exercise science, kinesiology, human biology, integrative physiology, and human performance. One of the notable curricular additions brought about by this evolution was the need to increase students' knowledge about relevant research, as both practitioners and consumers. It was recognized that one of the skills required in this broader curriculum is familiarity with the research process, thus allowing students to keep up with the constantly growing body of knowledge.

The human performance curriculum has evolved into the preliminary training for students interested in pursuing professions in physical and occupational therapy, in nursing, as personal physical trainers, as physician assistants, medicine, public health, and other allied health professions. Most professionals in these fields are required to use evidence-based research to make decisions. As a result, universities increasingly require undergraduate research experiences as part of experiential learning, even including research experiences as part of the core curricula. Additionally, human performance programs now allow undergraduate students to receive academic credit through participation in research activities. All these changes have resulted in the growth of courses such as Introduction to Research Methods and Introduction to Statistics in the human performance curriculum in undergraduate and master's programs.

We, the three authors of this textbook, have been a part of this evolution. Collectively we have taught thousands of students at the university level, with the majority of our time devoted to measurement, research methods, statistics, and computer-related courses. Even though we have taught at different institutions, our methods are remarkably similar. Over the years we have observed how students learn this material and have identified which instructional approaches work and which don't work.

Statistics is a combination of logic and mathematics, but we have found ways to minimize the role of complex mathematics while making statistics understandable and even enjoyable (yes, even enjoyable). In our courses, as well as in this textbook, our approach is to give students the ability to see how and why statistics work. Understandably, students will need to practice what they are learning and be active participants in conducting studies and interpreting statistical results. The textbook and its associated online components provide students with many ways and opportunities to do this.

THE TEXTBOOK'S ORGANIZATION

Statistics is a tool, and like any tool, it needs to be used correctly to be effective. Because statistics is a *research* tool, we initially present a chapter that explains the connections between research methods and statistics. For those students who have completed a course in research methods, this chapter can be omitted or used as a review. Chapter 2 introduces measurement and statistics, presents the importance of accurate measurement of variables to be included in a research project, and begins students on their statistics journey. Subsequent chapters are presented in an order similar to that of other statistics texts, with statistical concepts in each building on previous material. Unique to this book are the chapter on research methods (Ch. 1), a separate chapter (8) on inferential statistics issues (these are usually scattered throughout statistics textbooks), and chapters on advanced statistics (13) and on nonparametric statistics (14) that are more comprehensive than typically found in introductory statistics textbooks. Our approach to presenting information in this textbook is as follows: To begin, we explain concepts in great detail so that students can learn how statistics work. As students gain understanding, we are able to reduce the level and detail of explanation offered. By the time we cover complex statistics, discussions focus on when particular statistics would be used, some considerations about their use, and how they are interpreted.

Acknowledgments

Although we take full responsibility for the contents of *Introduction to Statistics in Human Performance*, we must acknowledge many individuals and groups of people who made this textbook possible.

First and foremost we thank our families, who have put up with Husband, Dad, or Grandpa (Grandpap) missing from some family occasions because he was "busy with the textbook". We tried to keep these to a minimum, but they still occurred more often than we would have liked.

Importantly, we thank the literally thousands of students who over the years have taught us as much as we taught them (not always about statistics). We especially thank Rubin J. Issac and Allison A. Morgan, who provided their insight by working through all of the questions, problems, and computer activities to check them for ambiguous writing and correct answers.

Finally, we thank the reviewers of early versions our manuscript and the anonymous reviewers of this second edition. Although we had each other to help shape ideas about how to present material, the old adage about two heads are better than one (in our case modified to many heads are better than two) is very true. The suggestions provided by the following reviewers resulted in a better presentation of the material, and we appreciate their help: Tim Anderson, California State University, Fresno; Jason Boyle, University of Texas at El Paso; Lee Brown, California State University, Fullerton; Jared Coburn, California State University, Fullerton; Jarold L. Cosby, Brock University; Shirley Darracott, Georgia Regents University; Craig A. Horswill, University of Illinois at Chicago; Mark Ricard, University of Texas at Arlington; Charlotte Sanborn, Texas Woman's University; and John Storsved, Eastern Illinois University.

Student Introduction

"Statistics". The word and the course often arouse trepidation in students. With *Introduction to Statistics in Human Performance*, we hope to alleviate this fear. As you see the big picture and grasp the underlying logic of descriptive and inferential statistics and hypothesis testing, we believe you will come to learn, appreciate, interpret, and use statistics in your professional life. Our goal for this textbook is to make statistics relevant and even enjoyable. Our examples, questions, and activities focus on human performance concepts that you are likely to encounter professionally. We hope you will enjoy learning the logic of statistical processes and appreciate how what you learn from this textbook can and will influence your life.

Statistics is a combination of logic and mathematics, but we have tried to minimize the complex mathematics and make statistics understandable. With a reasonable knowledge of arithmetic, simple algebra, and a little geometry, you will be well-prepared to understand the concepts in this textbook. This is not to suggest that you won't have to devote effort to this endeavor. Our approach is intended to help you see how and why statistics work, but you will need to be an active participant in this process. This textbook offers you many tools and opportunities to do this, and we'll present more about these below. Your learning will be enhanced if you take full advantage of these resources.

The primary use of statistics is as a research tool to help in making accurate decisions. You make decisions based on information in your everyday life, in your course work, and in your career. You collect data and then choose one of two (or several) alternative paths. To succeed in this process, you need accurate data, sound investigative skills, and the ability to interpret the data that you possess. After you reach a decision, you may find that the alternative you chose was incorrect. In this case, you gather additional data and modify your decision based on the new information. The skills you learn in this textbook will help you improve this process in your personal and professional lives.

In the media, you often hear or read the phrase "According to a new study. . . ." For example, did research actually show that creatine supplementation improves athletic performance? What can you really conclude from the reported research? Is it accurate? Could the researchers have been wrong in their data collection, analyses, or conclusions? The answer to each of these questions is yes. But with a solid understanding of the scientific method, research methods, and statistical procedures, you can make meaningful decisions about the accuracy of researchers' conclusions. It is wise to be skeptical when you first encounter research reports. By learning the concepts in this textbook, you will be well equipped to draw your own conclusions about the quality of research.

Statistics is a tool, and like any tool it must be used correctly to be effective. Using a wrench to pound a nail isn't very efficient. Because statistics is a *research* tool, the first chapter in this textbook explains the connections between research methods and statistics. (If you have completed a course in research methods, you can consider this chapter as a review.) Following Chapter 1, the order of topics presented is similar to that in most statistics texts. Just as adding and subtracting come before multiplication and division, statistical concepts build on those previously learned.

Our approach to presenting material in this textbook can be summarized this way: At the beginning we explain concepts in great detail, allowing you to learn how statistics are used. As we move on, because you have this foundational understanding, we can reduce the level of explanation. By the time we present complex statistics, we limit our discussion to when particular statistics would be used, some considerations about their use, and how they are interpreted. At that point we present virtually no mathematical considerations. As it turns out, by its general nature, statistics consistently involves determining the probability of the occurrence of a particular event under certain circumstances. The complexity evolves as the types and number of factors involved in the research experiment increase.

As we mentioned, this learning process is not intended to be a passive event—your learning and understanding will increase if, in addition to reading, you also take advantage of the tools we've provided throughout the textbook. These include:

Meet Two of the Authors for an Introduction to the Tools

1. **Review questions and practice problems at the end of each chapter.** For some of the questions and problems, we provide the answers in Appendix C so that you can check your progress in learning the chapter material. For others, we have omitted the answers to allow your instructor to use them as assignments. We highly recommend that you complete all the questions and problems as a way to monitor your grasp of each topic.
2. **Computer activities that will guide you in learning how to use SPSS/RStudio, computer statistical packages, which can do much of the required calculating.** As the statistics become increasingly complex, the mathematics involved are often not so much difficult as they are tedious. Used judiciously, software can save you much time without sacrificing your understanding of statistics. In addition, the format of these activities supports our goal of avoiding complicated mathematical instructions. As you think about the correct way to organize data to be entered in the software, you will also be learning about aspects of research design, such as types of measurement, levels of measurement, and number of levels. Simply entering data into SPSS/R (or any other statistical program) without a full understanding of the appropriateness of the analyses can result in invalid interpretations, but if you have a

strong understanding of the logic and nature of the statistical procedure, the use of software can help you immensely. In this second edition we include both SPSS and R, statistical packages used to calculate many of the statistics presented in the text. We alert you to the fact that these computer packages are constantly being updated and the version you are using may differ slightly in the commands and output obtained. This may result in you seeing computer output that is not exactly like that presented in the textbook or on the videos but the interpretation remains the same. Don't let these minor changes detract from the helpfulness of these resources.

3. **More than 80 video presentations to give you a joint visual and verbal description of specific points in the textbook.** A marginal icon, χ, in the text indicates that a related video is available for you to watch. The videos were not necessarily scripted; they will provide you with an experience much like you might have when you stop by your instructor's office to discuss a concept you don't understand. In the videos, we as authors talk to one another and use slide presentations to illustrate information for you. The videos provide additional examples, SPSS/RStudio instructions, analyses, and interpretation. The typical instructional video begins with a presentation of selected content from the chapter. This is often followed by an illustration using SPSS terminology, instructions, and output. With inclusion of R in this second edition the original videos are often followed by a new video designated by -R. These new -R videos will often begin with the identical selected content (including the mention of SPSS) but then will be followed by specific RStudio commands and RStudio output. We trust you will find them a helpful addition to the text.
4. **A companion website.** This textbook's student website, **www.routledge.com/9780815381204**, gives you easy access to the 80+ videos that accompany the textbook; you may also print out the related PowerPoint® slides and use them to take notes. In addition, the site provides important study tools such as a chapter-by-chapter key concept review, the textbook's Review Questions and Practice Problems, and the data files to be used with the SPSS/R computer activities.

We hope you will use this textbook, the questions and activities, the videos, and the student website to learn and become comfortable with the concepts and applications of statistics.

1 Research Methods

INTRODUCTION

For the human performance professional, statistical analyses and research studies play a useful and important role in everyday life. To see this, consider the physical therapist who works with a wide variety of patients, such as knee replacement recipients, stroke patients, patients with arm or shoulder injuries, patients with severe burns, and so forth. The therapist is interested in treating each patient with the modality that has been shown to be most effective for each individual's specific injury, surgery, or condition. Physical therapists review the scientific literature and make decisions regarding the best practice to help return each patient to full function

V.1.1
Introduction to Research Issues and Vocabulary

as quickly and efficiently as possible. The effectiveness of the various modalities is determined through research studies that adhere to the scientific process. For another example, think about a coach who has to decide which potential players to keep on the team and which to cut. In this decision process, what are the most important physical characteristics, skills, and abilities that the coach should consider? What psychological attributes are important, what leadership and motivational traits are crucial, and how can all of these characteristics, traits, and attributes be measured accurately? The coach who knows how to read the scientific literature critically and who can learn through experimentation will make sound decisions based on the available evidence. Professionals need to make **evidence-based** decisions, whether they be in a clinical, academic, athletic, or performance setting. Thus, we hope that you will become excited about conducting research, gathering data, and making decisions that are based on data. Such decisions will positively influence your clients, students, colleagues, and, ultimately, the world of human performance.

It is not unusual in human performance curricula to devote an entire semester to the study of research methods. The research methods course often covers topics such as defining the research question, reviewing the literature, describing the methods used, analyzing and interpreting the results, writing the conclusions, and disseminating the findings. Because this text is essentially about statistics, we will not discuss all of these topics, but, rather, we will focus on those aspects of the research process that directly relate to statistical procedures and decisions. Thus, as we build a foundation for learning how to use statistical procedures when making scientific decisions, we will introduce and distinguish between *research* and *statistics* terminology.

In some cases, the design of a research project will dictate what statistical procedures are appropriate; in other situations, the statistical procedures that may be used will necessitate the adoption of certain aspects of the research design. For an example of the first situation, if the variable of interest can be measured only in a categorical or ranking manner, then the use of a **nonparametric** statistical procedure (see Chapter 14) is called for. The second situation might occur when the assumptions required for a particular statistical procedure dictate how subjects will be grouped or the number of subjects necessary in each possible combination of treatments.

Therefore, to select efficient statistical procedures, some knowledge of research methods is required, and to select the best research design, knowledge of statistics is necessary. Thus, in this chapter we discuss the elements of research methods that affect the selection of appropriate statistical procedures.

THE NATURE OF RESEARCH

How do people "know" things? There are several ways. For one, individuals or groups have experiences, and through these experiences we learn about phenomena. Experiences might include personal observations or input from other people, and they might be rational in nature or, perhaps, influenced by superstitions. Experiences are limited, and they are sometimes misleading. Imagine that you have observed that every time you eat a particular meal before a sporting event, you perform well. That may or may not be a true relationship (that is, you might have made inaccurate observations or have come to an unjustified conclusion), and, if it is a true relationship, the relationship may or may not be causal. That is, the fact that two things tend to happen at the same time does not in itself necessitate that one causes the other.

Because of these limitations of learning through experience, humans have developed the research process as the preferred way of learning. The research process is *organized, logical, and scientific*. The results of sound research processes are reproducible and generalizable (that is, the findings and conclusions from a research study conducted on a sample of subjects can be extended, or generalized, to the population from which the subjects were selected). One element of the research process is the *scientific method*, which helps researchers learn about the relationship between a treatment (or an exposure) and the impact of that treatment on measured outcomes. A *treatment* is something that subjects are required to do (such as take a particular medication), and *exposure* refers to a condition or behavior in which a subject engages (such as a sedentary lifestyle). For example, one relationship that a researcher might explore is that between a sedentary lifestyle and negative health consequences, and whether this relationship is the same for both males and females. The research process helps scientists make decisions about relationships like this one. The research process is not infallible. Quite the opposite: much of the research process is based on probability, and drawing incorrect conclusions is always a possibility.

GAINING KNOWLEDGE

Theories

A **theory** is an educated supposition about a relationship among some natural phenomena. Generally, theories are derived through observation, experimentation, and reflective thinking. For example, Henry and Rogers (1960) proposed the memory drum theory, which has to do with how humans learn and execute motor skills. According to Henry's theory, unconscious neural patterns acquired from past experiences are stored in the central nervous system as a "memory storage drum". A competing theory, called schema theory, was advanced by Richard Schmidt (1975). Schema theory suggests that, instead of storing a memory for every previous movement, we store previously obtained information about the *relationships* involving joint actions and muscular contractions to produce a movement to be used to adapt to new situations. Additional theories have since been postulated by other researchers in the field of motor behavior, but the purpose here is not to debate which theory is correct. Rather, the purpose is to indicate that a theory is not a proven fact; instead, it is someone's supposition about the relationships among variables and the expectations that flow from such a supposition. When scientists conduct experiments, often the goal is to provide evidence to support or to refute a particular theory.

Hypotheses

A hypothesis is a prediction derived from a theory or a researcher's speculation regarding the likely outcome of an experiment. A **research hypothesis** might take a form like the following: "If subjects are exposed to a certain treatment, then there will be a measurable change in a particular variable". A research hypothesis might state that greater changes would be expected from exposure to treatment A than from exposure to treatment B—a supposition that, again, is based on measurement of an outcome variable. For example, an exercise scientist might hypothesize that two different strength-training regimens result in dissimilar modifications of muscle fiber characteristics.

The treatment that is manipulated by the researcher is known as the **independent variable.** The variable that is expected to change (the outcome) is called the **dependent variable.** Thus, the research hypothesis predicts the expected relationship between the independent variable and the dependent variable. Let's consider another example, this one based on current thinking regarding the relationship between physical activity and body weight. A general research hypothesis might state that "increases in physical activity are associated with decreases (or maintenance) of body weight". A specific research study could be conducted to examine this hypothesis. Depending on the researcher's interests, subjects could be assigned to one of several possible levels of the independent variable (physical activity), such as sedentary, light, moderate, and vigorous. After subjects' exposure to these treatments, the researcher can determine whether meaningful differences occurred among the average body weight (dependent variable) of the groups. This result would support or refute the hypothesis. Notice we did not say this result would *prove* or *disprove* the hypothesis.

In contrast to a research hypothesis, a **null hypothesis** states that the independent variable and the dependent variable are not related. That is, exposure to a particular treatment will not result in any predictable change in the dependent variable.

Inferential statistical tests are designed to examine the viability of the null hypothesis, *not* the research hypothesis. As will be explained in detail in later chapters, statistical tests examine the relationship between the independent and dependent variables. The relationship might take the form of differences among measurements or a correlation between the independent and dependent variables. The researcher tests the likelihood of obtaining such a difference or relation, assuming the null hypothesis is actually true. If the probability of obtaining such a relation or difference is found to be very unlikely, the researcher will reject the null hypothesis and accept the research hypothesis (sometimes called the **alternative hypothesis**). In some cases, the null hypothesis might be the research hypothesis. For example, if two physical therapy modalities (one expensive and the other inexpensive) are available for treating a particular malady, a researcher might hope that the null hypothesis is tenable.

V.1.2
How Statistics Fit into Research

These methods and interpretations are important to understanding hypothesis testing and much of the scientific literature. They provide the keys and logic that will help you comprehend all inferential statistical testing, as well as the many very advanced statistical explanations that research consumers (i.e., human performance professionals) encounter on a daily basis.

Levels of Knowledge

The word *science* is derived from the Latin word *scientia*, meaning "having knowledge". There are several hierarchical levels of scientific knowledge. The lowest level is *description;* then we move to *prediction*, then to *control*, and finally to *explanation*. Because cardiovascular disease is a serious illness in our society today, we will use it to illustrate these levels of knowledge.

Description Level of Knowledge

Heart disease, although decreasing in prevalence in recent years, still accounts for nearly a quarter of the annual deaths in the United States in both males and females. Several agencies (e.g., Centers for Disease Control and Prevention, American Heart Association) collect descriptive data on causes of death and compile the information according to many different categories (e.g., gender, age, ethnicity). This is an example of *descriptive research*, as no attempt is made to predict who will die from cardiovascular disease or to explain its causes. It simply describes the current state of affairs. This type of information, however, is vitally important, as it allows us to determine trends over time. Figure 1.1 provides descriptive data on causes of death in the United States in 2014. Another example of description research is the current research that has led to awareness of increases in the incidence of obesity.

FIGURE 1.1 Causes of death in the United States in 2014.

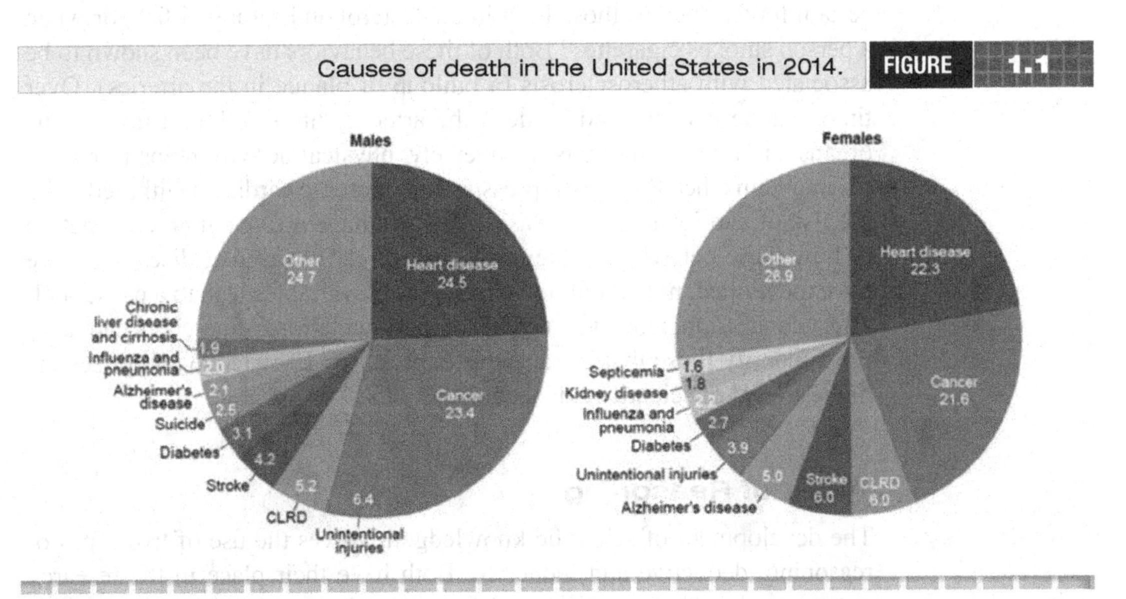

Source: Heron, M. (2016). Deaths: Leading causes for 2014. National vital statistics reports; vol 65 no 5. Hyattsville, MD: National Center for Health Statistics.

Prediction Level of Knowledge

The next level of knowledge occurs when relationships among variables are identified, providing the possibility of prediction. For example, many risk factors for cardiovascular disease, such as smoking, stress, lack of physical activity, sedentary behaviors, age, obesity, genetics, high blood pressure, and elevated levels of blood lipids, have been identified. By using correlational statistical procedures, researchers can combine measures of these and other variables to predict an individual's risk of susceptibility to cardiovascular disease. Although description can tell us what the current status is, prediction goes beyond this by allowing us to forecast what may happen in the future. Prediction is never perfect—there will always be some error in prediction—but it does allow us to move to the next level of knowledge, that of control.

Control Level of Knowledge

Although a correlation between variables does not necessarily indicate causation, we often use knowledge of relationships to attempt to control events. Not all factors are modifiable (e.g., gender, age, and genetics), but many are, and we can reduce (control) the risk of cardiovascular disease through proper diet, exercise, stress reduction, smoking cessation, and other behavior modifications. Modifications such as these may have beneficial effects, but to achieve a true understanding of how best to avoid cardiovascular disease, we must eventually get to the highest level of knowledge, that of explanation.

Explanation Level of Knowledge

Why does the risk of cardiovascular disease increase when a person ingests certain foods, such as those high in cholesterol and saturated fats, or when a person smokes cigarettes? Both of these behaviors have been shown to be associated with atherosclerosis (a buildup of plaque in the arteries). Over time, plaque narrows and hardens the arteries, limiting blood flow to the organs and other body parts. Conversely, physical activity helps to reduce (or maintain a healthy) blood pressure and increase cardiac health, reducing the risk of cardiovascular disease. Thus, we have arrived at an explanation of how these behaviors relate to the risk of cardiovascular disease. As we have mentioned, not all relationships between variables lead to a cause-and-effect conclusion. However, through experimentation, researchers can, over time, identify those that do, and this explanatory knowledge comprises the highest level of scientific advancement.

Types of Reasoning

The development of scientific knowledge involves the use of two types of reasoning, deductive and inductive. Both have their place in the research process, although statistical procedures are more strongly associated with **inductive reasoning** than with **deductive reasoning.**

Inductive Reasoning

Inductive reasoning moves from observations of specific events to predictions about general principles. For example, as the result of using a statistical procedure, researchers might determine that group A performed significantly better than group B on some task. ("Significantly better" means that there is a very low possibility that the difference between the groups can be attributed to chance.) If the experiment was conducted carefully with respect to certain issues that we will present in detail later in this chapter and throughout the text, the researchers can generalize this finding to the population from which the specific subjects (i.e., the sample) were selected. Some physicians recommend that older adults take a "baby aspirin" each day because they believe a relationship exists between a daily aspirin dose and cardiac risk. Interestingly, as the result of ongoing research based on the principles presented in this chapter, some in the medical field are questioning the effectiveness of this treatment for all segments of the population. Such discrepancies in research findings help move science forward.

Deductive Reasoning

Deductive reasoning is just the opposite of inductive reasoning, in that it proceeds from observations of general information to the explanation of specific events. For example, the development of theories involves deductive reasoning. After careful observation and reflection on a great deal of general information, an investigator might deduce and propose the manner in which various specific phenomena are related to each other. Consider a researcher who has observed that individuals living in a particular country perform well in long-distance running events and wonders about what could explain this phenomenon. The researcher also has observed that this country is at a high altitude, leading her to theorize that training or living at high altitude may be related to cardiorespiratory fitness. The researcher could then develop a specific research study to examine the relationship between altitude training and cardiorespiratory fitness.

Types of Research

There are many ways to categorize research into groups, and, on occasion, statistical procedures that are appropriate for one type of research may not be appropriate for another type. In this section, we describe the following categories of research: descriptive and experimental, basic and applied, qualitative and quantitative, longitudinal and cross-sectional, and human and animal.

Descriptive and Experimental Research

Descriptive research describes the current status of behavior and events, whereas experimental research seeks to determine cause-and-effect relationships. Descriptive research typically precedes experimental research. Surveys, observational techniques, historical investigations, and correlational assessments are the main tools of descriptive researchers. Figure 1.1, listing causes of death in the United States, is an example of descriptive research. **Experimental research,** on the other hand, has as its goal the exploration and determination of whether the correlations among variables are cause-and-effect relationships. Typically, this research examines different levels, categories, or values of an independent variable to determine whether they cause changes in a dependent variable. Experimental researchers are interested in knowing how changes in an independent (exposure) variable *X* affect a dependent (outcome) variable *Y*. Conducting a research study to determine the most effective muscle-strengthening activities to increase muscular power is an example of experimental research.

Basic and Applied Research

These words *basic* and *applied* reflect the two extremes of a continuum, and it is not always easy to draw a line between them. **Basic research** is theoretical in nature, and it seeks to explore fundamental principles. The researcher's approach is likely to be to "see what happens" rather than to expect a predicted outcome. **Applied research,** on the other hand, is generally conducted to answer a very precisely stated question. Whereas basic research has the goal of increasing the depth of understanding of some process, applied research usually seeks to determine an immediate solution to be applied to a present problem. An example of basic research is to study how much of an individual's body fat is subcutaneous. Applied research would use this basic knowledge in the development of skinfold equations to predict total-body fat from skinfold measurements.

Qualitative and Quantitative Research

The categories of qualitative and quantitative research form another continuum, where the line between them is not always clear. In general, **qualitative research** relies on observation of events as they occur, and its goal is to describe and qualify what occurs. It relies on subjective and observational information and does not rely as much on numerical information as does quantitative research. Qualitative research typically does not involve the manipulation of variables or the administration of various treatments. Terms often used for qualitative research include ethnography, phenomenology, and field research. At the other end of the continuum is **quantitative research,** in which the researcher typically manipulates and controls variables and strives for the elimination of all possible causes of an effect except the one identified in the research hypothesis.

Qualitative researchers might observe the ongoing, daily activities in which students engage in a typical elementary physical education class. Quantitative researchers, on the other hand, might gather specific data on the amount of moderate to vigorous physical activity that is occurring in an elementary classroom and test whether boys and girls engage in different amounts of physical activity.

Longitudinal and Cross-Sectional Research

Human performance research is often concerned with mapping changes that occur over time as a result of growth and maturation. Investigators in this area use both longitudinal and cross-sectional research in this pursuit. In **longitudinal research** studies, the same subjects are observed and measured repeatedly over a relatively long time (e.g., weeks, months, or even years). This type of research allows precise assessment of changes occurring over time, but it is time consuming, and longitudinal studies often face the possibility of losing subjects. In **cross-sectional research** studies, groups of subjects of differing ages are all observed or measured at the same point in time, and then the researchers compare the results that are obtained from the different age groups to arrive at their conclusions. Although this method is more efficient than longitudinal research, it does not permit direct observation of developmental changes, because different groups of subjects are being assessed. A cross-sectional study might report the differences in throwing patterns among children of ages 6, 7, 8, 9, and 10. A longitudinal study would follow a specific group of children starting at age 6 and study the changes in throwing patterns that occur in these individual children for a five-year period.

Human and Animal Research

The most efficient way to evaluate human performance is to use humans as subjects in experiments, that is, to conduct **human research.** However, due to a variety of factors (e.g., risk of physical, psychological, or emotional harm; time commitment required; effort and restrictions of daily activities; cost), the use of human subjects may not always be possible. Many of these problems can be overcome through the use of animals as subjects. In **animal research,** researchers can more easily control such factors as genetic differences; age; diet; and exercise duration, intensity, and mode. Of course, the main disadvantage of using animals has to do with external validity (discussed later in this chapter). Although other mammals have similar physiology to humans, many adaptations that occur in the animal model are difficult to transfer to humans (e.g., drug dosages and exercise tolerance). Physiologists might use animal subjects because the study design requires sacrificing the subjects to gain information about changes in muscle fiber content as a result of muscular strength training. Subsequently, researchers may use knowledge learned from animal model research to investigate these effects in humans.

Summary of Types of Research

Notice that the categories of research are not mutually exclusive. For example, experimental research can be basic or applied, longitudinal research can be qualitative or quantitative, and human research can be descriptive or experimental. The goals and objectives of the research will often dictate the research type that is most appropriate and efficient. The researcher's knowledge of available statistical techniques might also determine the type of research conducted.

THE SCIENTIFIC METHOD

The essence of the **scientific method** is to gather information objectively and make decisions based on evidence. These decisions are based on probability, so there is "room for error". However, careful planning, data collection, and interpretation can help the researcher make truthful statements about the observed phenomenon.

Scientists often develop research questions based on the following process:

1. They discover a theory that another scientist has proposed, or they observe a phenomenon or interaction that raises their interest. They "question" what they are seeing and ask, "Why?"
2. They propose research hypotheses to explain why the phenomenon or interaction occurs. (As mentioned above, a research hypothesis proposes a relationship between variables.) They also develop a null hypothesis stating that there is *no* relationship between the variables. Statistics will be used to support or reject this null hypothesis.
3. They design a method of obtaining data related to the phenomenon and the hypotheses.
4. They gather data.
5. They make a decision about the null hypothesis based on the available evidence (the data), either to accept or reject.
6. Their decision about the null hypothesis leads to a decision about the research hypothesis.
7. Based on the decision that is made, they refine or extend their research interests and cycle through the research process again.

The Data Collected

Fundamental to the scientific method is the ability to collect trustworthy data. The scientist must feel confident that the data obtained are reliable (consistent across time and researchers) and valid (truthfully reflecting the phenomenon being observed). Consider the researcher who determines that there is no relationship between the variables under investigation. The researcher could reach this conclusion for one of three reasons:

1. There truly is no relationship between the variables; the researcher has made an accurate decision.
2. The experiment was not properly designed to draw accurate conclusions; the researcher has made a design error.
3. The data were not sufficiently reliable and valid to identify the relationship; the researcher has made a measurement error.

Note that reasons 2 and 3 result from some type of experimenter error. The use of the scientific method and the study of research methods can help researchers avoid or, at least, minimize such errors.

Populations and Samples

A researcher must obtain data on a random sample that is representative of the **population** to which the researcher desires to generalize the study results. Populations vary in size from very large (e.g., adults over the age of 50) to somewhat smaller (adults over the age of 50 who are members of Mensa) or relatively small (adults over the age of 50 who are members of Mensa and live in a specific city). Similarly, **samples** can be very large (in the thousands) to quite small (a dozen). *Most important to the researcher is that the sample be representative of the population.* If it is not, the conclusions of research based on the sample cannot be generalized to the population.

To obtain a sample that is representative of the population, the researcher considers various types of sampling procedures. One such procedure is simple random sampling, in which every person or member of the population has an equal likelihood of being chosen. In **stratified sampling,** the researcher categorizes the population into segments based on specific characteristics, to ensure that the sample will accurately reflect the proportions of those characteristics in the population. For example, a researcher might stratify based on gender or ethnicity. Convenience sampling is often used in field research. **Convenience sampling** occurs when the researcher simply selects the first available subjects from a population. This is also called *accidental sampling*. The problem with convenience sampling is that the sample often does not accurately reflect the characteristics of the population to which the researcher desires to generalize the results. For example, recall the iconic photo of president-elect Harry Truman holding up the *Chicago Daily Tribune* with the headline *DEWEY DEFEATS TRUMAN*. The *Tribune's* conclusion was based on sampling that did not represent the electorate in 1948. Adequate sampling is vital to extrapolating conclusions drawn via the scientific method.

Variables

Variables are the "nuts and bolts" of the scientific process. Researchers are interested in the relationships between independent and dependent variables. Often, the independent variable is called variable *X*, and the dependent is called variable *Y*. Therefore, researchers are interested in what changes, if any, occur in *Y* (the affected dependent variable) when changes occur in *X* (the independent variable). The researcher must be certain that the variables of interest are accurately measured and that they truly reflect the phenomenon being studied. Thus, the researcher must describe and measure the independent and dependent variables accurately, so that the study's results may be interpreted accurately.

Variance

Variance is simply the fact that not all things are the same; there is variability in things, persons, and observations. Science, essentially, tries to explain why there are differences. Researchers use the scientific method to attempt

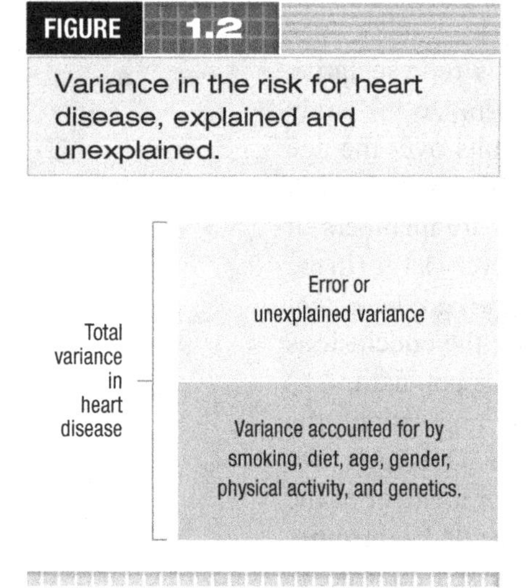

FIGURE 1.2 Variance in the risk for heart disease, explained and unexplained.

to identify the independent (*X*) variables that account for or explain the variation in the dependent (*Y*) variables. You might think of variance as the box illustrated in Figure 1.2. There is variance (i.e., variation) in the propensity for heart disease (the dependent variable). Some people have a greater propensity, others a lesser one, and some none. What are the independent variables associated with heart disease? Some variables believed to be related to heart disease are smoking, diet, age, gender, physical activity, and genetics. People have different values for these characteristics, such as either male or female for the variable of gender or a specific number for the variable of age. These differing values might be related to the propensity for heart disease. The unexplained variance in Figure 1.2 is "error"—it is that part of the variation in risk for heart disease that is unaccounted for by the variables that the researchers have studied. To help reduce the unaccounted for (error) variance, researchers consider what additional variables might be included in future experiments.

Summary of the Scientific Method

Scientists observe phenomena, develop hypotheses, gather data from samples, and arrive at conclusions, which may be generalized to a population. All of this is done according to sound procedures that can be replicated by other scientists. The researchers choose variables that will help them explain variation in the phenomena so that they can accurately describe (and manipulate, as necessary) the independent variables to account for the variance in the dependent variables. Each research effort incrementally is intended to reduce the amount of error in predicting the dependent variable.

Knowing and the Scientific Method

RESEARCH VALIDITY

Careful selection and measurement of the independent and dependent variables does not, in itself, guarantee a good experiment. In addition, issues related to the validity of the overall experiment are important. Experimental validity issues are broadly classified as internal or external.

Internal Validity

Internal validity refers to the technical quality of a study. It involves the certainty with which the results of an experiment can be attributed to the effect of the independent variable rather than some other, confounding variable(s). If internal validity is high, then the effect on the dependent variable can be attributed to the independent variable. If internal validity is low, then the observed effect of the independent variable on the dependent variable is brought into doubt. Internal validity ranges along a scale from very high to very low, and the exact level of internal validity of a particular study may not be readily apparent. Assessing internal validity is a subjective process, because there is no definitive way to measure internal validity. However, researchers can take steps to reduce threats to internal validity.

External Validity

External validity refers to the degree to which the findings of a study can be applied to other situations, subjects, or environments. If external validity is high, then the results apply to a wide range of situations or to larger populations. If external validity is low, then the findings apply only to a narrowly defined situation. As with internal validity, external validity ranges along a scale from very high to very low. Assessment of external validity is also subjective, because there is no exact way to quantify external validity. The researcher can provide evidence for external validity through inferential statistics such as significance levels, probabilities of various errors, and power. The discussion of these and other, related concepts make up the bulk of the rest of this text.

The term **generalizability** is used to describe how well the results of a study conducted on a sample of subjects will transfer to the population from which the subjects were selected. The term **transferability** is sometimes used when a consumer of research contemplates how well the results of a study can describe, explain, or predict the behaviors of individuals different from those in the study or in dissimilar situations and environments. In this text, we will use the term *generalizability* to include transferability.

Balancing Internal and External Validity

When designing an experiment, researchers balance the benefits of internal and external validity, because these concepts tend to be inversely related. An experiment can be designed to have very high internal validity with very strict experimental controls, but that same experiment might have low external validity (i.e., generalizability) because the experiment may bear little resemblance to the situation to which researchers desire to generalize the results. Designing a very generalizable experiment that suffers from low internal validity is also possible. If the internal validity is weak, the overall value of the research is questionable. Researchers need to reduce the negative consequences of studies conducted with poor internal or external validity. Let's examine factors affecting internal and external validity that researchers should consider.

Factors Threatening Internal Validity

Uncontrolled variables are the single largest threat to the internal validity of a study. Internal validity is reduced by uncontrolled factors such as subject attrition during the course of the experiment, uncontrolled events that occur during the experiment, and other factors detailed below. Threats to internal validity generally fall into two categories: threats concerning subjects and their behavior or treatment during the experiment, and threats related to experimental procedures or instrumentation.

Local History

Local history is defined as any unanticipated events occurring during a study that might alter the subjects' behaviors in an uncontrolled and unaccountable way. For example, students in a particular region may become more interested in physical activity because the Olympics are being held locally, whereas students in another area are not influenced by the Olympics. Researchers manage local history threats by using control groups and by maintaining equal conditions among groups.

Maturation of Subjects

Maturation of subjects is a threat to internal validity when change due to growth or development affects the subjects' characteristics. For example, did the middle school students get stronger because of the weight training class or just because they matured physically during the semester? Random or matched subject assignment procedures are the best ways to control for subject maturation; these procedures randomly distribute maturation among groups. A second technique is to use norms to compare changes in the subjects to normal maturation effects. However, norms for the characteristics being studied may not be available. Choosing an appropriate statistical model (some of which are described later in Chapter 13) is another technique for controlling for maturation effects.

Pretesting

Pretesting is a threat to internal validity because subjects might "learn" how to take the posttest, resulting in improvements that are not related to the treatment. An example of such a situation might occur in the use of a treadmill VO_2max test measuring oxygen consumption during exercise to assess cardiorespiratory fitness. A problem with the treadmill VO_2max test is that the procedures and equipment can be intimidating to the subjects. Thus, pretesting with a treadmill VO_2max test could by itself produce superior posttest results, simply because the subjects learn how the test works, and their apprehension levels are reduced during the posttest because they have been through the test before. Unfortunately, if, as a researcher, you don't conduct a pretest, you lose the assurance that the groups are randomly equivalent at baseline. Alternatively, if the subjects are randomly assigned to the experimental groups, then the assumption is that they are randomly equivalent on all variables *except* the independent variable.

Instrumentation

Instrumentation refers to the effect of human or equipment measurement errors during the data collection or data analysis phase of a study. Suppose, for example, that during a study measuring body composition, the skinfold calipers were dropped and damaged. The continued use of those calipers could result in inaccurate measurements. Calibrating equipment and carefully training research assistants are two good methods of controlling instrumentation threats.

Another example of instrumentation error is a change in the definition of the variable being measured. For example, changing the definition of sufficient physical activity for health benefits from five days per week for 30 minutes to 150 minutes per week can dramatically change the reported percentage of people achieving sufficient physical activity. Researchers need to compare the definitions of measured variables carefully with those reported in the literature.

Conducting a pilot study is a method of controlling for instrumentation threats. During a pilot study, testing procedures and equipment calibrations can be evaluated prior to data collection. The main drawbacks to pilot studies are that they can be time consuming and that they require resources such as supplies, time, and subjects. Pilot studies are, nonetheless, very important to conducting valid research activities.

Statistical Regression

Statistical regression is also referred to as *regression to the mean*. Regression is the statistical tendency for extreme scores to move toward the group mean when measured a second time. For example, subjects who are selected for a study because they are extremely unfit would tend to score better on a fitness test the next time, even though their fitness level is little improved. The measured improvement in fitness level might simply reflect the unreliability of scores from the initial test, resulting in movement toward the mean on subsequent testing occasions. Researchers can control the effect of statistical regression by using the most reliable measure available and by using a control group.

Differential Selection of Subjects

Differential selection of subjects occurs when the characteristics of the subjects are initially different in the treatment conditions, due to improper or unavoidable selection or assignment procedures. For example, an experiment that relies on volunteer subjects might be exposed to this threat because volunteers are, by their particular willingness to join the experiment, different from those who do not choose to join the experiment. Another example is the use of "intact groups" (groupings that exist prior to the research study). Two different intact fifth-grade physical education classes may differ in important but unknown ways. Random subject selection and matching are good methods for eliminating differential subject selection threats.

Experimental Mortality

Experimental mortality refers to the uneven loss of subjects from the various treatment categories during the course of the study because they drop out or fail to complete the treatment conditions. If one condition in an experiment is physically more demanding than another, leading to greater numbers of injuries or higher stress, subjects might choose to drop out in that group, while subjects do not drop out in a less demanding group. This is a problem only if there is *differential* mortality, meaning that subjects are lost from one group for a certain reason and not from another group, or lost from groups in different amounts. If the loss of subjects is random, then the problem is no longer that of experimental mortality but becomes a reduction in statistical power due to reduced sample size. (Statistical power is described later in this text.) The effect of experimental mortality can be limited by several methods, including: (1) providing incentives for subjects to complete the study, (2) determining the types of subjects that tend to drop out and removing similar subjects from other groups, (3) using random assignment methods, and (4) starting with more subjects than are needed.

Increasing Internal Validity

Researchers attempt to address internal validity issues by controlling several basic aspects of studies. One of these is the selection of an appropriate experimental design. Carefully conducting the experiment increases internal validity by reducing instrumentation errors, local history events, differential selection of subjects, and other threats.

The systematic elimination of potential confounding variables may be the most important factor in protecting the internal validity of a study. Although not all confounding variables are equally threatening to the validity of a study, the presence of any confounding variable always reduces the internal validity of a study to some degree. Perhaps the best recommendations to maintain high internal validity are: (1) use control groups, (2) randomly assign subjects to groups, and (3) be as careful as possible in making measurements.

V.1.4 **Internal Validity and Threats to It**

Factors Threatening External Validity

Issues in external validity concern how well the results of a study generalize to other situations, to other subjects, and to other environments.

Subjects

Behavioral research, including research in human performance, sport, and exercise, often relies on college students as subjects. Sears (1986) and Smart (1966) found that in psychological research, college students served as subjects in more than 70% of the studies reported in selected psychology journals. Besides the question of how well one set of college students represents all college students, the larger external validity issue is how well college students represent other segments of society. Recently, Hanel and Vione (2016) report the use of such subjects can be problematic as they often vary randomly from the population to which one desires to generalize.

Pretest Sensitization

Pretesting is a commonly used method of assessing group equivalence prior to the initiation of treatment trials. Pretesting is uncommon in real life, and it might limit the ability to generalize the results from a study to the population from which the subjects were selected. **Pretest sensitization** occurs when subjects become aware of the nature of the study and this influences their behavior during the treatments or on a posttest. For example, recording subjects' body weight prior to enrolling them in a weight-loss study might sensitize some individuals to their body weight, and these individuals might then modify their diets, physical activity, and other lifestyle behaviors.

The topic of pretest sensitization provides a good example of the tradeoffs confronting the researcher in designing a study. Pretesting serves to improve internal validity by assuring that groups are equivalent at the inception of a study, but it can reduce external validity because the population at large to which the researcher would like to generalize has not been exposed to a pretest. One way to reduce the problems posed by a pretest is to design the study so that half of the subjects are pretested and half are not. A second method is to replicate the study without a pretest. In either case, the number of subjects required is increased and additional time and resources must be invested in the study. If possible, it is better to rely on proper random sampling procedures to ensure initial group equivalence.

Expectancy

Expectancy refers to the unintentional effects on the study caused by the researcher's bias or prior beliefs about the study's outcome. This is also referred to as "the self-fulfilling prophecy". Researchers try to control for expectancy effects by using double-blind methods (where neither the subject nor the researcher is aware of the assignment to treatment conditions) or objective methods whenever possible. For example, the use of an accelerometer to measure physical activity is more objective than a self-report of physical activity behaviors. Perhaps the best control for expectancy effects is for the researcher to be aware of this issue and guard against it.

Hawthorne Effect

The term **Hawthorne effect** comes from a study involving the Hawthorne Electric Company, in which employees were shown to become more productive simply because they thought they were in an experiment and were being evaluated (Levitt & List, 2011). In essence, the subjects performed better because they knew they were subjects in a study. Another example of the Hawthorne effect might be a study where elementary grade children are given pedometers, which the researcher is using to validate that a treatment curriculum increases physical activity more than the traditional curriculum. However, the children become more active in their physical education classes simply because they have been given pedometers. A control group and unobtrusive measures are the two best means of limiting the Hawthorne effect.

Overgeneralizing

Overgeneralizing is the tendency to extend the conclusions from a study beyond the conditions of the original research. Experiments are based on a limited number of independent and dependent variables. The study by Quadagno, Faquin, Lim, Kuminka, and Moffat (1991) on menstrual cycle and athletic performance investigated two measures of swimming speed and two weight lifting exercises during three stages of the menstrual cycle in college-age women. The researchers found no reduction in athletic performance during menses. Can these findings be generalized to performance in other sports or to different populations? It will not be possible to generalize in these ways until further experiments are devised to examine a larger range of activities and additional samples of women.

Improving External Validity

Subject selection is an important factor in considerations of external validity. Defining the population of interest as completely as possible and then sampling this population properly are the two keys to assuring external validity.

Researchers need to describe their methods carefully, including operationally defining the outcome variables. They should provide complete specifications for any instrumentation used and define any terms that might be ambiguous. For example, some consider physical fitness to be defined completely by a VO_2max value. Others also include muscular strength and muscular endurance in their definition of physical fitness. Some include flexibility, and others do not. In addition, treatments should be described specifically, so that another researcher could duplicate the exact processes of the initial study.

External Validity and Threats to It

Replication is the best method for demonstrating external validity. Replication, either exact or conceptual, with different subjects, in a different environment, or under modified experimental conditions will help to establish the level of external validity. **Conceptual replication** is more common than exact replication. Conceptual replication occurs when a study is replicated with modifications, such as changing the independent variables or measuring the dependent variable differently. The goal is to demonstrate similar results under a broader range of conditions than the original research. For example, a laboratory study could be replicated in a field setting.

The Relationship Between Internal and External Validity

Judgments about the external validity of a study depend to some extent on judgments of the study's internal validity. If an experiment is found lacking in internal validity, examination of its external validity is not important. The results of the study are in doubt and, thus, its generalizability is of little practical value. On the other hand, studies that are high in external validity must necessarily have high internal validity.

EXPERIMENTAL DESIGN

Experiments are the tools that researchers use to draw conclusions. As discussed previously, scientists are interested in the relationships between independent variables (*X*, treatment or exposure) and dependent variables (*Y*, outcome). The researcher hypothesizes that manipulating the independent variable in some manner will influence the dependent variable. That is, the researcher hypothesizes that understanding the effects of changes in the value of *X* will increase the ability to account for variation in *Y*. Thus, scientific research is about identifying the relationship between *X* and *Y*. However, other variables, which the researcher may not have control over, could also influence the outcome. These other variables might confound the ability to interpret the relationship between *X* and *Y* and, hence, they are known as *confounding variables*.

Researchers want to control these extraneous variables and other influencing variables that might mask the true relationship between *X* and *Y*. For example, if a researcher is studying the relationship between physical activity behaviors *(X)* and health outcomes *(Y)*, the researcher might want to control (account or adjust) for differences in age, because it is well known that positive physical activity behaviors decrease with increasing age. Alternatively, the researcher might want to include age as a variable in the study. Deciding which variables should be included in the analysis and how they should be included is the domain of experimental design. Proper experimental design is essential to correct interpretation of study results.

To understand experimental design, a few additional, broad concepts are necessary. We will only introduce these concepts here; they are further described in specific sections throughout the remainder of this text. An important distinction is that of (1) between-groups designs versus (2) within-groups (repeated-measures) designs. In between-groups designs, each subject or observation is measured once and only once. Measurements are compared for the different groups. In within-groups designs, the subject or observation is measured multiple times, in multiple levels of an independent variable, and the measurements may be compared in multiple ways. This distinction is illustrated in Figure 1.3, where panel (a) reflects the fact that the subjects are grouped according to gender and measured only once. Panel (b) shows that both genders are measured repeatedly across four weeks.

A within-groups design provides the opportunity to study learning, decay, and carryover effects. However, order effects (i.e., the order in which the various levels of the independent variable are applied) could also have an impact on interpretation of the results. In this experimental design, all subjects receive all levels of the treatment; hence, it is important to rule out the order in which the various levels are presented as a possible explanation of the results. If there were three levels of the independent variable (A, B, and C), there would be six different orders in which the treatments could be presented (ABC, ACB, BAC, BCA, CAB, and CBA). The researcher could examine the order effect, if any, as an additional independent variable by analyzing the mean differences among these six levels. Alternatively, the order effect could be eliminated as an explanation of the results if an equal

FIGURE 1.3 Between-groups (a) and within-groups (b) experimental designs.

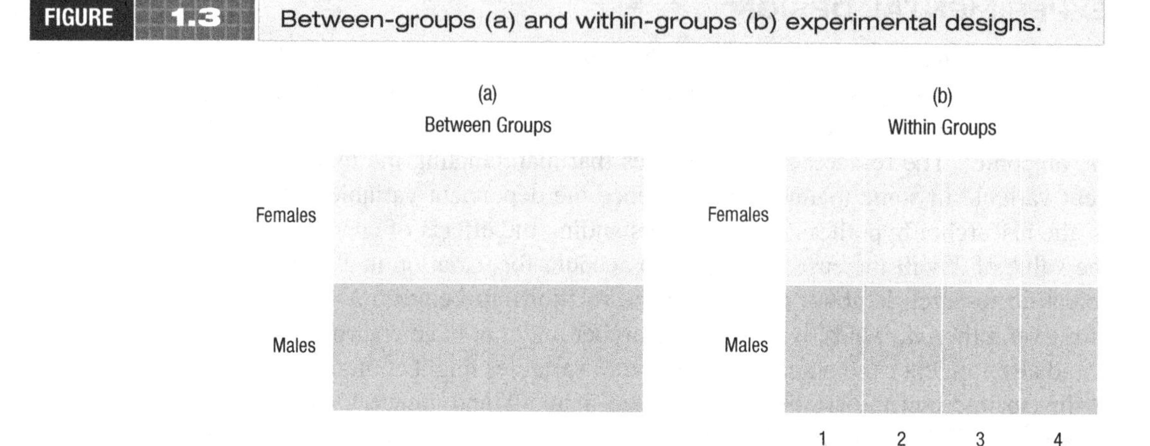

number of subjects completed each order of presentation. The researcher must design specific experimental study steps to account for order effects.

The between-groups design shown in Figure 1.3, panel (a), allows the researcher to identify any differences between the genders. The within-groups design, in panel (b), allows the researcher to identify any differences in genders, any differences in weekly measurements for each gender, and any changes in weekly measurements that occur for both genders. We will discuss experimental design more thoroughly in later chapters, because decisions about study design affect the selection of appropriate statistical tests.

SELECTION OF STATISTICAL PROCEDURE

With this foundational knowledge of research methods, we are ready to begin the study of statistics in human performance. In the remaining chapters, we will continue to use examples from human performance, kinesiology, exercise science, physical education, and related areas. We will guide you in choosing appropriate statistical procedures for each example and help you with the interpretation of the results. Many statistical procedures are introduced in this text. Collectively, they are provided to help you use hypothesis testing and the scientific method, which have been introduced in this chapter. The statistical procedures presented in the remaining chapters are listed in Appendix B. Appendix B consists of two tables (one for parametric and one for nonparametric) that indicate the correct statistical procedures to use depending on the number of variables (independent and dependent), the number of levels of each, and the levels of measurement involved. After studying this text, you will understand how you can use this information to identify which statistical procedure is appropriate for your research. Appendix B will become a valuable tool for you in your current and future work.

SUMMARY

We have introduced you to some fundamental aspects of research methodology. This chapter discussed how we gain knowledge, types of research activities, the scientific method, research validity, and experimental designs. These concepts are foundational to conducting research studies and selecting statistical procedures to evaluate their results. Choosing the appropriate statistical procedure is essential to conducting valid research, and to assist in this task Appendix B lists the statistical procedures presented in the remainder of this text. With this background knowledge, in the following chapters we present statistical concepts and procedures involved in research.

V.1.6 **Chapter 1 Summary: Concepts and Terms**

Review Questions and Practice Problems

PART A (ANSWERS PROVIDED IN APPENDIX C)

1. What is the general goal of all types of research?
   A. To contribute to the body of knowledge
   B. To publish research papers
   C. To develop theories
   D. To test hypotheses

2. Which research method can establish cause-and-effect relationships?
   A. Descriptive research
   B. Correlational research
   C. Longitudinal research
   D. Experimental research

3. Your research methods professor asks you to conduct a survey on cheating by college students. To save time, you set up a table in the student union and give the questionnaire to students who happen to stop by. What is this sampling method called?
   A. Simple random
   B. Convenience
   C. Quota
   D. Stratified

4. A researcher investigates the relationship between taking a vitamin C supplement (none, 500 mg, 1000 mg) and workers' (office, outdoors) frequency of colds. What is/are the independent variable(s)?
   A. Colds
   B. Vitamin C
   C. Colds and workers
   D. Vitamin C and workers

5. In the previous question, what is/are the dependent variable(s)?
   A. Colds
   B. Vitamin C
   C. Colds and workers
   D. Vitamin C and workers

PART B

For the situations in questions 6–10, select the most serious possible threat to *internal* validity. Mark each one as:

A. if the most serious possible threat is history
B. if the most serious possible threat is maturation
C. if the most serious possible threat is statistical regression

D. if the most serious possible threat is selection bias

E. if the most serious possible threat is experimental mortality

6. Students below the 15th percentile in fitness who volunteer to participate are placed in the experimental group. Non-volunteers at the same fitness level are placed in the control group.

7. During the experimental treatment subjects are exposed to a variety of special instructional experiences from sources outside of the experiment.

8. Low achievers drop out at a higher rate than other subjects from a treatment designed to increase fitness levels.

9. Subjects falling below the 15th percentile on a fitness test later achieve higher scores, but subjects initially scoring above the 85th percentile achieve lower scores when retested.

10. Subjects who attend a summer camp for three years in a row show yearly increases in physical strength.

11. Which level of knowledge requires the greatest depth of understanding of a topic?

    A. Description

    B. Prediction

    C. Control

    D. Explanation

12. Which research method works with a single group of subjects measured over an extended period of time?

    A. Qualitative

    B. Experimental

    C. Longitudinal

    D. Cross-sectional

13. What is the simplest level of knowledge, and what is the most difficult level of knowledge to achieve?

    A. Prediction, description

    B. Description, explanation

    C. Control, description

    D. Description, control

14. What is the goal of human performance research?

    A. To describe behaviors

    B. To predict behaviors

    C. To control behaviors

    D. To explain behaviors

    E. All of the above

15. What term is associated with describing how well the findings from a study generalize to another situation?

    A. Ecological validity

    B. Experimental validity

    C. External validity

    D. Internal validity

REFERENCES

Hanel, P. H. P., & Vione, K. C. (2016). Do student samples provide an accurate estimate of the general public? *PLos One*, *11*(12). doi:10.1371/journal.pone.0168354.

Henry, F. M., & Rogers, D. E. (1960). Increased response latency for complicated movements and a "memory drum" theory of neuromotor reaction. *Research Quarterly*, *31*(3), 338–458.

Heron, M. (2016). *Deaths: Leading causes for 2014. National vital statistics reports* (vol. 65, no. 5). Hyattsville, MD: National Center for Health Statistics.

Levitt, S. D., & List, J. A. (2011). Was there really a Hawthorne effect at the Hawthorne plant? An analysis of the original illumination experiments. *American Economic Journal: Applied Economics*, *3*(1), 224–238.

Quadagno, D., Faquin, L., Lim, G., Kuminka, W., & Moffat, R. (1991). The menstrual cycle: Does it affect athletic performance? *The Physician and Sports Medicine*, *19*(3), 121–124.

Schmidt, R. A. (1975). A schema theory of discrete motor skill learning. *Psychological Review*, *82*(4), 225–260.

Sears, D. O. (1986). College sophomores in the laboratory: Influences of a narrow data base on social psychology's view of human nature. *Journal of Personality and Social Psychology*, *51*(3), 515–530.

Smart, R. G. (1966). Subject selection bias in psychological research. *Canadian Psychologist*, *7a*(2), 115–121.

2 Introduction to Measurement and Statistics

INTRODUCTION

Statistical tests are tools used by researchers in virtually every field of scientific inquiry. They often are part of the scientific method. As described in Chapter 1, the scientific method has developed since the time of the ancient Greeks. It takes slightly different forms in various disciplines, but, as described earlier, its purpose is always to answer questions through organized observations and experiments. It generally serves to determine cause-and-effect relationships, and its most important features are that it is reproducible and verifiable.

The scientific method usually begins with the formulation of a question derived from one of many possible sources (observation, other research, curiosity, reading of relevant literature, or existing theories). As important as it is, we will not dwell on this step. Next, the researcher determines how to measure the variables involved in constructing a hypothesis about the question at hand. Because the measurement process is extremely important to statistics, we will spend a considerable amount of time on this topic later in this chapter.

In Chapter 1, we presented a few issues regarding the next two steps—hypothesis formulation and experimental design—especially as they relate to statistical tests. These two topics are very important and are the focus of research methods. As suggested in Chapter 1, the curriculum of many science fields often includes a separate course in the area of research methodology.

In our discussions of statistical tests, we make the assumption (sometimes dangerously) that the data resulting from experiments are accurate (that is, proper measurement procedures were used) and meaningful (that is, proper experimental designs were employed). The next steps—analyzing the data and drawing conclusions—are the main foci of this text. These steps almost always employ various statistical tests. Note that statistical tests do not provide proof of anything, nor are the ultimate decisions based on their results guaranteed to be absolutely correct. You might often hear commentators say, "A new research study has proven" some proposition. Frankly, such statements are wrong, because it is impossible to "prove" anything with statistical analysis. Statistical analysis can provide a great deal of evidence, but you always run some risk of your decision being incorrect. Statistical methods are simply tools to help the researcher analyze collected data and make reasonable decisions based on those data.

Often, the last step of the scientific method has to do with communication and dissemination of results. Although this is an important part of the process, we have little to say on this topic, since it is not within the scope of this text. Finally, you should be aware that the scientific method is an iterative process, as the conclusions from a study often result in as many or more new questions as answers. Let's now move to examining, in general, how researchers use statistics in making decisions.

STATISTICS AS A TOOL

The body of knowledge called statistics uses logic and mathematics to help researchers make decisions during and after experiments. The following scenario illustrates how statistics works.

A friend tells you that he is going to toss a coin in the air five times, catch it each time, and obtain five heads in a row. He proceeds to do exactly that. Your first reaction (based on logic) would probably be rather skeptical—you might doubt that everything was on the up and up, and you would be correct to do so, because (based on mathematics) the probability of this event occurring by chance is approximately 3 in 100; that is, in the long run, it would occur naturally 3 times out of 100 tries. The mathematical way to calculate this is:

$$(.5)^5 = 0.031$$

where .5 is the chance of each coin coming up heads, and 5 is the number of coin tosses.

You could even use logic to determine the rarity of obtaining five heads in a row by chance. Suppose there are 100 people standing in a room, and each of them flips a coin and notes the result. Logic suggests that 50 heads and 50 tails would be the result. Of course, this exact result would be unlikely to happen every time, but in the long run this would be the average outcome. If the 50 (or so) people who obtained tails sat down and the rest of the people remained standing and tossed their coins again, we would expect, on average, about 25 heads. If these 25 tossed their coins a third time, we would expect, on average, 12 or 13 to obtain heads. Let's say there were 12 heads. On the next (fourth) toss, we would reasonably assume six heads as the result, and when these six tossed their coins for the fifth time, we would expect three heads. Thus, after five tosses, only three of the original 100 people would still be standing; those three got five heads in a row. In this way, we can determine logically that the probability of obtaining five heads in a row (with an honest coin) is about 3%. Although this exact outcome would not be expected to occur every time, the long-run average would be 3 five-heads runs out of 100 five-toss trials.

You now have a decision to make. Was your friend's performance one of those 3 times out of 100 that this event could have happened by chance—or is something funny going on? Logically, most people would look for an answer other than chance. Perhaps your friend has perfected his coin-flipping technique to achieve a desired outcome, perhaps he was using a weighted coin, or perhaps he was using a coin that had been modified to have heads on both sides.

This illustration is representative of a common use of statistics in research. An investigator might give one group of subjects a medication believed to lower systolic blood pressure and give another group of subjects a placebo. As noted in Chapter 1, the hypothesis that the researcher tests statistically is called the null hypothesis, and in this case it would be phrased to hypothesize *no difference* in the mean systolic blood pressure of the two groups after administration of the medication over some length of time. The null hypothesis suggests that there is no relationship between the dependent or outcome variable (here, blood pressure) and the independent or exposure variable (here, the type of medication taken). Next, the researcher would measure each subject's systolic blood pressure and calculate the mean blood pressure for each of the two groups. Let's say that the researcher found the mean value to be 115 mmHg for the treatment group (those who received the experimental drug) and the mean value for the control group (those who received the placebo) to be 122 mmHg. Just as in the coin-toss illustration, the task is to determine whether the difference between the means can reasonably be attributed to chance, if there really is no difference in the effects of the administered medication. Is the difference of 7 mmHg too improbable to believe it could have happened solely by chance, if there is no medication effect? In this case, the researcher would use what is called an *independent t-test* to make this decision. You will learn more about this statistical test in Chapter 7.

This rather simple-sounding scenario gives you an idea of how statistics work. However, there are many issues, some of which were mentioned in Chapter 1, that the researcher needs to consider. Here are just a few examples:

- Whom do the subjects used in the study represent?
- How were the subjects selected?
- What other variables might cause blood pressure to change, and have these other variables been controlled or accounted for in any way?
- How many subjects should be used?
- How should blood pressure be measured?
- How accurate is the measurement used?
- To whom do the results of this study apply?

As you should recognize after reading Chapter 1, many of these questions relate to research methods and experimental design. Some of them relate to the field of statistics. The answers to these questions all affect the accuracy and usefulness of the results of the experiment. In this text, we will concentrate on the statistical issues, but, because they are overlapping, we will occasionally address research methods and experimental design issues, as well.

INTRODUCTION TO MEASUREMENT

Statistical procedures are generally performed on numerical data. The numbers involved are derived primarily from the measurement of some characteristic of the subjects in an experiment. Perhaps you are measuring the time to complete an event, the number of repetitions completed, or the amount of force generated. Measurement means assigning a symbol (usually a number), according to a set of rules, to a subject to reflect the amount of some trait or characteristic that the subject possesses. It is impossible to overestimate the importance of accurate measurement. If measurements are not **valid** (i.e., they actually assess what they are intended to assess) and **reliable** (i.e., they assess whatever they assess consistently), then statistical tests based on these measurements will not be useful in making meaningful decisions.

Levels of Measurement

Some measurements are more informative than others. Measurements have been categorized into four levels, according to the amount of information they convey. To see how measurements can convey differing amounts of information, consider that through observation you could rank a group of individuals according to their heights, or you could use a measuring tape to assess the height of each individual in centimeters. The amount of information available from the second procedure is greater than that obtained from the ranking by inspection. With the data from the ranking procedure, you would only be able to indicate that one individual has more (or less) height than another individual, but by using the measuring-tape data you could determine the exact amount by which two individuals differ in height. To illustrate further, you might group people into two groups: (1) those shorter than 6 ft and (2) those 6 ft (72 in, 183 cm) or taller. Without the exact measurements, you would now know only whether a person is in one group or the other. People in the "short" group are all shorter than those in the "tall" group, but you know nothing else about their heights. Organized from the least to the most information conveyed, the levels of measurement are called *nominal, ordinal, interval*, and *ratio*.

Nominal Measurement

Nominal measurement conveys a minimal amount of information and permits only the assessment of equality or difference. Nominal measurement typically uses descriptors to classify subjects into categories. Often, it is necessary to define the categories. For example, a person might be classified as a student-athlete or a non-student-athlete; or be classified as a freshman, a sophomore, a junior, or a senior; or, in a triathlon, each competitor might be classified as finished or did not finish. Treatment group is usually a nominal measure; each subject is typically in either the treatment group or the control group. Gender is a nominal variable with two levels, male and female. When numerals are used to categorize subjects (e.g., numbers on baseball uniforms), the mathematical properties of the numerals are irrelevant.

Ordinal Measurement

Ordinal measurement permits the ranking of subjects based on whether one has more or less of some characteristic than another. Finishers in a race might be ranked as first, second, third, and so forth. However, unless the actual time to finish the race is known for these individuals, it is not possible to determine how much time elapsed between the contestants' arrivals at the finish line. Ordinal measurements are often based on subjective judgment. For example, a person hired for employment is judged to have placed first among the applicants. Placement in many athletic events (gymnastics, springboard diving, ski jumping, dancing, and many others) is based on rankings involving subjective judgment. The difference between rank #1 and rank #2 is the same as the difference between rank #8 and rank #9—they are one rank apart—but the actual values associated with these ranks might be considerably different. The actual measured difference between rank #1 and rank #2 might be considerably different from the measured difference between #8 and #9 or any other two consecutive ranks.

Interval Measurement

The most important feature of **interval measurement** is the use of some sort of scale to quantify the amount of some characteristic possessed by the subjects being assessed. A simple example is the use of a thermometer to measure temperature in degrees Fahrenheit. The scale used in interval measurement permits us to make statements about the equality of intervals. For example, it is the same distance between 0 degrees and 40 degrees as it is between 40 degrees and 80 degrees on the Fahrenheit scale. However, it is not appropriate to state that 80 degrees is twice as hot as 40 degrees, because the zero point on the Fahrenheit scale is an arbitrary point and does not indicate the total absence of heat. Similarly, a boy who can do four chin-ups is not twice as strong as another boy who can do only two chin-ups, because the ability to do zero chin-ups does not indicate the absolute lack of any upper-body strength.

Ratio Measurement

Ratio measurement permits statements of comparison, such as *twice as much* or *one-third as much*. These are called ratios. This level of measurement is possible when zero on the scale that is being used indicates a total lack of the characteristic being measured. For example, a 6-ft-tall man is twice as tall as his 3-ft-tall son, because zero on the height scale means no height at all. Consider the example of strength in the previous paragraph. The strength of the boy who does four chin-ups is not twice as great as that of the boy who does two chin-ups, but the actual number that the first boy does *is* twice as many as the second boy's number. Thus, strength is measured on an interval scale, but the number of chin-ups is measured on a ratio scale. Time, distance, and force are typically measured with ratio measurements.

Comments About Levels of Measurement

The levels of measurement are hierarchical, so that it is possible to convert a higher level of measurement to any level below it, but it is not possible to convert a lower level to a higher level. An ordinal scale is also nominal. A ratio scale is also nominal, ordinal, and interval. Interval and ratio levels of measurement are classified as **continuous** or scalar measurements. For what is called **parametric statistics,** at least interval-level measurement is required. **Nonparametric statistics** are available for situations in which only nominal- or ordinal-level measurements are available. We will discuss parametric statistics in Chapters 6 through 13 and nonparametric statistics in Chapter 14.

A researcher should always choose to use the highest level of measurement available, because this increases the amount of information and the precision of the measurements. However, not everything can be measured with a ratio scale. For example, a physical therapist might be interested in examining the effectiveness of various treatments in reducing pain in a particular joint. How do you measure pain? Usually, this is accomplished by asking patients to rank their level of pain with a numerical symbol linked to a verbal description of the amount of pain that is present (e.g., 10 represents the worst possible pain, 5 represents a moderate degree of pain, 0 represents no pain, and so forth). This is an ordinal-level measurement. However, the same physical therapist uses a ratio scale when measuring the number of repetitions of an exercise that a patient did in week 1 and comparing that to the number of repetitions completed in week 4 of treatment. Another example of an ordinal scale is Borg's (1982) scale of perceived exertion, where subjects rate how hard they are exercising on a scale from 6 to 20. Alternatively, one could measure energy consumption in VO2, which is a ratio measure of cardiorespiratory fitness.

V.2.1 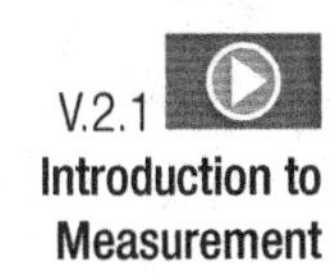
Introduction to Measurement

ACCURACY OF MEASUREMENT

The accuracy of measurements depends on their validity and reliability. The validity and reliability of measurement are essential in statistical analysis. If a measurement process results in data that are low in validity and reliability, applying statistical tests to them is a waste of time. Assume you are examining the relationship between two variables and, through the use of a statistical procedure, you conclude that the two variables do not appear to be related. This conclusion could result from the fact that your measurements of the two variables are not very good, or it might result from the fact that the two variables are not actually related. If your measurement process is not accurate, you cannot know which is true.

No measurement process is perfectly valid or reliable. There will always be some error associated with measurement. In fact, the examination of sources of error is fundamental to statistical reasoning.

Measurement Validity

When we say that a measurement is valid, we mean that the measurement process results in a true assessment of the attribute that is being assessed. In other words, you are measuring what you say you are measuring.

Measurement validity is not a yes-or-no proposition, but rather it exists on a continuum ranging from low to high. There are many ways to collect evidence to determine the amount of validity that a measurement process possesses. Some of these procedures rely on a high level of subjective judgment, whereas others rely on empirical evidence. Figure 2.1 may help you organize these types of validity evidence.

Relevance Validity

Two types of evidence for measurement validity that are determined exclusively through subjective judgment are *face validity* and *content validity*. Because both are concerned with whether or not the measurement process appears to be *relevant* to the attribute being assessed, we categorize face validity and content validity as types of *relevance validity*.

Face validity

We determine **face validity** by examining the measurement process to answer the question, "On the face of it, does it look like this procedure measures what it is supposed to measure?" For example, standing on one leg on

FIGURE 2.1 Measurement validity.

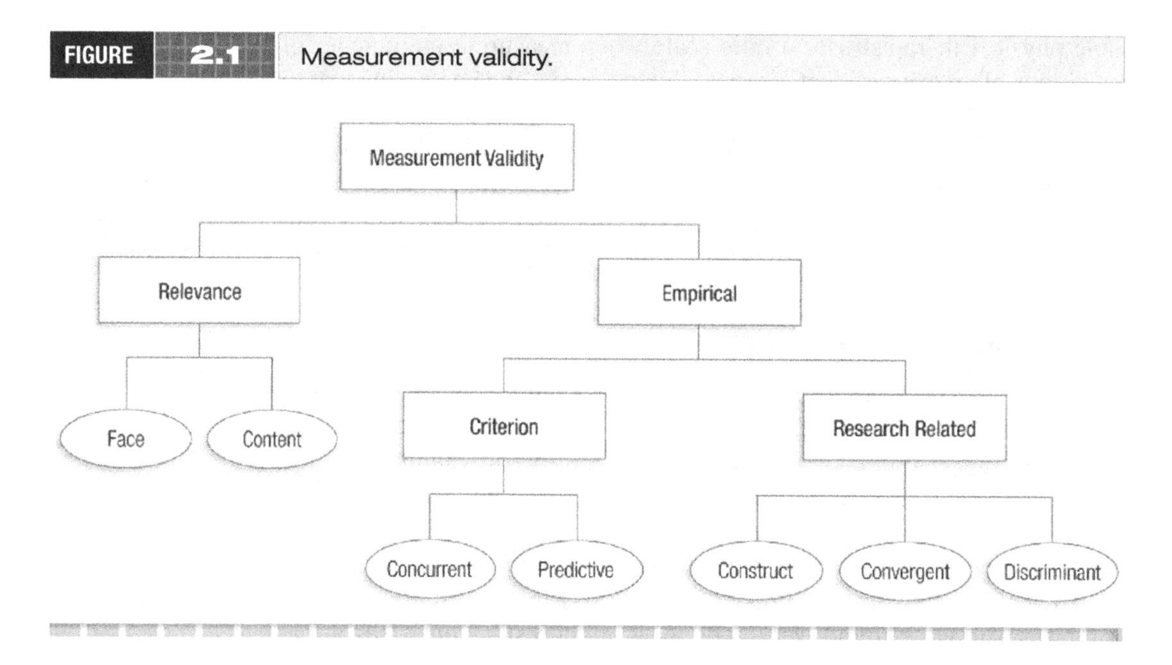

a narrow piece of wood would seem, on its face, to be a good way to measure balancing ability, since the task is practically the definition of the attribute being assessed. In this case, the measure has face validity. However, if the same task were being used to measure leg strength, even though there might be a positive relationship between balance and strength, the balancing task would probably be judged to be less relevant and have less face validity as a strength measurement. Face validity, while not insignificant, is by itself the weakest form of validity evidence, due to its totally subjective nature.

Content validity

We say a measure has **content validity** if the content of the attribute being assessed is adequately covered by the measurement process. Again, notice that this decision requires subjective judgment. Examining the content validity of a process is easier in some situations than in others. For example, determining whether the questions on a written test cover all the relevant educational outcomes and all the pertinent topics that the test is designed to assess is easier than deciding whether a certain questionnaire adequately covers a construct such as the self-esteem of regular exercisers. This is because the content of the self-esteem construct is difficult to define.

Empirical Validity

Several types of evidence for validity are based on the gathering and examination of data. Two of these, *concurrent validity* and *predictive validity*, are classified as *criterion-related validity* because they compare the results of a measurement process for which validity evidence is being sought to the results of another measurement process that is known to be valid (the criterion). Three other types of empirical validity evidence—*construct validity, convergent validity*, and *discriminant validity*—focus on whether the measurement process functions as it theoretically should, if it is valid, by comparing it to other measurements. Because this type of evidence is typically gathered through experimentation and the use of statistics, we will classify *construct validity, convergent validity*, and *discriminant validity* as *research-related validity.*

Criterion-Related Validity: Concurrent Validity

In exercise physiology, a test known to be an excellent assessment of cardiorespiratory fitness is treadmill assessment of VO_2max. It determines the number of milliliters of oxygen a person can utilize, per kilogram of body weight, per minute (ml/kg/min). However, this test requires expensive equipment and trained personnel for its administration, and it is time consuming. Suppose a researcher devised a questionnaire (much less expensive to administer) that is intended to measure subjects' cardiorespiratory fitness levels. To provide **concurrent validity** evidence for the questionnaire, the researcher could administer both the treadmill VO_2max test and the questionnaire to a group of subjects and determine the relationship of the resulting two sets of scores. Through the use of a correlation coefficient (see Chapter 5), it would be possible to determine how well the questionnaire results matched (or did not match) the criterion scores from the treadmill VO_2max test. This comparison would result in evidence of the questionnaire's concurrent validity or lack thereof. For example, Jurca et al. (2005) used criterion-related validity methods with common health indicators (i.e., gender, age, BMI, resting heart rate, and self-reported physical activity level) to estimate cardiorespiratory fitness.

Criterion-Related Validity: Predictive Validity

Predictive validity is the same as concurrent validity, except that the criterion is not available at the same time as the "new" measurement process is conducted. Instead, the criterion measure is obtained sometime in the future (even possibly the distant future). For example, let's imagine that you have invented a new type of microscope, and you need to hire employees to sell your new product. Naturally, you want to hire people who will sell lots of your microscopes, so you devise a written instrument designed to help you select such salespeople. You administer the instrument to 20 applicants, and you hire all of them, regardless of how well they scored on the instrument. A year later, you examine the relationship between the written instrument scores and the number of microscopes each of your salespeople has sold. If the correlation is high, it is evidence that the instrument has predictive validity, and then you could use it to discriminate among your next set of salespeople applicants. Another example of predictive validity is illustrated by Kestilä et al. (2015), who used measures of adversities and social-economic status obtained in the subjects' childhood to predict sedentary lifestyle behaviors when the individuals reached early adulthood. Notice that the criterion, sedentary behavior, was measured many years later.

Research-Related Validity: Construct Validity

Many times, measurements are required for attributes that are not directly observable (e.g., intelligence, motivation, creativity, and self-esteem). Researchers often measure these constructs by observing patterns in subjects' behavior. Imagine that you have devised a measurement procedure to assess kinesthetic perception (the sensation of position, movement, tension, etc. of body parts, perceived through nerve-end organs in muscles, spindles, tendons, and joints). To provide evidence of your instrument's **construct validity**, you might set up an experiment in which the instrument is administered to two groups of subjects. One group consists of people believed to possess a high degree of kinesthetic perception (e.g., dancers, springboard divers, and gymnasts), and the other group is made up of people who have had no experience in such activities. If the athletic group's mean score on the kinesthetic perception instrument is significantly higher than that of the other group, this would provide construct validity evidence for the instrument. With construct validity, you are essentially "constructing a case" for the validity of the instrument, test, or protocol based on empirical and statistical evidence.

Research-Related Validity: Convergent and Discriminant Validity

Both of these types of validity evidence rely on the use of correlation, but they do so in opposite ways. **Convergent validity** evidence exists when we find high correlations between the results of a measurement process and other measurements, where a positive relationship is expected. **Discriminant validity** evidence, on the other hand, exists when we find low correlations between the results of measurement processes that are not expected to assess the same attributes. For example, suppose we find moderate to high correlations among several different upper-body *muscular strength measures*. Similarly, we find moderate to high correlations among different measures of upper-body *muscular endurance measures*. These findings provide convergent validity evidence for the various measures of muscular strength and muscular endurance. However, the correlations among measures of muscular strength and measures of muscular endurance are found to be low to moderate. This finding provides discriminant validity evidence for the measures, because muscular strength and muscular endurance are two different constructs. Note that both convergent and discriminant validity are often used as evidence of construct validity.

V.2.2 **Measurement Validity**

Measurement Reliability

Reliability is a very important concept in statistical analysis. If you are going to gather data and make decisions on its basis, you must have confidence in the data you collect. This is where reliability (and validity) comes into play. Reliability is essentially *consistency of measurement*. That is, a measure is reliable if, upon repeated observations, the same value is obtained, observed, or recorded. This might not be the true value, but it is significant that you have, at least, observed the same measure on multiple occasions.

A variety of interpretations (and statistics) result in estimates of measurement reliability. These include:

Stability Reliability

Are measures consistent across time? For example, are self-reported physical activity measures consistent from month to month?

Equivalence Reliability

Do two different types of measures give the same results? For example, do forms A and B of a written examination result in consistent scores?

Internal Consistency

Are measures consistent within themselves? For example, on a scale assessing attitude toward physical activity, do the individual items all contribute to the total scale score?

Measurement Reliability

Measurement Objectivity

Objectivity is a special case of reliability. Objectivity is inter-rater reliability. That is, when two raters (or observers) perceive the same thing, if they record the same value for this observation, then the measure reflects objectivity. Think of a written test that you might take. A multiple-choice test is objective in nature. It generally reflects objectivity, in that two raters will nearly always obtain the same score for a given answer sheet. This is different from an essay test, which is much more subjective in nature and, thus, reflects less objectivity. Consider researchers who observe physical activity behaviors in a park setting. Multiple observers might be used, and it is important that they agree on their interpretations of the physical activities they record (for instance, what constitutes hiking versus walking). Two different observers seeing the identical activity should record it in the same way.

Note that you can obtain evidence for the reliability of a measurement only if the score is obtained on at least two occasions. In general, the reliability of measurement increases when something is measured a greater number of times and when the number of people who are observing is greater (assuming that the observers are not biased).

Reliability is often reported as $r_{xx'}$, indicating that you are reporting the reliability (r) for a measurement taken on two occasions (x and x'). Reliability ranges from 0 to 1, with 1 signifying perfect reliability and 0 indicating no reliability. Depending on the situation and the type of measurement obtained, researchers generally hope to achieve reliabilities of at least .70 and preferably .80 or higher. For some measures, even higher reliabilities are necessary (e.g., $\geq .90$). However, no measurement process is perfectly reliable. There is always some amount of error associated with any score. The amount of error is reflected in a statistic called the **standard error of measurement (SEM)**. The SEM reflects the degree to which an observed score will vary as the result of errors of measurement. In your research, you will want to take steps to increase measurement reliability and decrease measurement error.

To make sound decisions based on data that you collect, you must verify that the data are reliable and valid. With inaccurate data, any decisions based on those data will be questionable. One of the first things that you should do before conducting any statistical analysis is to verify the reliability and validity of the measures that you use.

The Relationship Between Measurement Validity and Reliability

Researchers desire to use measurement instruments that yield scores at the highest level of measurement possible and that are high in validity and reliability. It is possible for a measurement instrument to be highly reliable but low in validity. For example, if a researcher had a subject repeatedly step on a scale to determine her weight, with most scales, the results would be very consistent and, thus, reliable. If, however, the researcher attempted to use this same process to measure the subject's percent body fat, its validity would be questionable. This would be a case of obtaining a very consistent measurement of an incorrect attribute.

It is also possible for a measurement instrument to be low in both validity and reliability. In this instance, you would obtain inconsistent measurements of an incorrect attribute. Such measurements are virtually worthless. Applying statistical tests to such data do not improve them, and any resulting conclusions would be useless.

It is not possible for a measurement instrument to be low in reliability and high in validity. If the instrument is high in validity, this means that it measures something consistently. It would be impossible actually to measure some attribute (to have validity) with an instrument that produced inconsistent scores. Figure 2.2 illustrates the relationship between validity and reliability.

FIGURE 2.2 The relationship between validity and reliability.

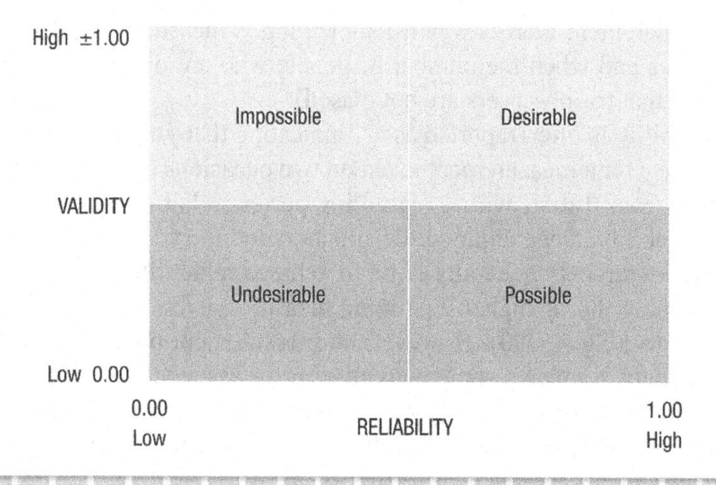

TYPES OF STATISTICS

Statistics may be descriptive, correlational, or inferential. In some sense, these terms are interchangeable, and they certainly overlap and are related to one another. It is often necessary to consider all three types of statistics in any decision-making process.

Descriptive Statistics

Descriptive statistics are measurements obtained on a sample that are used to describe the data set. Descriptive statistics indicate how alike or different the various measures are. If you plotted the measures on a line graph, are the data points all clustered together, or do they have a great deal of variability? How would you describe the shape of a bar graph of the data set—is it bell-shaped, or does it skew in one direction or another? Descriptive statistics answer these questions. There are many descriptive statistics that help you describe and interpret the data you have collected. We present a large selection of these in Chapter 3. As one example, notice how in Figure 2.3(a) the data are relatively bell-shaped, but in Figure 2.3(b) the data are skewed toward the right.

Correlational Statistics

Correlational statistics are often more interesting than simple descriptive statistics. Correlational statistics help you describe the *relationship* between two or more variables. For example, how you dress each day for class is a

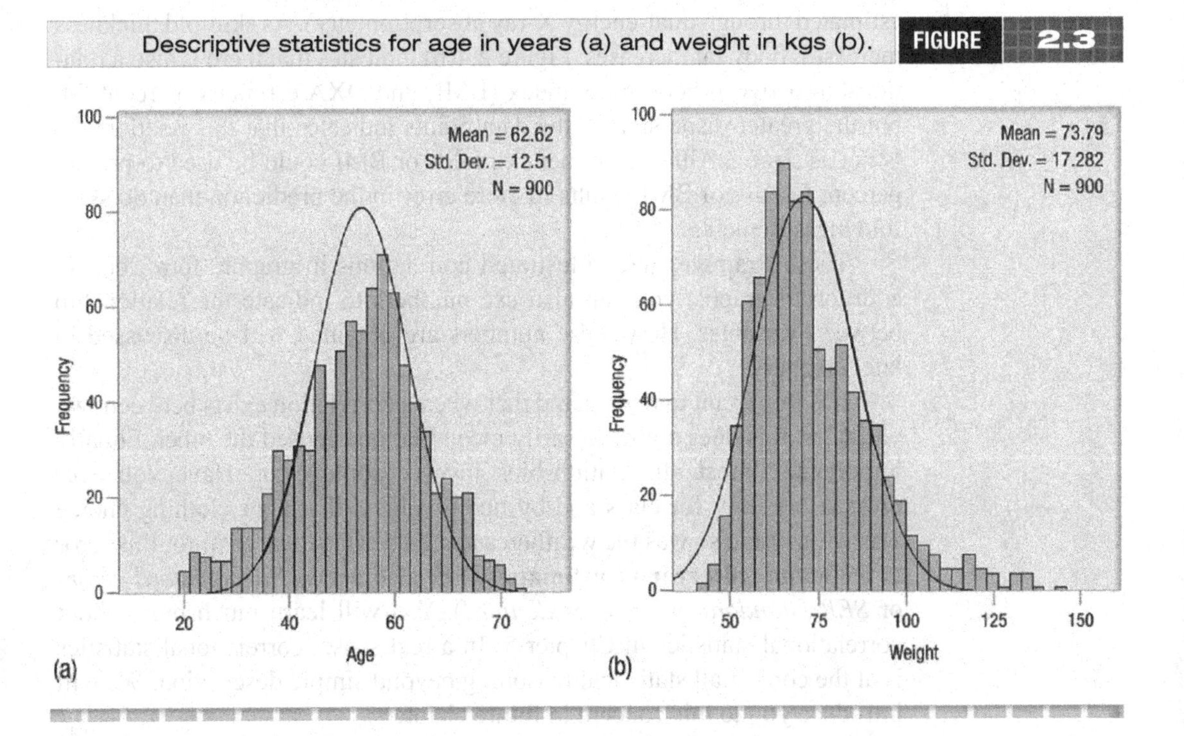

Descriptive statistics for age in years (a) and weight in kgs (b). FIGURE 2.3

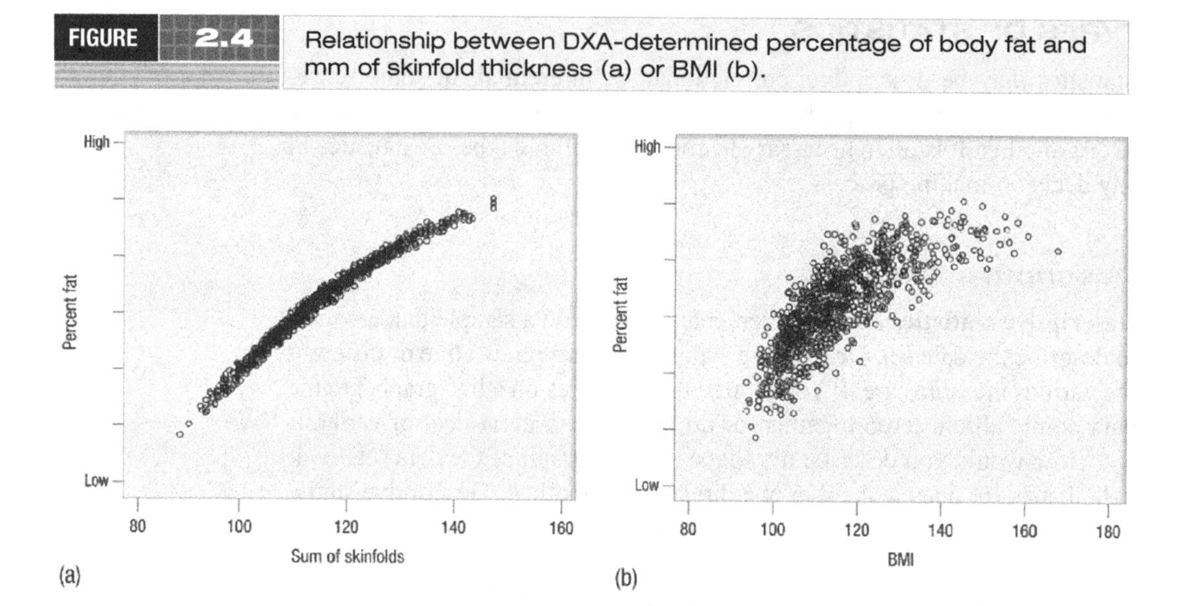

FIGURE 2.4 Relationship between DXA-determined percentage of body fat and mm of skinfold thickness (a) or BMI (b).

function of (that is, can be predicted from) the expected weather. There is a correlation between your clothing choices and the weather forecast. In a real sense, correlational statistics can also be viewed as descriptive in nature, because you can also illustrate the relationship with a figure or a graph. Consider the graphs in Figure 2.4(a) and (b). In Figure 2.4(a), the close concentration of the data points on the graph illustrates a strong relationship between skinfold thickness and DXA estimated percent fat (body fat estimated through dual-energy X-ray absorptiometry). As skinfold thickness increases, body fat increases. Figure 2.4(b) indicates that there is also a relationship between body mass index (BMI) and DXA estimated percent fat, but the greater dispersion of the data points indicates that this relationship is not as strong. Although either skinfolds or BMI could be used to predict percent fat, use of BMI results in more error in the prediction than do skinfold measurements.

These examples have illustrated correlations in graphic form, but, in addition to graphs, you can also use numbers to indicate the relationship between variables. How these numbers are obtained will be discussed in later chapters.

It is important to understand that when a correlation exists between two variables, this does not necessarily mean that one caused the other. Equally important, almost all relationships involve some error. Have you ever dressed one way for class and by noon realized that your clothing choice was wrong (and so was the weathercaster)? The statistical term for this error is the **standard error of estimate** (SEE, also called *SE [standard error]* or *SEP [standard error of prediction]*). You will learn much more about correlational statistics in Chapter 5. In a real sense, correlational statistics is at the core of all statistical reasoning beyond simple description. We will introduce you to this concept in future chapters.

Inferential Statistics

Inferential statistics are essential to scientific reasoning. Inference and its associated steps will be presented in nearly every subsequent chapter in this text. The concept of inference pertains to the ability to infer from a sample to the population from which it was taken. Consider political polling. Polling is conducted on a limited sample (often fewer than 1,000 people) and, as a result, inference is made to the population. Inference is what is accomplished through the Nielsen television ratings. A relatively small number of households is surveyed, and the researchers make estimates about how many people are watching a particular television program. On the basis of these estimates, the networks decide how much to charge advertisers for commercials. Super Bowl commercials cost a great deal because the television ratings suggest that millions of people watch the game.

Important considerations for inference are the random selection of subjects that are observed and their representativeness of the population from which they were obtained. As with prediction, inference always involves some amount of error. Researchers want to control the error and report it to readers. For example, when the results of political polling are presented, the margin of error is frequently specified. What is called the margin of error is technically the standard error of estimate (SEE) discussed above, which you will learn much about in future chapters.

V.2.4 **Categories of Statistics: Descriptive, Correlational, and Inferential**

USE OF SPSS AND R

Most of the data manipulation and analyses that you will do as you study this text will be done by computer. To present examples, we will be using Statistical Package for the Social Sciences (SPSS) and R (RStudio), very powerful statistical packages that are available worldwide. We will illustrate data formats and procedures with screen captures throughout the text. Beginning with Chapter 3, computer activities at the end of each chapter will provide opportunities for you to use the various procedures presented in each chapter. You could, of course, use other statistical packages (such as SAS) or even use MS Excel for some of the analyses. Following the Review Questions below, we provide an introduction to SPSS and instructions for downloading R and RStudio, and this chapter ends with an activity related to its content. We encourage you to read the introduction and complete the end-of-chapter activity before moving to Chapter 3, where you will begin to use SPSS or RStudio frequently.

One caveat: occasionally, the nomenclature used by SPSS and RStudio differs slightly from ours in minor ways such as capitalization and use of italics (e.g., T-test vs. *t*-test and B coefficient vs. *b* coefficient).

SUMMARY

As we begin our introduction to statistics, measurement accuracy is among our first concerns. The level of the measurements taken and the validity and reliability of measures will influence many decisions that researchers make.

Once the data are shown to be trustworthy (that is, they are reliable and valid), the researcher then considers the types of statistical procedures to use. Statistics are descriptive, correlational, or inferential, and these categories often overlap. The remaining chapters in this text focus on these types of statistics. This chapter also introduced you to SPSS and RStudio, powerful statistics packages that you will use throughout the remainder of the text.

Review Questions and Practice Problems

PART A (ANSWERS PROVIDED IN APPENDIX C)

1. Which one of the following attributes can be measured most precisely?
   A. Reaction time
   B. Self-confidence
   C. Physical fitness
   D. Muscle soreness

2. John has compared his number correct (25) with Susan's 50 correct answers on a written examination. What is the highest scale of measurement involved in this comparison?
   A. Nominal
   B. Ordinal
   C. Interval
   D. Ratio

3. What kind of evidence is most indicative of a *reliable* measuring instrument?
   A. That it actually measures what it is designed to measure
   B. That it measures whatever it does measure consistently
   C. That the scores it determines are the same no matter who administers it
   D. That the scores are relevant

4. Which of the following is a special condition (i.e., type) of measurement reliability?
   A. Efficiency
   B. Objectivity
   C. Relevance
   D. Validity

5. What is necessary for a measure to be valid?
   A. It must be measured on at least an interval level.
   B. It must be based on a large amount of data.
   C. It must contain little measurement error.
   D. It must measure what it is supposed to measure.

6. What is the possible range for a reliability coefficient?
   A. 0 to +1.00
   B. –1.00 to +1.00
   C. –1.00 to 0
   D. > .80
   E. It depends on test length.

PART B

7. Which level of measurement provides the *least* amount of information?
   A. Nominal
   B. Ordinal
   C. Interval
   D. Ratio

8. What type of research has as its main purpose the establishment of baselines?
   A. Descriptive
   B. Correlational
   C. Inferential
   D. Experimental

9. In human performance studies, what value of the reliability coefficient is usually considered the lower limit for a test to be considered reliable?
   A. .70
   B. .75
   C. .80
   D. .85

10. We sometimes rank measurements into categories of nominal, ordinal, interval, and ratio based on the amount of information they provide. What level of measurement does this categorization itself represent?
    A. Nominal
    B. Ordinal
    C. Interval
    D. Ratio

11. What type of measurement validity is normally assessed qualitatively?
    A. Content validity
    B. Concurrent validity
    C. Predictive validity
    D. Convergent validity

12. Which of the following is *not* considered research-related validity?
    A. Construct validity
    B. Concurrent validity
    C. Convergent validity
    D. Discriminant validity

COMPUTER ACTIVITIES

Learning the Basics

Using SPSS

Welcome to the world of Statistical Package for the Social Sciences (SPSS) programming. To augment your learning of statistical concepts in this text, we will provide SPSS computer activities that will allow you to apply your new knowledge. The following introduction is intended to familiarize you with the use of this software package, which combines spreadsheets, statistics, and graphic functions. You will need this application to analyze your data from various exercises and to complete activities. SPSS can be a little tricky in the beginning, but as you come to understand it, your assignments will go smoothly. Taking time to learn it now will save you hours in the long run.

Entering Data

To open the program, click on the SPSS icon on the desktop. This will open a new spreadsheet in SPSS. Exactly how SPSS will open on your computer may vary. Check with your instructor.

Move your cursor to the bottom of the spreadsheet and click on **Variable View**. Type "GENDER" (here, and in the following instructions, do not type in the quotation marks—type just what is inside of them. Also note that we will use All CAPS to name variables in computer files throughout the remainder of the textbook. This is true whether you are accessing Excel, csv, SPSS, or RStudio files) in the first box under the Name column to identify the first variable to be entered. Gender is measured nominally, so we will be using M and F (letters, rather than numbers) to identify gender. Thus, it is necessary to change the type of variable being entered from Numeric to String. To do this, click on the first box in the Type column. The word "Numeric" will appear. Click on the little box next to the word "Numeric" and a **Variable Type** screen will appear. Select **String** and click OK.

Next, move to the column called Label and click on the first box. Here you can label your variable. Type in "Gender of subject". Now go to the column called Measure and make sure it reads "Nominal". If it doesn't, click on the first box, and a pull-down menu will appear. Open the pull-down menu and click **Nominal**. Next, move the cursor to the bottom of the spreadsheet and click on **Data View**. You will now begin to enter the data for your first variable. Type M in the first box. Move to the next row to enter your second data point, and type F. Continue in this manner, entering the gender of the remaining eight subjects as M, F, F, F, M, M, M, F.

Go back to the **Variable View** screen. Click on the box under the Gender box and name your next variable "BODYWT". (*Note:* In SPSS, variable names cannot contain spaces or special characters and must begin with a letter.) In the Type column, the word "Numeric" should appear, which is correct for this variable. Go to the Label column to enter the full identification of this variable as "Body Weight, lb". Check to be sure Scale is listed in the Measure column. (This indicates that body weight is measured at either an interval or ratio level.) Now, go back to the **Data View** screen to enter the data for the body weight variable. Go to the data entry boxes and insert, in this order, the following body weights: 170, 120, 140, 129, 100, 107, 190, 149, 116, 90.

Click on **Variable View** to move to the next input column, and title it "HEIGHT". Be sure it is numeric, and label it "Height, in". Confirm that the measure is Scale. Go to **Data View** and enter the following heights: 68, 60, 64, 62, 58, 59, 70, 65, 65, 56. Notice that each row represents a different subject and each column represents a different variable.

Calculating New Variables

Suppose you wish to express your body weight data in kilograms rather than pounds. To do this, you have to divide each data point in the second column by 2.2. Let's have the computer do this for us. Go up to the main toolbar, click on **Transform**, and select **Compute Variable**. This will open a **Compute Variable** screen. In the Target Variable box, type "WEIGHTKG" and click on the blank box under the phrase "Numeric Expression". To enter the formula, double-click on **Body Weight, lb**. This will insert BODYWT in the formula box. Move the cursor to the end of BODYWT and type "/2.2" (without the quote signs). Alternatively, you can also use your mouse and cursor to select these values from the calculator that appears on the screen. Now, click OK and confirm that your new variable has been inserted in the WEIGHTKG column. If you have done this correctly, you should see 77.27 in the first box, 54.55 in the second box, and so forth.

Saving Your Spreadsheet

To save your spreadsheet now or at any time, select **Save As. . .** from the **File** menu on the top menu bar. Give your spreadsheet a distinctive name. As in any other application, you must be sure to save your spreadsheet to some location. You can save it to the desktop temporarily (if placed on the desktop, it might be erased when you log out), or you can save it to your flash drive or a cloud service, for example. When you are certain that you are saving to wherever it is you want, click **Save**.

Creating Bar Graphs

Suppose you are interested in examining the relationship between body weight and gender. That is, do males and females differ in their body weights? Similarly, we might be interested in the difference in height between males and females. To analyze weight and height as a function of gender, click on **Graphs** on the menu bar, select **Legacy Dialog**, and then select **Bar**. Now, select **Clustered** and **Summaries of separate variables**, and click on **Define**. A window entitled **Define Clustered Bar** will appear. Because we are interested in weight and height, we want to assign weight and height as variables to the **Bars Represent** box on the right side of the window. You assign these variables by highlighting them in the left side of the box at the left and using the arrow to move them into the box on the right. Move Body Weight, lb into the box. When you transfer height and weight to the box, they will appear as MEAN(BODYWT) and MEAN(HEIGHT). Next, highlight GENDER and move it into the **Category Axis** window by using the arrow. There are many other options to choose from, but we will do only this much for now. Click OK. This will result in an untitled view of a bar chart displaying the mean heights and weights of the male and female subjects.

To change the look of the bar graph, you can alter its shape, configuration, and other features. For example, to change the *Y*-axis, double-click in the middle of the graph. A new window labeled **Chart Editor** will appear. Double-click on the numbers of the *Y*-axis and make any appropriate changes to the *Y*-axis in the box that pops up. For example, change the number of major increments by selecting **Scale** on the top menu of the **Properties** box and changing 50 in the **Major Increment** box to 25. Experiment with other possible modifications.

Modifying and Transferring Figures to a Microsoft Word Document

After you have created a graph, you can paste it into a word processing document, such as one you might prepare for a report. Go back and create the original bar graph, if you have not saved it to this point. Assume that you wish to paste the bar chart into a report but also wish to remove the text at the top of the figure and add a title more to your liking before you paste it into your document.

First, in SPSS, create a box around your graph by clicking anywhere on the graph. In the **Edit** menu on the SPSS toolbar, select **Copy**. Look at the **Copy Special** option. **Copy Special** allows you to copy and then paste in a variety of formats (e.g., jpg, bmp, eps, tif, png). Choose the format that best meets your needs. Now open a document file. Place the cursor on the document page where you want the graph to appear. Now, select **Paste** (or **Paste Special**). The graph will appear on your page. Try different copy and paste formats to see what occurs, and experiment with ways of editing your figures. To add an appropriate title, simply type it above the graph. You will need to move the graph down on the page, if you put it at the top originally.

Using R and RStudio

Installing R and RStudio

R is a powerful statistical computing and graphics package. The **R** software environment is free and can be run on any common operating system platform. Details about downloading and installing **R** can be found at the *R Project for Statistical Computing* website: www.r-project.org

RStudio is an integrated development environment (IDE) for **R**. **RStudio** provides an optimal environment for beginners and advanced users of **R**.

Steps for Installing R:

1. Download the base distribution of **R** for your operating system from: www.r-project.org.

   On the left side of the page under **Download,** click on the link labeled CRAN.

   CRAN stands for the **C**omprehensive **R** **A**rchive **N**etwork.

   Choose a CRAN Mirror for download.

   For example, under **USA**, choose one of the available links.

   Under the **Download and Install R** section, click the link for your operating system and follow the instructions.

2. To install **R**, simply double-click the downloaded file and follow the instructions.

Steps for Installing RStudio:

1. Download **RStudio** for your operating system from: www.rstudio.com.

   Under the **Products** menu, click on **RStudio.**

   Click on **Desktop.**

Click on DOWNLOAD RSTUDIO DESKTOP button under the **Open Source Edition.**

Click on your operating system under **Installers for Supported Platforms.**

2. To install **RStudio**, double-click the downloaded file and follow the instructions.

Now, you should be able to run ***R*** *through* ***RStudio (which we recommend you do throughout the remainder of the textbook if you are using RStudio rather than SPSS)***

Using RStudio to complete the Computer Activities presented above.

Open RStudio.

Install any necessary R packages.

When we installed R, we installed "base" R which includes the vast majority of functions and tools that we will need for our analyses. However, there are situations where we will need to install an additional package to carry out a specific function.

For Chapter 2, we do not require any additional packages.

Set up your RStudio workspace.

In the RStudio menu bar, click on **File → New File → R Script**

On the first line of the R Script (upper left pane of RStudio), type

#Chapter 2

When you have done that, you should see the following in your new R Script

1 #Chapter 2

[Note that any line with a preceding "#" is considered a "comment" and is ignored by R.]

1. Create a data frame in R by directly typing data into RStudio.

In your R Script that you created, type the following to create a character vector of gender codes for ten subjects:

- GENDER <- c("M","F","M","F","F","F","M","M","M","F")

[Note that in the typed text above, we combine (c) the character gender codes into a vector and assign this vector to the object (GENDER) using the symbol (<-).]

When you've finished typing, highlight the entire text that you've typed and click on the **Run** button.

This will create the object GENDER containing the vector of gender codes as can be seen in the upper right pane of RStudio under the "Environment" tab.

2. Next create two numerical vectors for the 10 subjects that contain body weight (WEIGHT_POUNDS) and height (HEIGHT_INCHES):
   - WEIGHT_POUNDS <- c(170,120,140,129,100,107,190,149,116,90)
   - HEIGHT_INCHES <- c(68,60,64,62,58,59,70,65,65,56)
3. Next create a data frame for these 10 subjects that incorporates gender, body weight and height using the data.frame function as follows:
   - mydata <- data.frame(GENDER,WEIGHT_POUNDS,HEIGHT_INCHES)

   Be sure to highlight the text you've typed and click on **Run** to execute. You should now see the three objects listed under the "Environment" tab in RStudio as well as the combined data frame (mydata).

Calculating New Variables

In your R Script, type the following to create a new variable that displays the weight in kilograms, instead of pounds.

- mydata$WEIGHT_KILOS <- mydata$WEIGHT_POUNDS/2.2

Highlight what you've just typed and click on **Run**

[Note that in the typed text above, the "$" directs R to a particular variable within a data frame. In this case, we are creating a new variable that will reflect body weight in kilograms (WEIGHT_KILOS) within the data frame (mydata) which is assigned (<-) a value that is calculated by dividing (/) body weight in pounds (WEIGHT_POUNDS) by 2.2.]

Click on mydata under the "Environment" tab in the upper right pane of RStudio to view the full data set that now includes four variables from 10 subjects (observations).

Saving the Data Frame

To save your data frame for future use, type the following into your R Script:

- save(mydata, file="mydata.RData")

Highlight what you've just typed and click **Run**

Creating Box Plot

To create a simple boxplot to compare body weight of males and females, type the following:

- boxplot(WEIGHT_KILOS~GENDER,
- data=mydata,
- main="Boxplot",

- xlab="Gender",
- ylab="Body Weight (kg)")

Highlight what you've just typed and hit **Run**

[Note that in the typed text above, we are generating a boxplot of body weight in kilograms (WEIGHT_KILOS) as a function of gender (GENDER) and then we provide a title (main) and label the x and y axes (xlab and ylab).]

The boxplot will be displayed in the lower right pane of RStudio under the "Plots" tab.

Exporting a Graph

To export a graph to be used in another application (i.e. word processor, presentation software, etc.), under the "Plots" tab (lower right pane), click on "Export" where you can then select to save the plot/graph as an image or as a PDF. Once you've created a digital file of the plot/graph you can use it to insert into other applications.

Save Your Work

In the RStudio menu bar, click on **File** → **Save** to save your R Script that you created for this chapter. You can name it whatever you like ("Chapter2") and save it anywhere on your computer that is convenient. RStudio automatically adds the file extension "R" such that the full name saved to your computer would be "Chapter2.R".

When you exit out of RStudio, it will ask if you'd like to save the workspace image. If you elect to save your workspace before closing, when you open RStudio again, you will see your workspace saved as it was right before you exited from RStudio. It is generally good practice to save the work you've done, even when you can recreate all the steps performed through the R Script that you saved.

Review

You should be able to perform the following tasks comfortably now:

- enter data (either numerical or word-based data)
- locate descriptive statistics for each set of data
- create new variables
- create a **histogram** (bar graph)
- change the graph in various ways
- transfer graphs into a Word document

This concludes the tutorial for using the SPSS and RStudio statistics and graphics packages. You may save your work to your own flash drive as you would in any application, by selecting **Save** in the **File** menu. You might wish to try other analyses or graphical functions of these statistical packages. They can compute literally dozens of parametric and nonparametric statistics. Browse when you have the time and the inclination. You can become more proficient with SPSS and RStudio as you practice, practice, practice. Additional resources include, *Discovering Statistics Using IBM SPSS Statistics*, (Field, 2013) *Getting Started with R: An Introduction for Biologist*, (Beckerman, Childs, & Petchey, 2017), and *Statistical Analysis with R for Dummies* (Schummer, 2017).

Important Note About Entering Data into SPSS and RStudio

V.2.5
SPSS Introduction

V.2.5R
RStudio Introduction

In the above Computer Activity, you were given explicit directions for how to enter the data into both the SPSS and RStudio packages. You may continue to enter data in this manner throughout the remainder of the text but sometimes it may be advantageous to import a data file that has already been created elsewhere. Instructions for how to do this will be given in the Computer Activities in Chapter 3.

REFERENCES

Beckerman, A. P., Childs, D. Z., & Petchey, O. L. (2017). *Getting started with R: An introduction for biologists* (2nd ed.). Oxford: Oxford University Press.

Borg, G. A. (1982). Psychophysical bases of perceived exertion. *Medicine and Science in Sports and Exercise*, *14*, 377–381.

Field, A. (2013). *Discovering statistics using IBM SPSS statistics*. London: Sage Publications.

Jurca, R., Jackson, A. S., LaMonte, M. J., Morrow, J. R. Jr., Blair, S. N., Wareham, N. J., . . . Laukkanen, R. (2005). Assessing cardiorespiratory fitness without performing exercise testing. *American Journal of Preventive Medicine*, *29*(3), 185–193.

Kestilä, L., Mäki-Opas, T., Kunst, A. E., Borodulin, K., Rahkonen, O., & Prättälä, R. (2015). Childhood adversities and socioeconomic position as predictors of leisure-time physical inactivity in early adulthood. *Journal of Physical Activity and Health*, *12*, 193–199.

Schummer, J. (2017). *Statistical Analysis with R for Dummies*. Hoboken, NJ: Wiley & Sons.

3 Descriptive Statistics

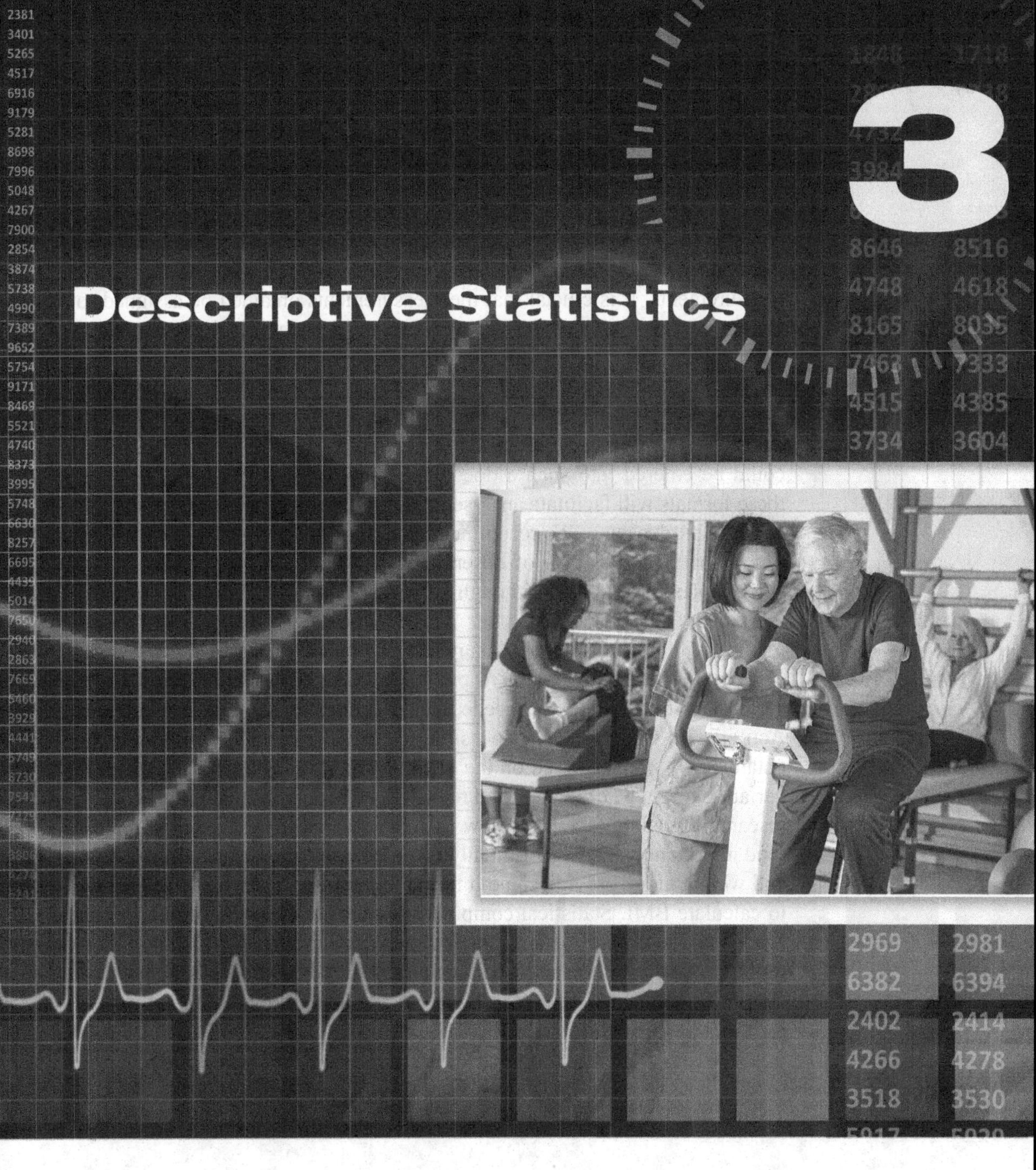

INTRODUCTION

Once you have collected a set of data, you will need to examine it and describe its characteristics. This is necessary to determine which statistical tests are appropriate for later use. In this chapter, we examine three aspects of describing data. These are: (1) the frequency with which various scores occur and what a graph of the score distribution looks like, (2) where the center of the distribution lies, and (3) how widely the scores in the distribution are scattered.

RECORDING AND DESCRIBING DATA

Before examining several statistics used to describe data, we need to mention a few issues regarding the collection and organization of the data. When collecting data from an experiment, it is very important that you consider, in advance, the process to be used to prepare an organized plan to record the measurements. It is crucial that the measurements obtained be assigned to the correct subjects, that all relevant conditions under which the measurements are obtained be recorded, and that any circumstances that might affect the accuracy of the data be noted.

Equally important is the actual organizational structure of the recorded data. Think of a spreadsheet such as that presented in Figure 3.1. Your data could be entered into such a document in a variety of formats. Some of these formats will facilitate analyses and decisions, while other formats can be quite burdensome. We recommend that you think of this spreadsheet as a data matrix, with rows of observations (usually subjects) and columns of variables (measures). Although other formats are certainly acceptable, thinking of and presenting your data consistently in this recommended method will help you conduct many statistical tests effectively.

In most statistical software programs, these columns of data can be sorted and manipulated as needed. As we pointed out in the Computer Activities in Chapter 2, it is possible to have the computer program calculate new variables based on the original raw data. These calculated new variables might be of interest and related to the questions that motivated the research. Often, in fact, these new calculated variables are of more interest (and importance) than the originally recorded raw data. For example, height and weight alone provide some information, but together they can be used to calculate BMI. Statistical computer programs, such as SPSS and R, can produce various graphs from data arranged in this spreadsheet fashion. Figure 3.2 is an SPSS Data View window of displayed data. Similarly, in R it is common practice to construct a table comparable to Figure 3.2 and import the data with RStudio.

FIGURE 3.1 A generic spreadsheet to organize data.

SUBJECT	TYPE	AGE	VARIABLE1	VARIABLE2	VARIABLE3	...	VARIABLEP
#1							
#2							
#3							
#4							
...							
N							

SPSS Data View. **FIGURE 3.2**

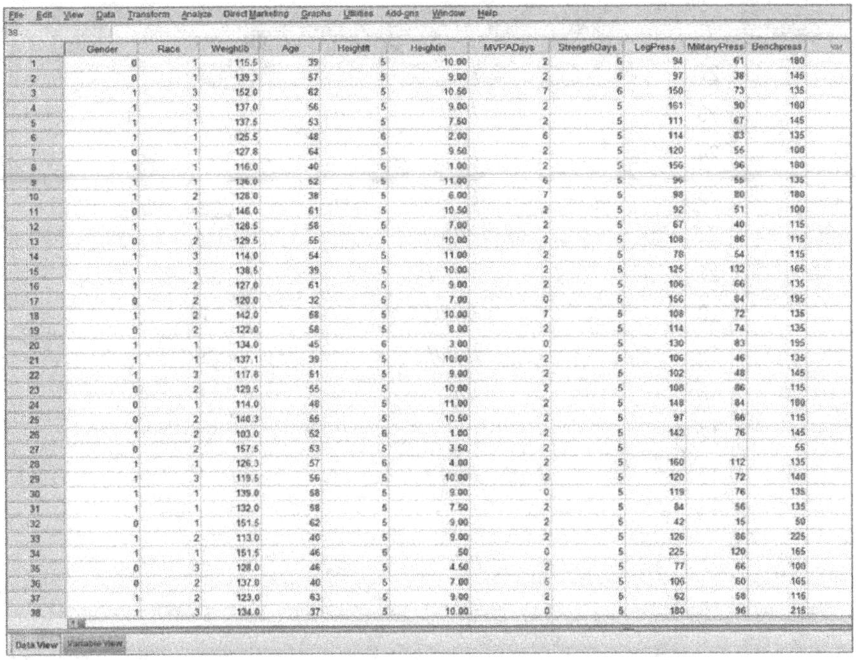

File Edit View Data Transform Analyze Direct Marketing Graphs Utilities Add-ons Window Help

38 :

	Gender	Race	Weightlb	Age	Heightft	Heightin	MVPADays	StrengthDays	LegPress	MilitaryPress	Benchpress	var
1	0	1	115.5	39	5	10.00	2	6	94	61	180	
2	0	1	139.3	57	5	9.00	2	6	97	38	145	
3	1	3	152.0	62	5	10.50	7	6	150	73	135	
4	1	3	137.0	56	5	9.00	2	5	161	90	180	
5	1	1	137.5	53	5	7.50	2	5	111	67	145	
6	1	1	125.5	48	6	2.00	6	5	114	83	135	
7	0	1	127.8	64	5	9.50	2	5	120	56	100	
8	1	1	116.0	40	6	1.00	2	5	156	96	180	
9	1	1	136.0	62	5	11.00	6	5	96	55	135	
10	1	2	128.0	38	5	6.00	7	5	98	80	180	
11	0	1	146.0	61	5	10.50	2	5	92	51	100	
12	1	1	128.5	58	5	7.00	2	5	67	40	115	
13	0	2	129.5	55	5	10.00	2	5	108	86	115	
14	1	3	114.0	54	5	11.00	2	5	78	54	115	
15	1	3	138.5	39	5	10.00	2	5	125	132	165	
16	1	2	127.0	61	5	9.00	2	5	106	66	135	
17	0	2	120.0	32	5	7.00	0	5	156	84	195	
18	1	2	142.0	68	5	10.00	7	5	108	72	135	
19	0	2	122.0	58	5	8.00	2	5	114	74	135	
20	1	1	134.0	45	6	3.00	0	5	130	83	195	
21	1	1	137.1	39	5	10.00	2	5	106	46	135	
22	1	3	117.8	61	5	9.00	2	5	102	48	145	
23	0	2	129.5	55	5	10.00	2	5	108	86	115	
24	0	1	114.0	48	5	11.00	2	5	148	84	180	
25	0	2	140.3	66	5	10.50	2	5	97	66	115	
26	1	2	103.0	52	6	1.00	2	5	142	76	145	
27	0	2	157.5	53	5	3.50	2	5			55	
28	1	1	126.3	57	6	4.00	2	5	160	112	135	
29	1	3	119.5	56	5	10.00	2	5	120	72	140	
30	1	1	139.0	68	5	9.00	0	5	119	76	135	
31	1	1	132.0	58	5	7.50	2	5	84	56	135	
32	0	1	151.5	62	5	9.00	2	5	42	15	50	
33	1	2	113.0	40	5	9.00	2	5	126	86	225	
34	1	1	151.5	46	6	.50	0	5	225	120	165	
35	0	3	128.0	46	5	4.50	2	5	77	66	100	
36	0	2	137.0	40	5	7.00	5	5	106	60	165	
37	1	2	123.0	63	5	9.00	2	5	62	58	116	
38	1	3	134.0	37	5	10.00	0	5	180	96	215	

Data View Variable View

Visualizing Distributions

Before submitting data for statistical analysis, it is important to examine the shape of the distribution. In the past, this was done through visual inspection of the raw data. Data were arranged in hierarchical order to make it easier to observe the highest and lowest values and to see where scores tended to be the most numerous. With a large data set, this process was cumbersome. If the range (high score minus low score) of scores was small, but the *N* (number of scores) was large, then a simple frequency distribution was constructed to reduce the data to one page for easier inspection. This was a manual tabulation of the frequency occurring at each possible score between the high and low score, using a hash mark to count each occurrence Figure 3.3 illustrates a simple frequency distribution.

If the *N* was large and the range was also large, a grouped frequency distribution was employed. The researcher divided the range into as many intervals as desired and tabulated the scores within each interval. Usually, the number of intervals chosen was between 15 and 25, so that, again, the results would fit on one page. The values resulting from tabulating the scores in each interval produced the grouped frequency distribution. Figure 3.4 gives an example of a grouped frequency distribution (panel a) and the resulting graphic representation (panel b), called a **histogram**. Note that when score values are grouped, information about the exact value of each score is lost.

FIGURE 3.3 Simple frequency distribution.

SCORE	TALLY	FREQUENCY (F)
27	/	1
25	//	2
24	/	1
23	///	3
22	////	4
21	///// //	7
20	///// ///// //	12
19	///// ////	9
18	/////	5
17	/	1
16	/	1
13	/	1
11	//	2
10	/	1

FIGURE 3.4 Grouped frequency distribution (a) and histogram (b).

Grouped Frequency Distribution

INTERVAL	FREQUENCY OF OBSERVATIONS
90–99	7
80–89	18
70–79	35
60–69	51
50–59	60
40–49	42
30–39	28
20–29	15
10–19	9

(a)

Histogram Based on Grouped Frequency Distribution

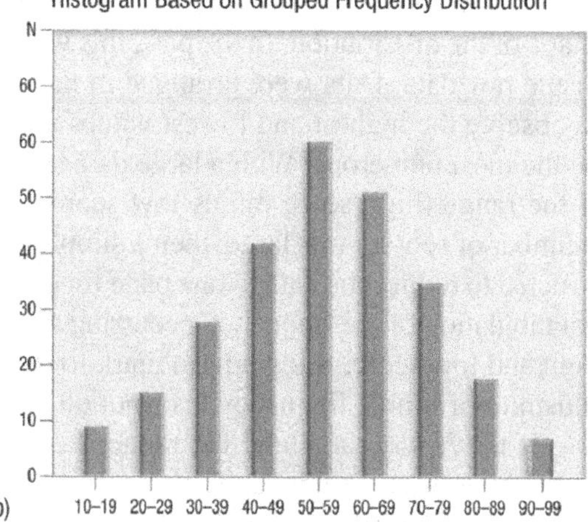

(b)

By observing the shape of a distribution, you can describe the data you have obtained. Imagine a line drawn to connect the tops of the bars in the graph in Figure 3.4b and notice the bell shape of this line. When data form a bell shape, it is said to resemble a “normal” distribution.

When data are organized into frequency distributions, we can use equations to calculate estimates of some of the descriptive statistics presented later in this chapter. However, computer software has made these procedures obsolete, as it is now possible to obtain precise values of these descriptive statistics from the raw data. As shown above, we can also use computer software to construct graphs of these frequency distributions to allow visual examination of the distributions. Various statistics programs (e.g., SPSS, R, SAS, Excel) have done away with much of the tedium involved in examining distributions. Visual examination of the data, however, is still important, allowing us to determine whether the distributions conform to expectations often required for certain statistical tests and to notice any extreme data points (called *outliers*), which might represent measurement errors.

Skewness and Kurtosis

V.3.1 **SPSS Calculating and Examining Skewness and Kurtosis**

V.3.1R **RStudio Calculating and Examining Skewness and Kurtosis**

Statistics users have agreed on words and statistical tools to describe the shapes of distributions. Earlier, we mentioned one example, the normal (bell-shaped) curve. Chapter 4 is devoted to the characteristics of the normal curve because of its critical importance in statistics. Two additional words we use to describe the shape of a distribution are *skewness* and *kurtosis*.

Skewness

Skewness refers to the degree to which a distribution departs from symmetry around its center. Is there a tail on one side or the other? If most of the scores are relatively low, but there are some extremely high scores, the tail will lie to the right, and this distribution is labeled as *positively* skewed. Alternatively, if most of the scores are relatively high, but there a few extremely low scores, the tail will lie to the left, and this distribution is called *negatively* skewed. Skewness is illustrated in Figure 3.5.

Normal (a), positively skewed (b) and negatively skewed (c) distributions. FIGURE 3.5

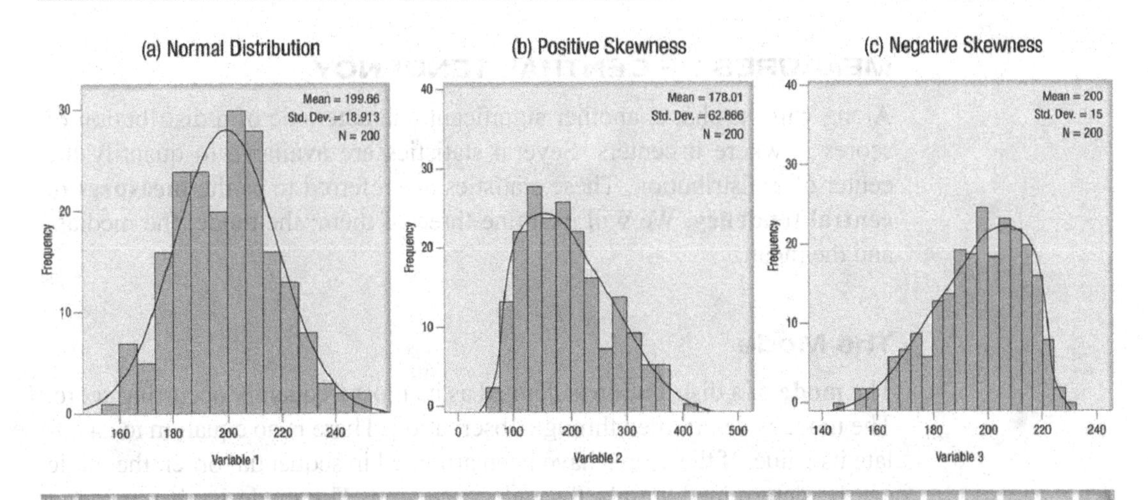

FIGURE 3.6 Mesokurtic (a), leptokurtic (b) and platykurtic (c) distributions.

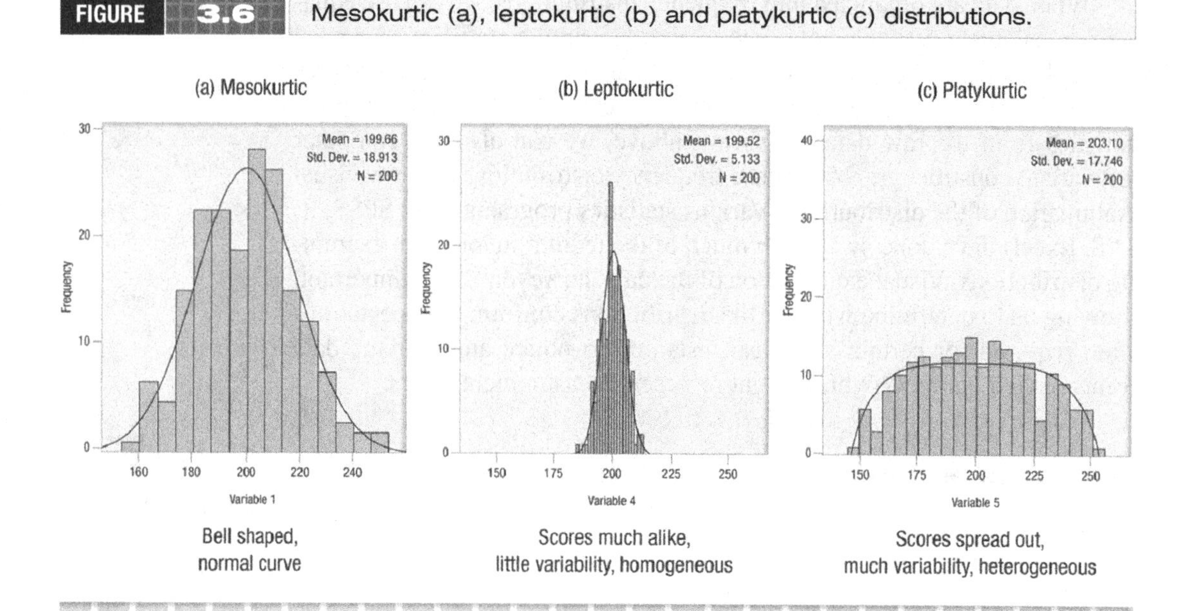

Kurtosis

Kurtosis is another shape characteristic of a distribution. Names for specific types of kurtosis are **mesokurtic, leptokurtic,** and **platykurtic**. The word *mesokurtic* is used to describe the height of a normal curve. Leptokurtic curves are more peaked, with scores that are more homogeneous (less scattered) than in a normal distribution. Platykurtic distributions are flatter, with more observations in the tails of the distribution when compared to a normal distribution. Figure 3.6 illustrates mesokurtic, leptokurtic, and platykurtic curves.

As you will learn in Chapter 4, statistical equations are available to quantify the amount that a distribution differs from the normal distribution in terms of skewness and kurtosis.

MEASURES OF CENTRAL TENDENCY

Along with its shape, another significant characteristic of a distribution of scores is where it centers. Several statistics are available to quantify the center of a distribution. These statistics are referred to as the **measures of central tendency**. We will examine three of them: the mode, the median, and the mean.

The Mode

The **mode** of a distribution is defined as its most frequently occurring score. The mode is determined through observation. There is no equation to calculate its value. If the scores have been arranged in sequential order, the mode can be determined simply through inspection. The mode is also apparent when the data are put into a frequency distribution. In a simple frequency distribution, the mode is the score with the highest frequency; in a grouped frequency distribution, it is the midpoint of the interval having the highest frequency. The mode in Figure 3.3 is 20, and in Figure 3.4(a) the mode is 54.5. The mode is typically near the center of a large and reasonably normally shaped distribution. In distributions with small *N*s, the mode may not be very representative of the center of the distribution. The mode is not a very useful statistic, although it can be interesting. For example, at the time of the printing of this textbook, the modal surname in the United States is Smith, and the modal first names for males and females are James and Mary, respectively. You will sometimes encounter distributions that are multimodal. For example, *bimodal distributions* have two values that are observed more often than other scores. Figure 3.7 illustrates a bimodal distribution.

FIGURE 3.7

Bimodal distribution.

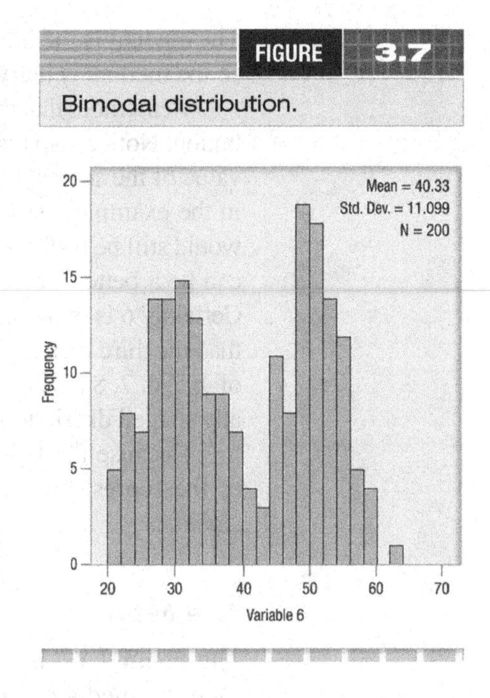

The Median

The **median** of a distribution is defined as the 50th percentile. A **percentile** represents the percentage of observations that fall at or below a given value. Hence, the median is the point in a distribution where 50% of the scores occur above it and 50% occur below it. If the scores are arranged in sequential order, the median can be located simply by counting. The location of the score that is the median can also be found by using the equation

$$\frac{N+1}{2} \qquad \text{Eq. 3.1}$$

where

N = the total number of scores in the distribution.

If there is an odd number of scores in a distribution, the median score is a whole number, but when the number of scores is even, the median is halfway between two scores.

Consider a simple distribution: 4, 5, 6, 7, 8. Here, $N = 5$.

$$\frac{N+1}{2} = \frac{5+1}{2} = \frac{6}{2} = 3$$

Thus, the median is the third score, or 6.

Now consider this distribution: 4, 5, 6, 7, 8, 9.

$$\frac{N+1}{2} = \frac{6+1}{2} = \frac{7}{2} = 3.5$$

The median is the 3.5th score. The third score is 6, and the fourth score is 7, so the median is halfway between them: 6.5.

Remember that the median is a point, not necessarily a score, in the distribution. Notice also that the value of the median is not influenced by the actual value of the scores, but rather by their location. For example, if the last score in the example distribution of 4, 5, 6, 7, 8 was 80 instead of 8, the median would still be 6. Thus, for some distributions with unusual shapes, the median can be a better indicator of the center than other central tendency statistics. Certainly 6 is more representative of the majority of scores in this distribution than the third measure of central tendency, the mean. As we will see, the mean of 4, 5, 6, 7, 80 is 20.4, which is not representative of the center of this unusually shaped distribution.

Because the value of the median is not influenced by the actual value of the scores, but by their location, the median is considered an *ordinal statistic.*

The Mean

The **mean** $(\bar{X})$ is defined as the arithmetic average of a set of scores. Like the median, it is a point (integer or decimal), rather than an actual score. Instead of dividing the data in half, as with the median, the mean is found at what we might call the center of gravity of the data. The mean is probably the most commonly used measure of central tendency, because it is sensitive to the actual value of every score in the distribution. It is a continuous statistic, and, thus, it is used with interval- and ratio-level measurements.

The value of the mean is obtained by adding up all of the scores in a distribution and dividing this sum by the number of scores *(N)* in the distribution. The equation for determining the value of the mean is

$$\bar{X} = \frac{\sigma X_i}{N} \qquad \text{Eq. 3.2}$$

where
$\bar{X}$ = the mean,
X_i = each score in the distribution, and
N = the number of scores in the distribution.

The symbol σ indicates that we sum the variable that follows. For example, the mean of the distribution of 4, 5, 6, 7, 8 is 30/5 = 6. The mean is influenced by the value of every score in the distribution. If even one score in a distribution changes in value, the mean also changes. For the distribution 4, 5, 6, 7, 9, the mean is 6.2.

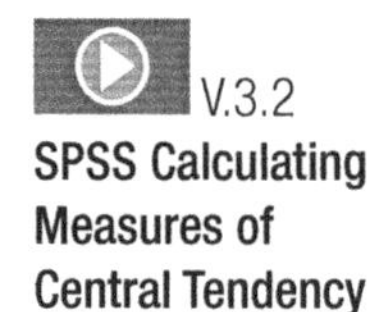
V.3.2
SPSS Calculating Measures of Central Tendency

Comparisons Among the Mode, Median, and Mean

The word *average* can refer to any of these three measures of central tendency. In fact, with a truly normal distribution, the value of all three statistics

TABLE 3.1 Measures of central tendency, their advantages and disadvantages, and when to use them.

MEASURE OF CENTRAL TENDENCY	ADVANTAGES	DISADVANTAGES	USE
Mode	▪ It is easy to determine. ▪ If data are close to being normally distributed, it describes the center of the distribution.	▪ It is an unstable statistic. ▪ It is a terminal statistic (cannot be used in further calculations). ▪ It disregards extreme scores.	▪ With nominal level measurements ▪ When the data are relatively normally distributed ▪ When it is desired to know the score value that occurs most often
Median	▪ It indicates the most typical or most representative score in a distribution. ▪ Its value is not affected by extreme scores.	▪ It is not sensitive to the value of the scores, only to their position.	▪ When the data are ordinal ▪ When the middle score is needed ▪ When the most representative score in the distribution is desired ▪ When the data are skewed due to extreme scores
Mean	▪ It considers all available information about the data. ▪ It can be used in further calculations. ▪ It can be used with interval- and ratio-level measurements.	▪ It requires calculation, as opposed to observation or inspection. ▪ It is not descriptive of the center of a very skewed distribution. ▪ It is not appropriate for nominal- or ordinal-level measurements.	▪ When the values of all the data need to be considered ▪ When the measure of central tendency will be used in further calculations ▪ When the data are interval- or ratio-level measurements and are relatively normally distributed

is identical. However, to avoid ambiguity, it is best to use the name of the statistic being employed rather than the word *average*, as we will demonstrate later. The advantages, disadvantages, and usage of each measure of central tendency are presented in Table 3.1.

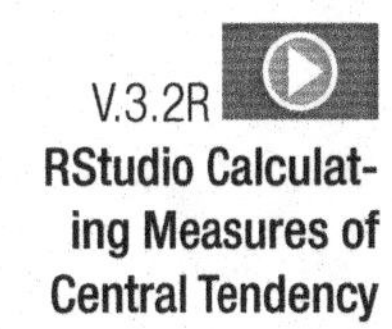

V.3.2R **RStudio Calculating Measures of Central Tendency**

The Danger of Using the Word Average

Seven recent (fictional) exercise science graduates from the University of West Florida reportedly "averaged" more than $1 million per year in salary for their first job upon graduation. Let's examine the salaries at which they were hired. Six were hired as exercise physiologists by health care organizations. Three received a starting salary of $45,000, and the other three received $46,000, $47,000, and $50,000. The seventh graduate signed a one-year professional basketball contract for $7 million. What does the word *average* mean in this situation?

If you choose to interpret *average* as the mode, the answer is $45,000, the most frequently observed salary. If you choose to use the mean to represent the average, the answer is $1,039,714.29. This value is not very representative of most of the salaries. In fact, only one of the seven graduates is earning more than that amount. It would be most appropriate, in this case, to use the median as your choice for the average, as $46,000 is a more typical or representative value of the salaries than the other two "averages". Half of the salaries are above this point and half below; thus, it is the "typical" salary in this distribution.

MEASURES OF VARIABILITY

Knowing the center of a distribution is helpful, but this information is inadequate for describing a distribution completely. For example, two students could both have a 2.0 grade point average, but suppose one student achieved this average by nearly always earning a C in each class taken, while another student received grades ranging from A to F. The difference in these two grade records is in the dispersion, or spread, of the grades. We need a way to describe the variability of the scores in a distribution. We will present four **measures of variability,** or ways to quantify the variability in a distribution of scores.

Range

We have already presented one of the measures of variability, the **range**, mentioned previously in this chapter. The range is usually defined as the high score minus the low score.

$$R_e = H - L$$ Eq. 3.3

where
R_e = the exclusive range,
H = the highest score, and
L = the lowest score.

This is known as the *exclusive range*. It is defined as the distance from the lowest to the highest score. In contrast, the *inclusive range* recognizes both end scores as included in the value of the range. The equation for the inclusive range is

$$R_i = (H - L) + 1$$ Eq. 3.4

where
R_i = the inclusive range,
H = the highest score, and
L = the lowest score.

If the high score and low score in a distribution are 12 and 7, the exclusive range is $12 - 7 = 5$. The inclusive range is $(12 - 7) + 1 = 6$.

Consider the possible scores that could exist in a set of data that ranges from 7 to 12: the scores could be 7, 8, 9, 10, 11, or 12. Notice that six different scores are possible, a number equal to the inclusive range.

The exclusive range is more commonly used than the inclusive range, and, for our purposes, the exclusive range is adequate for describing the variability of a distribution. Obviously, when the value of the range is large, this indicates a more variable distribution than when the range is a small value.

Unfortunately, the range is not a very stable indicator of the variability of a set of scores, because its value is dependent on only two scores in the entire distribution, namely, the lowest and the highest scores. If either of these two scores is extreme, this would create a misperception of the distribution's variability. As a measure of variability, the range is similar to the mode as a measure of central tendency: both are easy to determine but not all that informative.

Semi-Interquartile Range

To overcome the instability issue of the range, we can use the **semi-interquartile range** as an indicator of the variability of a distribution. Defined as half the distance between the 25th and the 75th percentiles, the semi-interquartile range will have a larger value for a more diverse or scattered (heterogeneous) distribution and a smaller value for a more compact (homogeneous) distribution. Its calculation requires determining the scores that represent the 25th and the 75th percentiles, finding the difference between them, and then dividing this value by 2. For example, if the 25th percentile in a distribution is 120 and the 75th percentile is 160, the value of the semi-interquartile range is 20.

If you have constructed a frequency distribution from the data or generated one with a statistical computer package, either can be used to determine the scores representing the 25th and 75th percentiles. These score values can also be found by arranging the data in numerical order, multiplying N by .25, and counting up from the bottom score to the product of this multiplication. The same procedure, multiplying N by .75, gives a way to determine the score representing the 75th percentile.

The reasoning here is that, because they are in the middle of the distribution, the 25th and 75th percentiles are more stable than the highest and lowest scores, and, thus, are more stable indicators of variability than the range. This is true, but the semi-interquartile range still depends on only two scores in the distribution.

Population Variance

A **population** refers to the total possible scores available. This is different than a sample. A sample is drawn from the population. For example, your class could be considered a population, and you could draw a sample from your class. This subset of the population would consist of fewer students than the population.

The **variance**, another statistic that quantifies the variability in a distribution, uses all of the information in the data: every score is involved in determining its value. To find the variance of a population, we determine how far every score in the distribution is from the mean of the distribution. This distance, called a *deviation*, is usually represented by x. To calculate the variance, we square all of the deviations, sum them, and divide this sum by N.

$$\sigma^2 = \frac{\sigma(X_i - \mu)^2}{N} = \frac{\sigma x^2}{N} \qquad \text{Eq. 3.5}$$

where

σ^2 = the variance of the population,
X_i = each score in the population,
μ = the mean of the population,
x^2 = the deviation, and
N = the number of scores in the population.

Can you see that the variance is the average squared distance of the values from the mean of the population? Notice that the mean is subtracted from each score, the result is squared, and then these squares are summed. The resultant sum (called the *sum of squares*) is divided by the number of scores. Thus, the variance is the average of the squared deviations from the mean. This definition may seem a bit wordy to you but understanding it will help you as you study later chapters. Notice also that the Greek symbols σ^2 and μ are used to represent the variance and mean of the population, respectively. The significance of this will be explained shortly.

The value of the variance will be larger for distributions that are more variable (heterogeneous), because the distances between the scores and the mean (the deviations) will be larger than they will be in a more compact (homogeneous) set of scores. Again, stated verbally, the variance is the average of the squared deviations from the mean. Can you see how leptokurtic distributions will have variances that are generally smaller in value than platykurtic distributions? Let's calculate the variance for a very small population of scores, just to see how the equation works. We will calculate the variance for the distribution 1, 2, 3, 4, 5. The mean of this distribution is 3, and $N = 5$.

	X	μ	x	x^2
	1	3	−2	4
	2	3	−1	1
	3	3	0	0
	4	3	1	1
	5	3	2	4
Σ				10

$$\sigma^2 = \frac{\sigma x^2}{N}$$

$$\sigma^2 = \frac{10}{5} = 2$$

Notice that if the distribution had been 1, 1, 2, 2, 3, 3, 4, 4, 5, 5, the sum of the squared deviations would have been 20, but this new distribution isn't any more variable than the original distribution. When we divide 20 by the new N, 10, the variance is still 2.

Population Standard Deviation

Another measure of variability is the **standard deviation** of a distribution. The standard deviation is simply the square root of the variance. For the distribution 1, 2, 3, 4, 5, the standard deviation is $\sqrt{2}$ or approximately 1.4. If you have calculated a variance, you then take the square root of the variance to obtain the standard deviation. The equation for the standard deviation of a population is:

$$\sigma = \sqrt{\frac{\sigma(X_i - \mu)^2}{N}}$$ Eq. 3.6

where

σ = the standard deviation of a population,
X_i = each score in the population,
μ = the mean of the population, and
N = the number of scores in the population.

It is important to understand that the standard deviation reflects a *linear measure* of variability. The variance is a "squared" measure of variability. The concepts of a square for variance and a line for standard deviation should become clearer as we progress. We will be using measures of variability, particularly the standard deviation and variance, throughout many of the remaining chapters in this book.

Calculating Variance and Standard Deviation

Typically, a researcher would use the raw score formula presented here to calculate the population variance:

$$\sigma^2 = \frac{\sigma X_i^2 - \frac{(\sigma X_i)^2}{N}}{N}$$ Eq. 3.7

where

σ^2 = the population variance
σX_i^2 = the sum of the squared scores
$(\sigma X_i)^2$ = the sum of the scores, the quantity squared, and
N = the number of scores.

Taking the square root of the variance gives the standard deviation.

$$\sigma = \sqrt{\frac{\sigma X_i^2 - \frac{\sigma X_i)^2}{N}}{N}} \qquad \text{Eq. 3.8}$$

Sample Variance and Standard Deviation

In Chapter 2 we introduced the concept of inferential statistics, where the focus is on estimating information about a population from data obtained from a subset of that population, called a sample. The subjects in the sample are obtained by randomly selecting them from the population, and, by definition, the sample is smaller in number (most of the time *much* smaller) than the population. It is not reasonable to assume that the variability of a sample will be as large as the variability of the population, because it is unlikely that a sample will contain the extreme scores of the population. Therefore, a correction is required to estimate the population variability from the sample variability. Importantly, all the equations for the variance and standard deviation presented thus far have been for populations.

To make the following calculations easier to understand, we need to present a few definitions. The mean of a population is represented by μ, the variance by σ^2, and the standard deviation by σ. These facts about a population are called **parameters.** The mean of a sample is represented by $\bar{X}$, the variance by s^2, and the standard deviation by s. These facts about a sample are called **statistics** or **estimates.**

In the numerator of the equation for the variance of a population (Eq. 3.5), we subtract the population mean (μ) from each score. It can be shown mathematically that summing the squared deviations from the mean will give a smaller result than if any other value than the mean is subtracted from each score. In the distribution 1, 2, 3, 4, 5, the sum of the squared deviations is 10. No value other than the mean (3), used in the variance equation, will give a result less than 10.

When we are working with a sample, we typically do not know the population mean. To determine the variance of the sample, our only recourse is to use $\bar{X}$ (the mean of the sample) in our calculation, as this is the best estimate of μ available to us. The use of $\bar{X}$ will produce the smallest possible numerator (sum of squares) to be used in the equation to calculate the variance of the sample. Thus, by using $\bar{X}$, we would underestimate the variance in the population. If we did know and use μ, it would result in a numerator that would always be larger than $\bar{X}$ gives (except in the case where $\bar{X}$ exactly equaled μ). A solution to this problem is to divide the numerator in the sample variance equation by something smaller than N, thus increasing the value of the calculated sample variance. As it happens, the precise value to divide the numerator by is $N - 1$.

The value $N - 1$ is called the **degrees of freedom**. This concept has to do with the number of values that are free to vary. When no limitations are applied to the data, as in a population, all values are free to vary. When we impose the limitation that the mean of the sample is expected to be equal to the mean of the population, all of the values are no longer free to vary. For example, if the mean of five scores is 10, the sum of the scores must be 50. If the first four scores are 15, 7, 9, and 11, the final score *must* be 8. So, of the five scores, four are "free to vary", but one is not. Thus, the degrees of freedom, in this case, are $N - 1$, or $5 - 1 = 4$. These considerations result in the equations for the sample variance and standard deviation being slightly different from those for the population variance and standard deviation. The equation for the sample variance is:

$$s^2 = \frac{\sigma(X_i - \bar{X})^2}{N-1} = \frac{\sigma x^2}{N-1}$$

Eq. 3.9

where
s^2 = the sample variance,
X_i = each score in the sample,
$\bar{X}$ = the mean of the sample,
x = the deviation, and
N = the number of scores in the sample.

The equation for the sample standard deviation is:

$$s = \sqrt{\frac{\sigma(X_i - \bar{X})^2}{N-1}} = \sqrt{\frac{\sigma x^2}{N-1}}$$

Eq. 3.10

where
s = the sample standard deviation,
X_i = each score in the sample
$\bar{X}$ = the mean of the sample, and
N = the number of scores in the sample.

Raw Score Equations for Sample Variance and Standard Deviation

Equations 3.9 and 3.10 presented above to calculate the variance and standard deviation of a sample are the definitional equations. Because they require the individual determination of each deviation, they are difficult to use. Instead, we can use algebraically equivalent equations that require only manipulations of raw scores to calculate these statistics. The raw score equation for the variance of a sample is:

V.3.3 **Variance and Standard Deviation as Measures of Variability**

$$s^2 = \frac{\sigma X^2 - \frac{(\sigma X)^2}{N}}{N-1}$$

Eq. 3.11

where

s^2 = the sample variance,
ΣX^2 = the sum of the squared scores,
$(\Sigma X)^2$ = the sum of the scores, the quantity squared, and
N = the number of scores in the sample.

The equation for s (standard deviation of a sample) is the square root of s^2.

$$s = \sqrt{\frac{\sigma X^2 - \frac{(\sigma X)^2}{N}}{N-1}}$$

Eq. 3.12

Let's return to the simple distribution that we used to illustrate the definitional equation—1, 2, 3, 4, 5 but now consider it to be a sample rather than a population—and determine the sample variance and standard deviation by using the definitional and raw score equations.

	X	μ	x	x^2	X^2
	1	3	−2	4	1
	2	3	−1	1	4
	3	3	0	0	9
	4	3	1	1	16
	5	3	2	4	25
$\Sigma =$	15			10	55

Definitional Equation

$$s^2 = \frac{\sigma (X_i - \bar{X})^2}{N-1}$$

$$s^2 = \frac{10}{4} = 2.5$$

Raw Score Equation

$$s^2 = \frac{\sigma X^2 - \frac{(\sigma X)^2}{N}}{N-1}$$

$$s^2 = \frac{55 - \frac{(15)^2}{5}}{5-1}$$

$$s^2 = \frac{55 - 45}{4} = 2.5$$

V.3.4
SPSS Various Equations to Calculate Variance and Standard Deviation

Notice that the value obtained with the definitional formula is equal to that obtained with the raw score formula. Of course, the sample standard deviation would also be identical, because in both equations it is $\sqrt{2.5} \approx 1.58$.

Notice, also, that when we refer to the variance and standard deviation of a population, we use Greek letters (σ^2 and σ), and when we refer to the same statistics for a sample, we use Roman letters (s^2 and s). This is also true for the population and sample means (μ and $\bar{X}$, respectively).

We have presented many definitional and calculating equations in this chapter. We did this purposefully, to illustrate where the values "come from". Although we realize that people normally use computer programs to calculate these values, we believe that knowing the derivations of these equations helps with understanding what these statistics mean.

V.3.4R
RStudio Various Equations to Calculate Variance and Standard Deviation

SUMMARY

In this chapter, we presented information about three important characteristics of data: the shape of the distribution, measures of central tendency, and measures of variability. All of these descriptive statistics will play important roles as you learn to use statistical tests.

Review Questions and Practice Problems

PART A (ANSWERS PROVIDED IN APPENDIX C)

1. What is another name for a statistic?

   A. Estimate
   B. Parameter
   C. Population fact
   D. Variable

For the types of data listed below, questions 2–5, indicate how you might best organize the data to study the distribution. Which of the following data organization methods would you use in each case? Choose A, B, or C.

   A. Rank order
   B. Simple frequency distribution
   C. Grouped frequency distribution

2. Small range, little data
3. Small range, lots of data
4. Large range, little data
5. Large range, lots of data
6. What is the "average" of the following scores? 10, 9, 9, 7, 3, 2

   A. 6.67
   B. 8.0
   C. 9
   D. All of the above

7. What type of curve would probably result if you plotted the grip strength of males from age 20 to 60?

   A. Normal
   B. Skewed negatively
   C. Skewed positively
   D. Symmetrical

PART B

Use the following data to calculate answers for questions 8 through 18.

12, 11, 10, 10, 9, 9, 9, 8, 8, 7, 6

8. What is the value of the mode?
9. The median is in which position in this distribution?
10. What is the value of the median?
11. What is the value of the mean?
12. Change the score of 12 to 22. What are the new values of the mode, median, and mean?
13. Using the original data, what is the range?
14. What is the value of the semi-interquartile range?
15. Using the definitional formulas, calculate the variance and standard deviation

of these data. (Assume they represent a population.)

16. What are the variance and standard deviation if the data are considered to be from a sample?
17. Use the raw score formula to calculate the standard deviation for a sample.
18. How many degrees of freedom are associated with these data?

COMPUTER ACTIVITIES

Descriptive Statistics

ACTIVITY 1 Examining Data

Suppose a group of researchers administered a 15-item questionnaire that asked subjects to rate how often they engage in various types of physical activity on a five-point scale. From the questionnaire, the researchers obtained the 65 scores in Table 3.2, which will be used as data for this activity.

Using SPSS

A. Call the data column "TESTSCR" and label it "Scores". Remember that these are numerical data. The last column in the Variable View box should be labeled Scale.

B. Go to the SPSS Data View window to enter all 65 scores.

Or as mentioned at the end of Chapter 2, an alternative to entering data directly into SPSS is to first have your data in a different spreadsheet (e.g., Microsoft Excel). The benefit of this is that many data sets already exist in spreadsheet formats and you can simply read them into SPSS and conduct your analysis. Follow these steps.

1. Open an Excel spreadsheet.
2. Use the first ROW to name your variables. Do not use special characters and do not include spaces. Name as many or few variables as you need (or use a table from the textbook).
3. Enter all of the data in the remaining cells. (Alternatively, you can download Table_3_2_Data from the textbook website.)
4. Save the file with a. xls file extension.

   Your data are now in Excel format but can be easily read into SPSS with the following commands:
5. Open SPSS.
6. Go to the **File** menu and scroll to **Open** and then over to **Data**. You will see a screen similar to Figure 3.8.

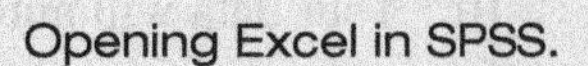

Opening Excel in SPSS. **FIGURE 3.8**

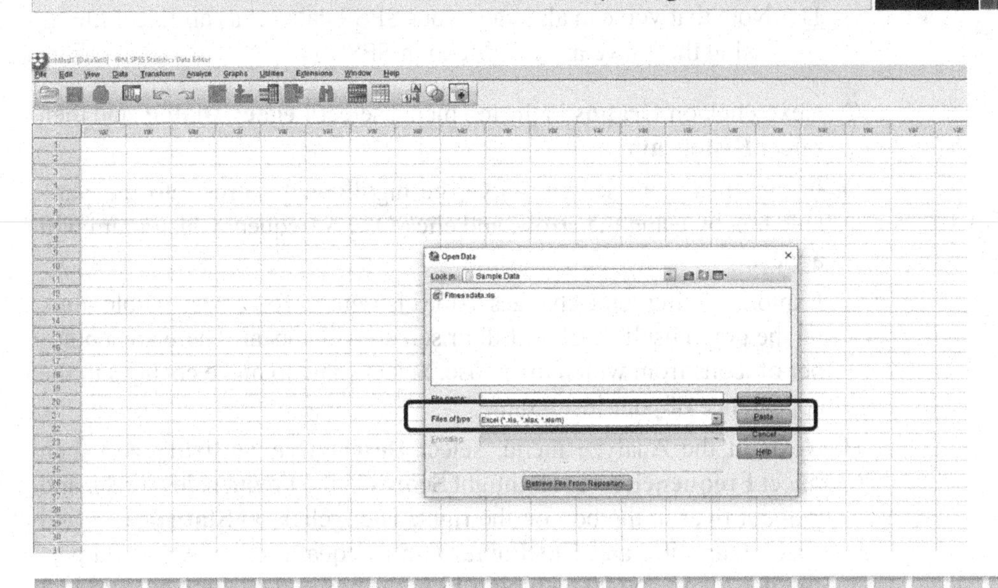

7. Within the "Open File" window that appears, use the downward arrow to indicate Excel (*.xls)
8. Locate the file you want to open and click on OPEN.
9. You will see a box asking if you want to "Read variables names from the first row of data". Click on the square box next to this statement. Recall that the first row of your Excel sheet has the variable names. See Figure 3.9.

Reading variable names from Excel. **FIGURE 3.9**

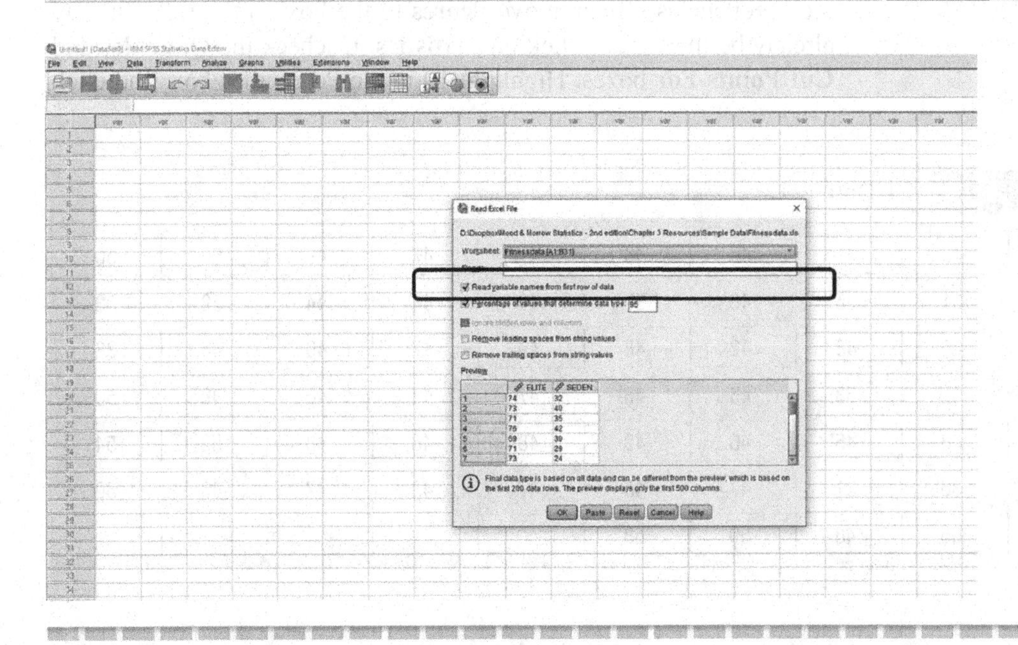

10. Your data will automatically be read into SPSS.
11. Note that you can also save your SPSS data set as an Excel file using the "Save as . . .". Menu in SPSS

C. Next, click on **Graphs** in the top menu, select **Legacy Dialog**, and then select **Histogram**.

D. In the Histogram screen, move the highlighted scores into the Variable box by using the arrow, and click OK. A frequency histogram will appear.

E. Explore making some changes to the histogram. To do this, double-click on the graph itself. A Chart Editor screen will appear. There are a number of icons from which to choose. Select some to make changes in the appearance of your histogram.

H. Next, in the Analyze menu, select **Descriptive Statistics** and then select **Frequencies. . . .** Highlight **Scores** in the Frequencies screen and move it over to the box on the right. Then, click on **Statistics. . .** and check **Quartiles** and **Cut Points For** 10 equal groups. Also, add percentiles for 10 and 90 by typing 10 in the box and clicking on **Add** and then typing 90 in the box and clicking on **Add** again. Click **Continue.** Click OK.

I. Use the two tables that appear to answer these questions:

1. What percentage of the subjects scored exactly 50 on the questionnaire?
2. What percentage of the subjects scored 42 on the questionnaire?
3. What questionnaire scores correspond to the 10th, 25th, 50th, 75th, and 90th percentiles, respectively?

J. Finally, go to **Analyze→Descriptive Statistics→Frequencies**. In the Frequencies screen, move **Scores** to the box on the right (it may already be there), and click on **Statistics**. Uncheck the **Quartiles** and **Cut Points For** boxes. Highlight 10 in the Percentiles box, and click

TABLE 3.2 Data set.

49	45	41	49	46	47	47	49	50	50
39	51	35	48	37	46	44	44	52	53
49	40	48	42	48	49	47	49	44	38
51	47	53	45	48	47	43	46	49	50
48	48	45	46	49	48	46	48	52	54
52	36	51	47	45	47	43	47	46	50
44	55	48	50	53					

on **Remove**. Highlight 90 in the Percentiles box, and click on **Remove**. This will remove the previous commands so that we can now insert new ones. Enter 0 and 100 in the Percentiles box. Click on the boxes for **Standard Deviation, Variance, SE mean, Skewness, Kurtosis, Mean**, and **Median.** Click on Continue and then on OK. Answer the following questions:

4. What are the values for the standard deviation, variance, and standard error of the mean?
5. Describe the data in terms of skewness and kurtosis.
6. What is the value of the mean?
7. What is the value of the median?
8. Do your answers to questions 6 and 7 agree or disagree with your observation about the skewness from above? Why or why not?
9. What score equates with a percentile rank of 100?
10. What score equates with a percentile rank of 0?

Using R

Use the following directions to answer items 1 through 10.

1. Open RStudio.
2. Install any necessary R packages.

When we installed R, we installed "base" R which includes the vast majority of functions and tools that we will need for our analyses. However, there are situations where we will need to install an additional package to carry out a specific function.

For Chapter 3, we will need to install the "moments" package to calculate skewness and kurtosis.

In the lower right pane of the RStudio, click on the "Packages" tab and then click on the "Install" button.

In the "Packages" field, type in "moments" and click "Install".

[Note that from this point forward, the package "moments" will be installed and you do not have to install it again.]

3. Set up your RStudio workspace.

In the RStudio menu bar, click on **File → New File → R Script**

Enter the following two lines in your R Script (upper left pane of RStudio):

- #Chapter 3
- library(moments)

Highlight these two lines that you've typed and click **Run**.

[Note that any line with a preceding "#" is considered a "comment" and is ignored by R.]
[Note that the second line is where we tell R that we will require the functions that are included in the "moments" package.]
[Note that any output generated will be displayed in the Console found in the lower left pane of RStudio.]

As mentioned at the end of Chapter 2 you may want to create a data file elsewhere and then import that file into RStudio. Here are the instructions for creating and then importing a data file.

A. Open a Microsoft Excel spreadsheet.

Open Microsoft Excel and create a "Blank Workbook".

B. Enter the data found in Table 3.2 under Column A of your spreadsheet.

First, let's provide a variable name for the values that you are about to enter. To do so, in the first row of the first column, enter the variable name: "TESTSCR".

Next, we will start to enter the actual values into the spreadsheet.

The second cell (row) in column A is "49", the third cell (row) in column A is "45" and so on.
After you've entered a value into a cell (row), click on the Return key to move down to the next cell (row). See Figure 3.10.

C. Once you've entered all of the data from Table 3.2 (there should be a total of 66 rows including the variable name in the first row), save the sheet as "TESTSCR":

FIGURE 3.10 First set of 10 values in Microsoft Excel.

	A	B
1	TESTSCR	
2	49	
3	39	
4	49	
5	51	
6	48	
7	52	
8	44	
9	45	
10	51	

In the Excel toolbar, click on **File → Save** and then save as "testscr".

Now we can export our data stored in the spreadsheet as a text file. In this situation, we only have one variable (row) of data that we have entered into our spreadsheet. We could have also entered a second row of values (a second variable) that corresponds to the same observation or research subject. When you have more than one variable (row), we have to delimit the values in the file in some way. A standard delimiter for text files is the comma ",". When using commas to delimit variables (rows) of data, the data are stored as comma separated values (csv) and the file is typically referred to as a comma separated value file, or csv file. The vast majority of statistical software packages can import csv files including SPSS and R.

D. Export spreadsheet file as a csv file:

In the Mac Excel toolbar, click on **File → Save** As and in the **File Format** pull-down menu, select, "Comma-Separated

Values". [The PC version will appear as "Save as type" and CSV (Comma delimited).]

4. Import data stored in a comma separated value (csv) file.

   In the upper right pane of RStudio, click on the "Environment" tab.

   Click on "Import Dataset" and select, "From Text (base) . . ."

   Navigate to the file, "testscr" on your computer and click "Open".

   On the Import Data screen, keep the default selections and click "Import".

   You can view the data in spreadsheet form in the upper left pane of RStudio.

   Alternatively, you can view the data frame by typing the name of the data frame directly into the console (lower left pane of RStudio) and hitting the "Return" key.

   - testscr

   You should also see the testscr data frame listed under the "Environment" tab (upper right pane of RStudio).

5. Generate a histogram of the Scores stored in the data frame testscr. First, we must specify which column (variable) we are interested in. For this example, there is only one column of data (TESTSCR) in the data frame (testscr). Nevertheless, we would direct R to the scores column by typing, "testscr$TESTSCR". The "$" directs R to a particular variable within a data frame. We can generate a histogram on the TESTSCR column by typing the following in your R Script:

   - hist(testscr$TESTSCR, main='Histogram', col='gray'. xlab='Score'.)

   When you've finished typing, highlight the entire function you've just typed in and click on **Run**.

   [Note that your histogram will now appear in the lower right pane of RStudio under the "Plots" tab.]

6. We can customize the histogram in various ways. For example, including adding additional "breaks" for a more refined scale.

   - hist(testscr$TESTSCR, main='Histogram', col='gray', xlab='Score', breaks=15)

   [Note that a new histogram will appear in the lower right pane of RStudio under the "Plots" tab.]

   [Note that you can cycle between the two histograms by clicking on the directional arrows in the lower right pane of RStudio under the "Plots" tab.]

7. Generate some descriptive statistics and a frequency distribution of the Scores stored in the data frame testscr by typing the following into your R Script:

- summary(testscr$TESTSCR)
- Frequencytable <- table(testscr$TESTSCR)
- Frequencytable

Highlight what you've typed and click **Run** to execute.

In the console (lower left pane of RStudio under "Console" tab) you will see the output generated by the command you just typed.

To calculate percentages, simply divide the frequency by N (65 in this case).

8. To view the mean, variance or standard deviation, type the following in your R Script:

- mean(testscr$TESTSCR)
- var(testscr$TESTSCR)
- sd(testscr$TESTSCR)

Highlight what you've typed and click **Run** to execute.

Output will be generated in the Console (lower left pane of RStudio).

9. To generate the 10th and 90th percentile, type the following in your R Script:

- quantile(testscr$TESTSCR, c(0.10, 0.90))

Highlight what you've above and click **Run** to execute.

Output will be generated in the Console (lower left pane of RStudio).

10. To generate the standard error, we could type the following in your R Script:

- sd(testscr$TESTSCR)/sqrt(length(testscr$TESTSCR))

[Note that we created a formula to calculate the standard error based upon the standard deviation (sd) divided by the square root (sqrt) of the number of observations (length) in the data set.]

Highlight what you've typed and click **Run** to execute.

Output will be generated in the Console (lower left pane of RStudio).

11. To estimate the skewness and kurtosis, type the following in your R Script:

- skewness(testscr$TESTSCR)
- kurtosis(testscr$TESTSCR)

Highlight what you've typed and click **Run** to execute.

Output will be generated in the Console (lower left pane of RStudio).

ACTIVITY 2 Calculating Descriptive Statistics

Suppose a group of researchers was interested in comparing the VO2MAX of elite bicycle riders and sedentary individuals, and they gathered the data shown in Table 3.3 on the VO2MAX values of the two groups of subjects.

TABLE 3.3

Data set.

ELITE	SEDENTARY
74	32
73	40
71	35
76	42
69	30
71	29
73	24
68	35
75	41
74	28
70	30
78	43
71	35
71	40
68	32
74	35
75	29
69	41
73	32
74	40
70	30
76	28
73	35
67	30
74	41
67	27
70	42
73	32
69	29
74	35

Using SPSS

A. In SPSS Variable View change the number of decimal places if desired.

B. Under Graphs in the top toolbar, select **Legacy Dialogs** and then select **Histogram**. Enter the data for the elite bicycle riders in the Variable box to be used for the graph. Click OK.

C. Double click on the histogram to bring up the "Chart Editor". Locate and click on the icon to superimpose the normal curve on the histogram. The icon is near the right of the Chart Editor menu bars.

D. Click on the graph, copy it, and paste it to MS Word. Give the graph an appropriate name.

E. Go to **Analyze→Descriptive Statistics→Descriptives**. Move the ELETE values into the Variable box. Click on **Options** and select all boxes (not bubbles). Click **Continue** and then OK.

11. What is the highest value?
12. What is the lowest value?
13. What is the range?
14. What is the mean?
15. What is the standard deviation?
16. What is the sum of the 30 values?
17. What is the variance?
18. What is the skewness?
19. What is the kurtosis?

F. Determine the same descriptive statistics for the SEDENTARY individuals.

G. Manipulate your Data View screen by using cut and paste to put all the data in one column. Determine the same descriptive statistics for the combined data.

20. If you graphed the combined data, what word would best describe the shape of the curve?

21. Describe in words the main differences between the two distributions (elite vs. sedentary), basing your comparison on appropriate descriptive statistics.

Using R

Use the following directions to answer items 11 through 21.

1. Load the R data file "Table_3_3_Data" into RStudio.

   On the lower right pane of RStudio, click on the "Files" tab.

   Navigate to where "Table_3_3_Data" is saved on your computer.

   Click on "Table_3_3_Data".

   Click "Yes" at the prompt that confirms that you'd like to load the R data file into RStudio.

2. Enter and Run the following commands to generate descriptive statistics for the elite and sedentary individuals:

   - #Chapter 3 – Computer Activity #2
   - #Read in Table_3_3_Data
   - library(moments)
   - hist(Table_3_3_Data$ELITE, main='Histogram', col='gray', xlab='Score')
   - summary(Table_3_3_Data$ELITE)
   - summary(Table_3_3_Data$SEDENTARY)
   - Frequencytable <- table(Table_3_3_Data$ELITE)
   - Frequencytable
   - Frequencytable <- table(Table_3_3_Data$SEDENTARY)
   - Frequencytable
   - range(Table_3_3_Data$ELITE)
   - range(Table_3_3_Data$SEDENTARY)
   - mean(Table_3_3_Data$ELITE)
   - mean(Table_3_3_Data$SEDENTARY)
   - sd(Table_3_3_Data$ELITE)
   - sd(Table_3_3_Data$SEDENTARY)
   - sum(Table_3_3_Data$ELITE)
   - sum(Table_3_3_Data$SEDENTARY)
   - var(Table_3_3_Data$ELITE)
   - var(Table_3_3_Data$SEDENTARY)
   - skewness(Table_3_3_Data$ELITE)
   - skewness(Table_3_3_Data$SEDENTARY)

- kurtosis(Table_3_3_Data$ELITE)
- kurtosis(Table_3_3_Data$SEDENTARY)

3. You need to combine both the elite and sedentary individuals into a data frame or data file (e.g., Excel or CSV) consisting of one column of data to obtain descriptive statistics on the variable VO2MAX for all 60 of the individuals. Then adjust the RStudio commands above using your new table name and VO2MAX as the variable name to obtain the combined results.

4

The Normal Curve and Standard Scores

INTRODUCTION

Centuries ago, it was observed that if a variable (say, height) was repeatedly measured and the frequencies of the various measurements obtained were plotted, the outcome was a bell-shaped curve, such as the one shown earlier in panel (a) of Figure 3.5. Since a particular person has only one true height, the variations among repeated assessments must result from measurement error. Because the bell-shaped curve was obtained with many different variables (e.g., weight, area, volume) and many different objects or subjects, researchers recognized that this curve (then known as the "curve of errors") must have some important underlying properties. Thus, began the study of the fundamentally most important distribution in statistics. It is called the *normal distribution.*

Later, it was discovered that the curve of errors occurred in situations other than measurement. The normal distribution also appeared to be related to events whose outcomes depended on chance. For example, if several individuals on several occasions each throw 10 coins in the air, let them land, count the number of heads (from 0 to 10), combine their data, and then graph it, the results will resemble the normal distribution. The occurrences of 0 heads or 10 heads will be relatively few, and the most common occurrence will be five heads and five tails. In fact, according to the laws of chance, if there were 10,000 total coin tosses of ten coins, about ten of them would turn up all heads, about ten of them would be all tails, and about 2,500 would be five heads and five tails. The discovery of the relationship between chance and the normal distribution is fundamental to the field of statistics.

The coin-tossing data would produce a histogram that looks somewhat like the normal curve. That is, if the tops of the bars were connected by a line, the shape of the line would be a curve resembling the shape of the normal curve (a bell shape). However, because the data would be discrete (i.e., there would be no data between, say, four heads and five heads, because it isn't possible to obtain 4.5 heads), the curve would not be continuous. Therefore, it would not, technically, be termed a normal curve. The normal curve exists only for continuous data.

THE NORMAL CURVE

The equation for the continuous normal curve was determined by Abraham DeMoivre, a French mathematician who lived in England about 300 years ago. DeMoivre was interested in the laws of chance, because he made part of his living through gambling. Working from the writings of several other mathematicians, DeMoivre published the equation for the normal curve sometime in the 1730s. The equation is:

$$f(X) = \frac{e^{-\frac{(X-\mu)^2}{2\sigma^2}}}{\sigma\sqrt{2\pi}} \qquad \text{Eq.4.1}$$

where

$f(X)$ = the height of the curve for any value substituted for X,

π = the mathematical constant, approximately 3.1416,

e = the mathematical constant, approximately 2.7183,

σ = the standard deviation of the distribution of X, and

μ = the mean of the distribution of X.

Substituting any value for the variable X in the equation (along with the constants and the mean and standard deviation of the distribution of X) will yield the height of the curve at that value of X. To produce the entire normal curve would involve substituting all possible values of X into the equation. Luckily, we will not need to work with this specific equation. Because it is so important to the field of statistics, mathematicians have produced tables that summarize this equation. These will be illustrated and explained later in this chapter.

In addition to applications involving measurement error and the laws of chance, researchers have determined over the years that the normal curve also fits data collected and examined from a wide variety of natural and social sciences. Applications for the normal curve have been found in chemistry, biology, anthropology, physics, meteorology, sociology, psychology, engineering, education, and, of course, human performance, among other disciplines. Measures of many physical and mental traits tend to be normally distributed. The fact that data from so many diverse sources conform to the normal distribution caused early measurement scholars to believe they had discovered a "law of nature". Unfortunately, this was not the case. Although the normal curve is certainly ubiquitous, all data will not conform to this shape. For example, VO_2max expressed in ml/kg/min is quite normally distributed; however, pull-ups are positively skewed, because many people can do very few pull-ups, and a few people can do very many.

When we use the normal curve, we are implicitly assuming that an infinite number of observations are available and that these are measured on a continuous (interval or ratio level of measurement) scale. In reality, however, even though a researcher might have a very large number of observations, the number remains finite. Thus, it is unlikely that any set of data will conform exactly to the normal curve, although often the difference is so slight that it can be assumed to be normal. In addition, as stated earlier, many variables are *not* normally distributed. For example, do you know why the median is almost always used in discussions of the "average" house price or the "average" family income? It is because these distributions are typically positively skewed (i.e., not normal). Now, consider the reported number of pedometer steps per week. This distribution, presented in Figure 4.1, is positively skewed because many people do not take a large number of steps per week (mean ≈ 45,000), but some take a very large number of steps (>100,000), resulting in a positively skewed distribution. As indicated in Chapter 3, with non-normal distributions, researchers typically report the median as the measure of central tendency rather than the mean. Recall the example of salaries of University of West Florida exercise science graduates in Chapter 3.

Although he was not the first to use the normal distribution, Carl Friedrich Gauss, a mathematician and astronomer, advanced its use when he published a paper in 1809 in which he applied the normal distribution to astronomical data. Since that time, the normal distribution has also often been called the Gaussian distribution.

As an example of one of the many early (nineteenth century) applications of the normal curve, Sir Francis Galton, a noted English scholar, published an article in 1876 in which he reported the heights of 799 fourteen-year-old boys, from various public schools. A compilation of his raw data

FIGURE 4.1 Positively skewed distribution of pedometer steps per week.

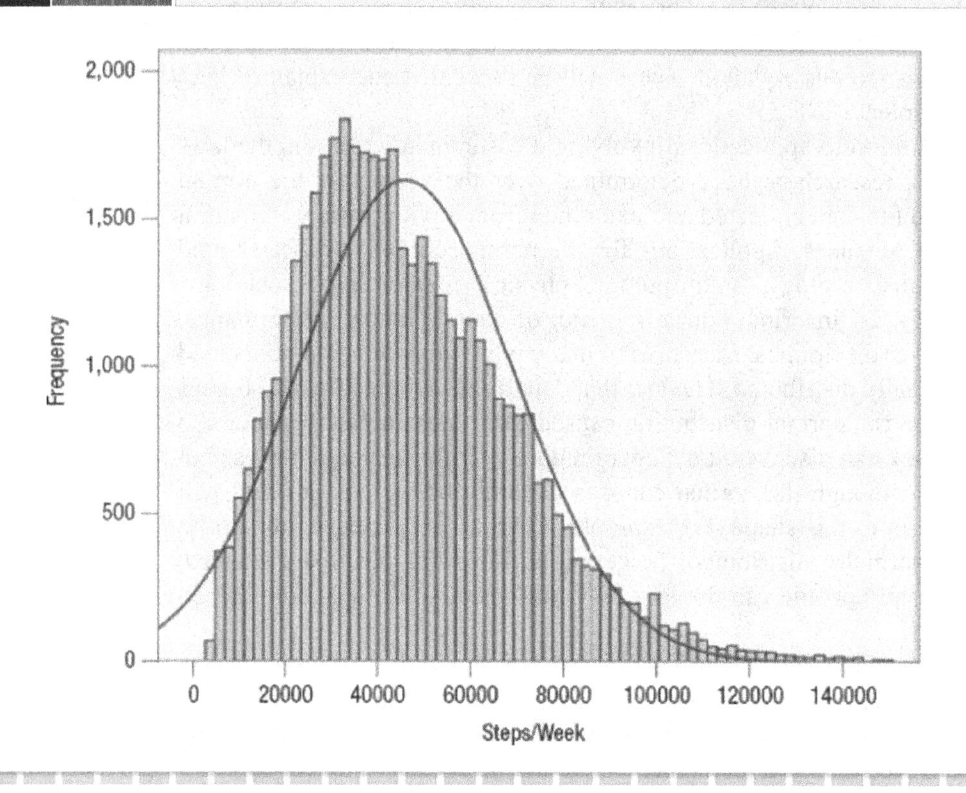

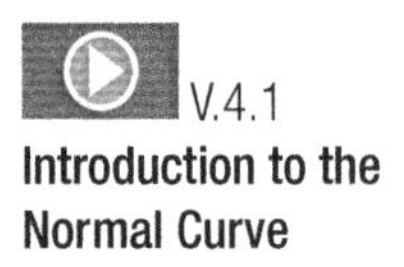
V.4.1
Introduction to the Normal Curve

is graphed in Figure 4.2. Notice the similarity to the shape of the normal distribution (refer back to Figure 3.5a). Also, notice that if Galton had measured and graphed the heights of boys *and girls* aged 14, in those same public schools, the results would not have been normally distributed, because the data would almost certainly have been bimodal (refer to Figure 3.7).

The *z*-Score

Comparisons among sets of data, even when they all conform to the shape described by the normal curve, are difficult when the units of measurement are different and the data have differing descriptive values (e.g., means and standard deviations). To overcome this problem, it is necessary to *standardize* the distributions. To accomplish this, we calculate standardized scores, called ***z*-scores,** which have a mean of 0 and a standard deviation of 1. Let's demonstrate this concept (assume the data are normally distributed).

Suppose that you took two fitness test items and were able to perform 55 sit-ups and 43 modified pull-ups. On which of the two tests did you perform relatively better? How does your performance on each of these tests compare to your classmates' results? What information do you need to answer these questions? If you learned that the mean scores for your

Heights of 799 English public school boys aged 14. FIGURE 4.2

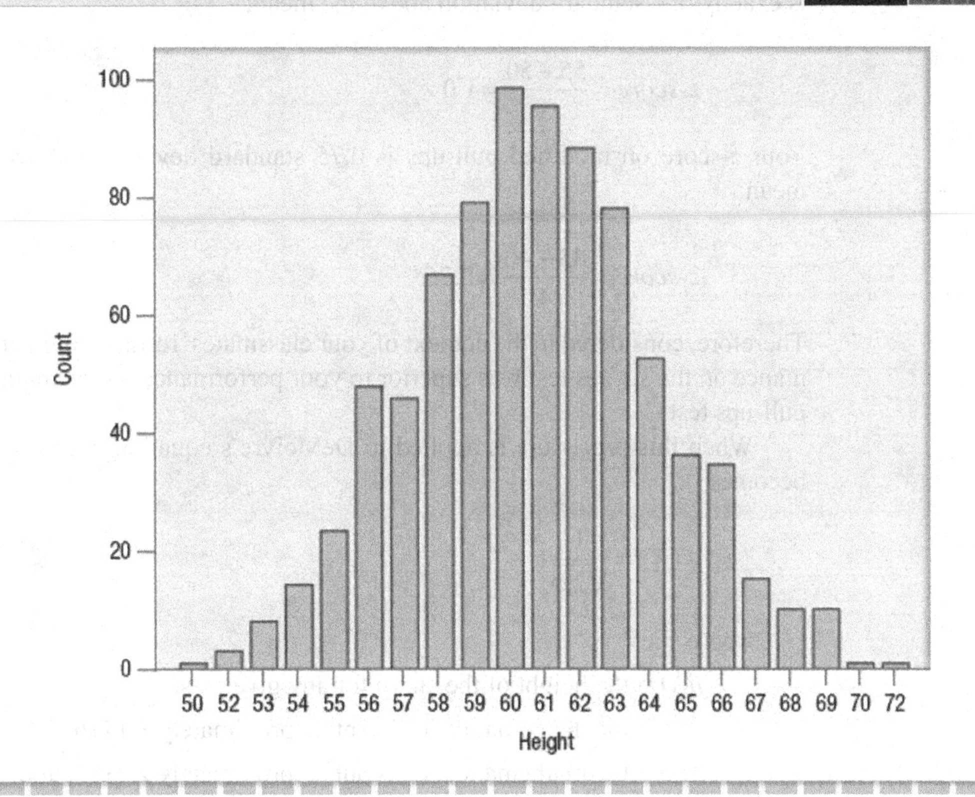

class were 50 and 40 for the two tests, respectively, you would know that you were above the mean on both tests. However, you still would not know which test result represents your better performance, because the distributions of the two sets of test scores would probably have differing amounts of variability. To overcome all of these issues and finally answer the questions, you need to express the scores in a standard way, so that you can compare them. If you learned that the standard deviations for the two tests were 5 and 4, respectively, now you can determine where your scores occur in each of the two distributions. You would calculate how far your scores are from the respective means of each distribution, measured in standard deviation units rather than raw score units. As mentioned, these standard deviation units are referred to as *z*-scores. The equation for *z*-scores is:

$$z\text{-}score = \frac{X - \overline{X}}{s} \quad \text{Eq. 4.2}$$

where

X = an observed score,

$\overline{X}$ = the mean of the distribution of scores, and

s = the standard deviation of the distribution of scores.

When we examine the fitness test results, we find that your z-score on sit-ups is exactly one standard deviation above the mean:

$$z\text{-}score = \frac{55 - 50}{5} = 1.0$$

Your z-score on modified pull-ups is 0.75 standard deviations above the mean:

$$z\text{-}score = \frac{43 - 40}{4} = 0.75$$

Therefore, considered in the context of your classmates' results, your performance on the sit-ups test was superior to your performance on the modified pull-ups test.

When this procedure is applied to DeMoivre's equation, the equation becomes

$$f(X) = \frac{1}{\sqrt{2\pi}} e^{-\left(\frac{z^2}{2}\right)} \qquad \text{Eq. 4.3}$$

where

$f(X)$ = the height of the curve for any given z-score,

π = the mathematical constant, approximately 3.1416,

e = the mathematical constant, approximately 2.7183, and

z = the z-score.

The resulting curve is called the *standard normal curve*. This curve is so fundamentally important to the area of statistics that all reasonable values of z have been substituted into the equation, and the results have been put into table format for convenient use. See Appendix A.1.

CHARACTERISTICS OF THE NORMAL CURVE

Much of the usefulness of the standard normal curve in statistics comes from its characteristics. It is unimodal, symmetrical, continuous, mesokurtic, and **asymptotic** (that is, it never touches the X-axis, except at infinity). The normal curve has only one high point (it is unimodal), and this peak is where the mode, median, and mean of the distribution all are located. Thus, in a normal distribution, the three measures of central tendency we discussed in Chapter 3 are all equal in value. When we say that the normal curve is symmetrical, we mean that a line drawn vertically from its peak to the X-axis bisects the curve into two identical halves. The fact that the normal curve is continuous and asymptotic indicates that the curve has some height value for any possible z-score.

V.4.2
Using the Normal Curve

To consider symmetry and height, skewness and kurtosis of a distribution of scores (described in the previous chapter) can be determined mathematically. For the normal curve, a skewness index of 0 indicates symmetry (i.e., no skewness), a positive value indicates positive skewness (a longer tail to the right), and a negative value indicates negative skewness (a longer tail to the left). For kurtosis, an index of 0 indicates a normal curve (mesokurtic), a positive index indicates a leptokurtic curve, and a negative index indicates a platykurtic curve. Because this procedure is relatively laborious to do by hand, most statistical computer programs will perform these calculations.

The area under the standard normal curve is defined to equal 1.0 (or 100%). This allows us to compare distributions to one another and to determine much information about scores coming from distributions that have differing characteristics, as will be demonstrated in the rest of this chapter.

USE OF THE *Z*-TABLE

Take a few minutes to examine Figure 4.3, which illustrates the normal distribution.

Now, turn your attention back to the *z*-table (Appendix A.1). Let's see how Figure 4.3 and the *z*-table can help us interpret information about the normal curve. The accompanying box offers key points to keep in mind.

The normal distribution. FIGURE 4.3

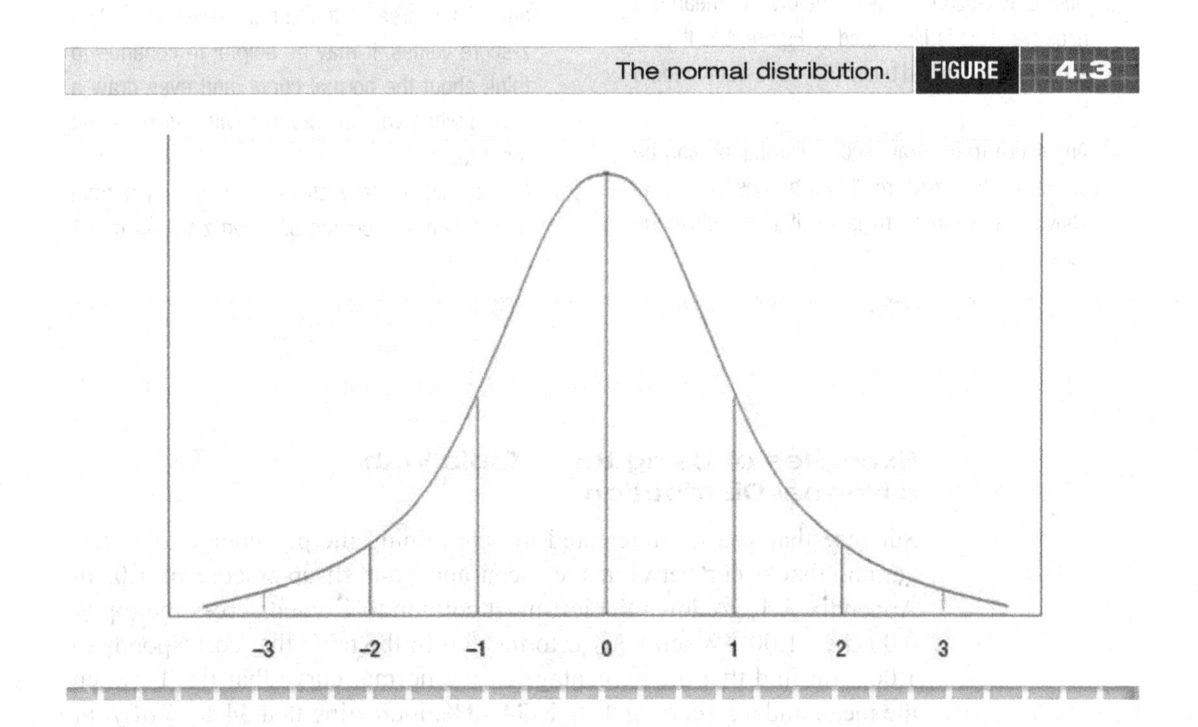

Key Points in Using the z-Table to Interpret Information from the Normal Curve

1. The normal distribution is symmetrical: it is the same on each side of the mean (see Figure 4.3).
2. The z-table in Appendix A.1 includes only one side of the distribution because, as explained in detail below, the value associated with a particular z-score located above the mean is identical to the value for that same z-score located below the mean, differing only in sign (+ or –). Other presentations of the z-table may differ slightly from the one in this textbook, but every z-table will provide identical information, regardless of the presentation style.
3. Remember that the z-score represents a raw score's distance from the mean of the distribution, measured in standard deviation units. Therefore, a score equal to the mean would convert to a z-score of 0. Recall the example of sit-ups and modified pull-ups tests earlier in the chapter, where the mean for sit-ups for the class was 50. If you had scored 50 on sit-ups, your z-score would be 0.
4. Note that z-scores that fall below the mean are negative. This is illustrated in Figure 4.3. If your score on sit-ups had been 45, your z-score would be –1.
5. Any score from a raw score distribution can be converted to a z-score. It will be positive if it is above the mean or negative if it is below the mean.
6. Now, turn your attention to the z-table in Appendix A.1. Note the following:
   a. The values in the first column are z-scores, in whole numbers and tenths.
   b. The numbers across the top give hundredths for the corresponding z-score in the first column.
   c. Each number in the body of the z-table is a percentage. It is the percentage of the area of the normal-curve distribution that lies between the mean and the specified z-score.
7. Again, notice that the z-score is exactly the same as distance that a score lies from the mean of the distribution, measured in standard deviations.
8. Recall that the total area under the normal curve represents 100% of the observations. Half of them are above the mean and half of them are below the mean.
9. You can use the z-table to determine the percentage of the area of the curve between any two z-score values. It may be helpful to continue to think about the normal curve (and even draw a figure) when you are working with z-scores and the z-table.
10. Notice that virtually all the scores in a normal distribution are located between z-values of –3 and +3.

Examples of Using the *z*-Table With a Normal Distribution

Suppose that you are interested in determining the percentage of observations that occur between the mean and your sit-up *z*-score of 1.0. In Appendix A.1, go down the left-most column to 1.0 and across the top to 0.00 (i.e., 1.00). When you go to the cell in the table that corresponds to 1.00, you find that the percentage of the normal curve that lies between the mean and a *z*-score of 1.00 is 34.13%, indicating that 34.13% of your classmates were able to perform between 50 and 55 sit-ups. Similarly, we can find the value between the mean and a *z*-score of 2.0; it is 47.72%. You can use this procedure for any *z*-score value. Use the table to confirm that the percentage of the normal curve between the mean and a *z*-score of 1.46 is 42.79%.

Let's consider *z*-scores to the left of the mean (negative *z*-scores). Notice that there are no positive or negative signs in the *z*-table. That is because the curve is symmetrical, and the values to the left of the mean are identical to those on the right (i.e., positive). Thus, 34.13% of observations fall between the mean and a *z*-score of −1.00, equal to the number between the mean and a *z*-score of 1.00. This information can be used in a wide variety of ways. For example, 68.26% of observations in a normal curve are between *z*-scores of +1.00 and −1.00 (34.13% + 34.13% = 68.26%), indicating that approximately 68% of your class would be able to perform between 45 and 55 sit-ups.

What percentage of observations fall between +1.96 standard deviations and − 1.96 standard deviations? When we consult the table, we find that the answer is 95% (47.50% + 47.50%). This means that the remaining 5% (100% − 95%) of the observations in a normal curve fall outside ±1.96 standard deviations. This 5% will become increasingly important to us in subsequent chapters.

We can also use the *z*-table to calculate a percentile for any single observation. Recall that a percentile represents the percentage of observations that fall at or below a given point. For an easy example, let's look at the mean. The mean of the normal distribution is the middle of the distribution, so 50% of the observations are below a *z*-score of 0.00 (or the middle of the distribution). Recall that the median (50th percentile) and the mean (and the mode) are all equal to each other in a normal distribution.

Now consider how you would calculate the percentile for any observation identified above the mean. You would determine the percentage of observations falling between the mean and the given point and *add* that to 50%, because there are another 50% of the observations below the mean. Confirm that if the distribution of sit-up scores in your class is normal, your score of 55 (*z*-score = 1.0) is equivalent to the 84.13th percentile.

Obtaining the percentile for a score below the mean requires a slightly different procedure. Recall that the *z*-table includes only positive numbers and that the reference point is zero (0.00). Suppose that you were able to perform only 45 sit-ups on the fitness test. Your performance would result in a *z*-score of −1.0. What is the percentile associated with this observation? Consider that 50% of the observations occur from the mean to negative infinity. Between the mean and a −1.00 *z*-score, there are 34.13% of the observations. Hence, the percentile associated with a *z*-score of −1.00 is 15.87 (50.00% − 34.13% = 15.87%). For practice, confirm that the percentile associated with a *z*-score of −0.47 is 31.92.

By using Appendix A.1, you can describe and calculate many other facts about a normal distribution. For example, you could determine the percentage of scores that lie between any two *z*-scores. As an illustration of this, let's say you wanted to know the percentage of observations between

FIGURE 4.4 Standard scores and the normal distribution.

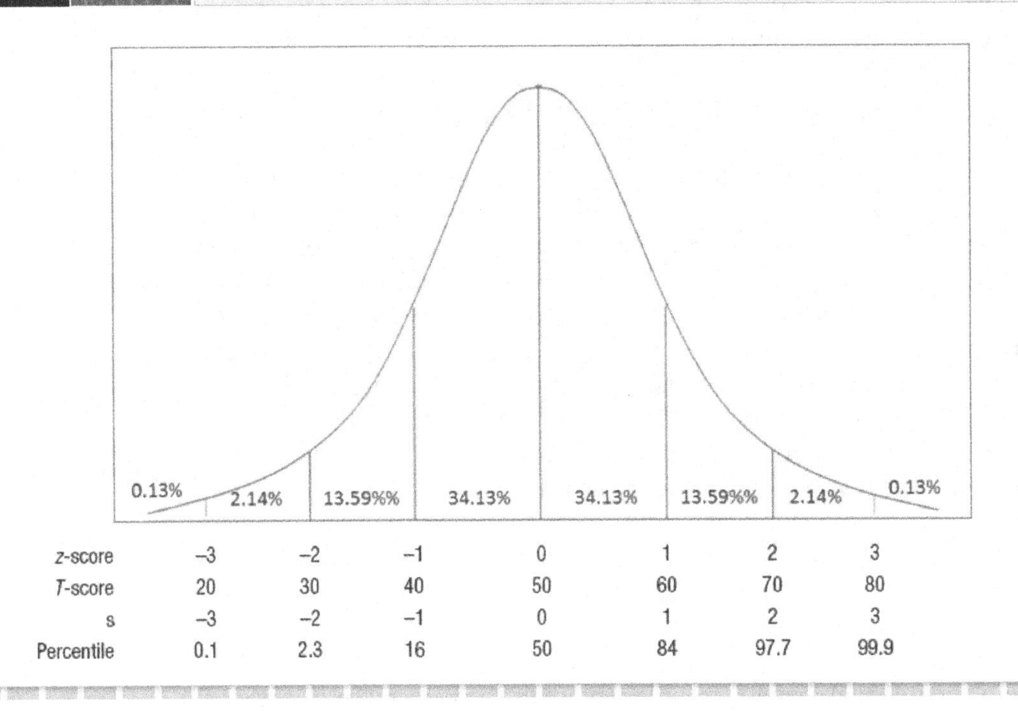

z-scores of 1.10 and 1.89. You would first determine the percentage between the mean and 1.89 (47.06%) and then do the same for 1.10 (36.43%). You would find the area between these two points by subtraction: 47.06% – 36.43% = 10.63%.

You can also use the *z*-table in the reverse direction to determine what *z*-scores would capture a particular percentage of the raw scores. For example, between what two *z*-scores does the middle 50% of the scores in a normal distribution lie? To determine this, find the closest value to 25% in the body of the *z*-table and determine the corresponding *z*-score (0.67). Therefore, the middle 50% of the scores lie between ± 0.67.

Try some other examples until you become familiar with the use and interpretation of *z*-scores, the normal distribution, and the *z*-table. Drawing a picture can be helpful as you determine a strategy for solving each example. Then, use the *z*-table to calculate the z-score. Remember that the *z*-table "starts" at the mean and that the numbers in the body of the *z*-table are the percentages of observations that lie between the mean and a given *z*-score (i.e., a given number of standard deviations). Look at Figure 4.4 to see a graphic depiction of the percentages of the normal curve between the mean score and various *z*-scores.

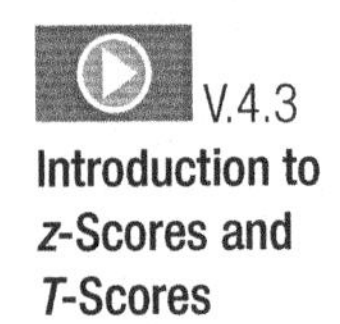
V.4.3 **Introduction to *z*-Scores and *T*-Scores**

STANDARD SCORES AND THEIR INTERPRETATION

When raw scores are converted to **standard scores,** it is easier to interpret and compare them to each other than when they are in raw score form. It is interesting to note that, regardless of the mean, standard deviation, and shape of any raw score distribution, if it is converted to *z*-scores, the *shape* of the resulting *z*-score distribution will be the same as the original distribution, but the *mean* and the *standard deviation* of the z-scores will always be 0 and 1, respectively. *z*-scores represent only one (undoubtedly, the most important one) of many possible "standardization" procedures.

Another standard score is the ***T*-score.** The equation to calculate a *T*-score is:

$$T - score = 50 + 10z$$ Eq. 4.4

where

T–score = the *T*-score to be determined, and

z = the z-score of the *X* value to be converted to a *T*-score.

The mean of any set of scores converted to *T*-scores is *always* 50, and the standard deviation is *always* 10. The main advantage of converting to a standard score is that it becomes easier to interpret and compare scores, because standard scores have a common unit of measurement and are equal in variability.

Many other standard scores have been developed over the years. Refer to Figure 4.4 to see how *z*-scores, *T*-scores, standard deviation units, and percentiles relate to one another.

Standard scores are a very important tool for researchers who are determining a composite score. For example, consider the constellation of risk factors associated with cardiovascular disease. Among others, these risk factors include the levels of blood pressure, triglycerides, cholesterol, and glucose, as well as waist circumference. It would not be appropriate to calculate an overall "risk factor score" by simply summing a person's values for these five measurements. These assessments have different units of measurement (e.g., mmHg, mg/dl, cm) and thus cannot be summed. An alternative is to calculate a *z*-score for each of the variables and sum the five *z*-scores to obtain an indication of the overall risk for an individual. Note that in this example, for each of these variables a *lower* value indicates less risk than a higher value. Thus, when we calculate the *z*-score for each of the five measurements, it is necessary to change the numerator of the *z*-score formula slightly:

$$z\text{-}score = \frac{\bar{X} - X}{s}$$ Eq. 4.5

Compare this equation to Eq. 4.2 and notice that the $\bar{X}$ and *X* have reversed positions: this changes the sign to reflect the fact that low scores are desirable. Alternatively, in situations where a low score is desirable, you could simply multiply the *z*-score for negatively scored items by −1.00, to change the sign.

It is most important that you consider how the *z*-scores are to be calculated in each situation you encounter, which will depend on whether a high or low score is a positive result. In this example, we could sum the *z*-scores for the risk factors and obtain an overall indication of the individual's risk (obviously, this will be relative to the risk of others on these measures). Tanha and colleagues (2011) followed exactly this procedure to create a composite risk factor profile for cardiovascular disease (CVD) in school-aged children. They found that, in their subjects, *low* amounts of moderate to vigorous physical activity and vigorous physical activity were related to *higher* CVD composite risk factor scores.

SUMMARY

Many observed characteristics are normally distributed. However, comparisons among data distributions, even when they all conform to the shape described by the normal curve, are difficult when the units of measurement are different, and the data have differing descriptive statistics values. Standardizing the distributions is accomplished through the use of z-scores. Because the normal curve is continuous and asymptotic, there is some height of the curve (value) for any possible z-score. The normal curve is so fundamentally important to the area of statistics that all reasonable values of z have been substituted into the equations, and the results have been put into table format, producing the z-table. The ability to use and interpret data from the z-table is fundamental to understanding statistical inference. Hence, it is important that you understand the z-table and practice its use and interpretation. A clear understanding of the z-score concept and use of the z-table will serve you well as we dig deeper into statistical inference procedures.

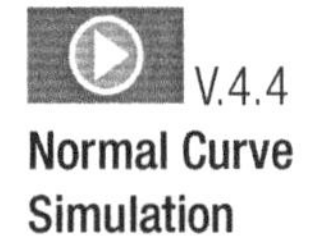
V.4.4
Normal Curve Simulation

Review Questions and Practice Problems

PART A (ANSWERS PROVIDED IN APPENDIX C)

1. Which of the following would *least* likely result in normally distributed data?

   A. Age of 5,000 individuals when suffering their first heart attack
   B. Heights of 5,000 ten-year-old boys
   C. Weights of 5,000 preschool children
   D. Percent fat of 5,000 sixteen-year-old girls

2. Which of the following most likely represents the *worst* performance?

   A. Percentile of 49
   B. T – score of 39
   C. z – score of – 0.8

3. If the mean of a distribution is 500 and the *variance* is 400, what score represents the 16th percentile?

   A. 470
   B. 480
   C. 490
   D. Impossible to determine from what is given
   E. None of the above

4. What advantage do T-scores and z-scores have over raw scores?

   A. There is no real advantage, but people use them because they are "standard" in nature.
   B. They can be added because of their standardized characteristics.

C. The mean is 50 and the standard deviation is equal to 1 for each.

D. The values are controllable.

E. They represent the normal curve, whereas raw scores do not.

PART B

For problems 5 through 8, assume the data are normally distributed, with a mean of 75, a standard deviation of 9, and an $N = 36$.

5. What is the *z*-score for a raw score of 80?
6. What is the T-score for a raw score of 87?
7. What is the percentile for a raw score of 82?
8. Karen's T-score was 65. What is her *z*-score?

A. – 2.0

B. – 1.5

C. – 0.5

D. Need more information

E. None of the above

Answer problems 9 and 10 by referring to the grouped frequency distribution in Table 4.1

TABLE 4.1

Grouped frequency distribution.

INTERVAL	FREQUENCY	PERCENTILE
148–152	5	100.0
143–147	7	95.0
138–142	18	88.0
133–137	20	70.0
128–132	10	50.0
123–127	15	40.0
118–122	9	25.0
113–117	8	16.0
108–112	6	8.0
103–107	2	2.0

9. In which interval would you probably find a person one standard deviation below the mean?

A. 123 – 127

B. 118 – 122

C. 113 – 117

D. 108 – 112

E. It is impossible to determine.

10. In which interval is a *T*-score of 60 located?

A. 143 – 147

B. 138 – 142

C. 133 – 137

D. 128 – 132

E. None of the above

11. Stanines are standard scores that have a mean of 5 and a standard deviation of 2. Suppose Ralph scored 75 on a spelling test that had a mean of 50 and a standard deviation of 12.5. What is Ralph's stanine score?

A. 2

B. 9

C. 70

D. 80

12. What is the standard deviation of the weight scores, given the information below? Assume that a student who is 70 inches tall and weighs 155 lbs will be at the same percentile point in each distribution (that is, assume normality).

Height	Weight
$\overline{X} = 64$ in	$\overline{X} = 140$ lbs
$s = 4$ in	$s = ?$ lbs

A. 1.5 lbs

B. 6.0 lbs

C. 10.0 lbs

D. 15.0 lbs

COMPUTER ACTIVITIES

Using Standard Scores

Table 4.2 shows test scores from four physical fitness knowledge examinations administered to twelve students in a physical fitness class.

1. What level of measurement do these data represent (i.e., nominal, ordinal, interval, ratio)?
2. What is the mean score for each examination?
3. What is the median score for each examination?
4. What is the mode for examination 2?
5. What is the range for each examination?
6. What is the standard deviation for each examination?

Using SPSS

A. Enter or import the data and perform the operations necessary to answer questions 1 through 6.

B. Obtain a composite score (COMPSCR) for each student, by having the program add together each student's four exam scores. To do this, go to the top menu bar and click on **Transform→Compute. . . .** A Compute Variable box will appear, where you can type in a formula for the program to calculate. First, enter "COMPSCR" in the Target Variable box. Next, move the cursor to the Numeric Expression box (click once), and then double-click on EXAM1 (listed in the box on the left) to move it to the Numeric Expression box. Go to the keyboard on the screen and click on the + sign. Double-click on EXAM2 and repeat the above command. Move EXAM3 and EXAM4 in a similar manner. (Do not put a + sign after EXAM4.) Click OK. You should now have the composite score (COMPSCR) for each student, which is the sum of the four "raw" scores.

C. Have SPSS calculate a *z*-score for each EXAM and COMPSCR. Obtain descriptive statistics via **Analyze→Descriptive Statistics→Descriptives.** Move all four exam scores and COMPSCR into the Variables window on the right-hand side of the Descriptives window. Be certain to click the **Save standardized values as variables** box in the Descriptives window. Click OK. Note that you have new variables for each exam. These are the *z*-scores.

D. The next steps involve converting each exam score distribution into a *T*-score. Click on **Transform→Compute.** Move TSCORE1 into the Target Variable Box. The general formula for this and the next three

Test scores. TABLE 4.2

STUDENT	EXAM 1	EXAM 2	EXAM 3	EXAM 4
1	98	30	87	7
2	94	85	82	40
3	96	79	83	9
4	97	38	84	15
5	97	62	85	39
6	95	42	84	34
7	95	37	86	29
8	96	59	83	23
9	98	62	84	12
10	95	40	85	19
11	96	53	86	38
12	97	33	83	25

columns is: 50 + 10**zscore*. Repeat this procedure for the remaining three exam score distributions. Remember to use the appropriate *z*-score for each exam.

E. Obtain a composite *T*-score for each student by summing their *T*-scores on each of the four examinations. Use **Transform→Compute.** Enter TOTALT as the target variable and insert a formula in the Numeric Expression window to sum the four *T*-scores.

F. On a sheet of paper, write the two composite score columns next to each other. In the first column, list the original composite scores (COMPSCR). In the second column, list the composite *T*-scores. Rank the scores in each column from 1 to 12, assigning 1 to the highest score and 12 to the lowest score. Assume you are going to assign grades to the students as follows: A to rank 1; B to rank 2; B− to rank 3; C+ to rank 4; C to ranks 5, 6, 7, and 8; C− to rank 9; D+ to rank 10; D to rank 11; and F to rank 12. Answer the following questions:

7. Are the rankings the same for the two composite distributions?
8. How many grades are different between the two distributions?
9. Determine what is unusual about student 1's performances on the tests.
10. What is strange about student 2's performances? How do the performances of these two students differ, and how are they the same?
11. What grades would be assigned to these two students if the total of the raw scores were used to determine grades?
12. What grades would be assigned to these two students if the total of the *T*-scores were used to determine grades?

13. Why are your answers to questions 11 and 12 different?
14. Based on your answers to questions 9 and 10, where should these two students' total performances place them (approximately) in the class?
15. What are the means and standard deviations of the four *T*-score distributions, to one decimal place?

Using R

Use the following directions to answer items 1 through 15.

1. Open RStudio.
2. Install any necessary RStudio packages.

   For Chapter 4, we do not require any additional packages.
3. Set up your RStudio workspace.

   In the RStudio menu bar, click on **File → New File → R Script**.

   On the first line of the R Script (upper left pane of RStudio), type

   - #Chapter 4
4. Enter the variable names (STUDENT, EXAM1, EXAM2, EXAM3, and EXAM4) and values found in Table 4.2 into a spreadsheet and create a csv file named "testscores" to be imported into R.
5. Import "testscores" into R (see Chapter 3).

   You should see the testscores data frame listed under the "Environment" tab (upper right pane of RStudio).
6. Enter the following RScript commands to obtain descriptive statistics to answer items 1 through 6:

   - #mean
   - mean(testscores$EXAM1)
   - mean(testscores$EXAM2)
   - mean(testscores$EXAM3)
   - mean(testscores$EXAM4)
   - #median
   - median(testscores$EXAM1)
   - median(testscores$EXAM2)
   - median(testscores$EXAM3)
   - median(testscores$EXAM4)
   - #range
   - range(testscores$EXAM1)
   - range(testscores$EXAM2)
   - range(testscores$EXAM3)
   - range(testscores$EXAM4)
   - #sd
   - sd(testscores$EXAM1)
   - sd(testscores$EXAM2)

- sd(testscores$EXAM3)
- sd(testscores$EXAM4)

Use the following RScript commands to obtain the data to answer questions 7 through 15. Data files will be created and located in the Environment pane of RStudio.

7. Create a composite score (COMPSCR) for each student by adding together each student's four exam scores:

   - testscores$COMPSCR <- testscores$EXAM1+testscores$EXAM2+ testscores$EXAM3+testscores$EXAM4

   When you've finished typing, highlight the entire function you've just typed in and click on **Run**.

   You should now have a composite variable created based upon the sum of the four exam scores.

8. Next, we will calculate a *z*-score for COMPSCR using the scale function. We will store the *z*-scores in the object "COMPSCRZ".

   - COMPSCRZ <- scale(testscores$COMPSCR)

Highlight what you've typed and click **Run** to execute.

9. Now calculate a *z*-score for each exam using the scale function. We will store *z*-scores for each exam in a new object as follows:

   - EXAM1Z <- scale(testscores$EXAM1)
   - EXAM2Z <- scale(testscores$EXAM2)
   - EXAM3Z <- scale(testscores$EXAM3)
   - EXAM4Z <- scale(testscores$EXAM4)

   Highlight what you've typed and click **Run** to execute

10. The next steps involve converting the *z*-scores to *T*-scores. We can do this by applying the formula: *50 + (10*z-score)*.

   - EXAM1T <- 50 + (10*EXAM1Z)
   - EXAM2T <- 50 + (10*EXAM2Z)
   - EXAM3T <- 50 + (10*EXAM3Z)
   - EXAM4T <- 50 + (10*EXAM4Z)

   Highlight what you've typed and click **Run** to execute

11. We can also create a composite *T*-score for each student as well.

   TOTALT <– EXAM1T+EXAM2T+EXAM3T+EXAM4T

   Highlight what you've typed and click **Run** to execute

12. Use the following RScript commands to obtain descriptive statistics for the T-scores

   - #means for T Scores
   - mean(EXAM1T)

- mean(EXAM2T)
- mean(EXAM3T)
- mean(EXAM4T)
- #sd for T Scores
- sd(EXAM1T)
- sd(EXAM2T)
- sd(EXAM3T)
- sd(EXAM4T)
- #Get Mode from frequencies
- frequencytable <-table(testscores$EXAM2)
- frequencytable

REFERENCES

Galton, F. (1876). On the height and weight of boys aged 14, in town and country public schools. *Bulletins de la Société d'Anthropologie de Paris*, *IX*(4), *X*(1).

Tanha, T., Wollmer, P., Thorsson, O., Karlsson, M. K., Lindén, C., Andersen, L. B., & Dencker, M. (2011). Lack of physical activity in young children is related to higher composite risk factor score for cardiovascular disease. *Acta Paediatrica*, *100*(5), 717–721. doi:10.1111/j.1651-2227.2011.02226.x.

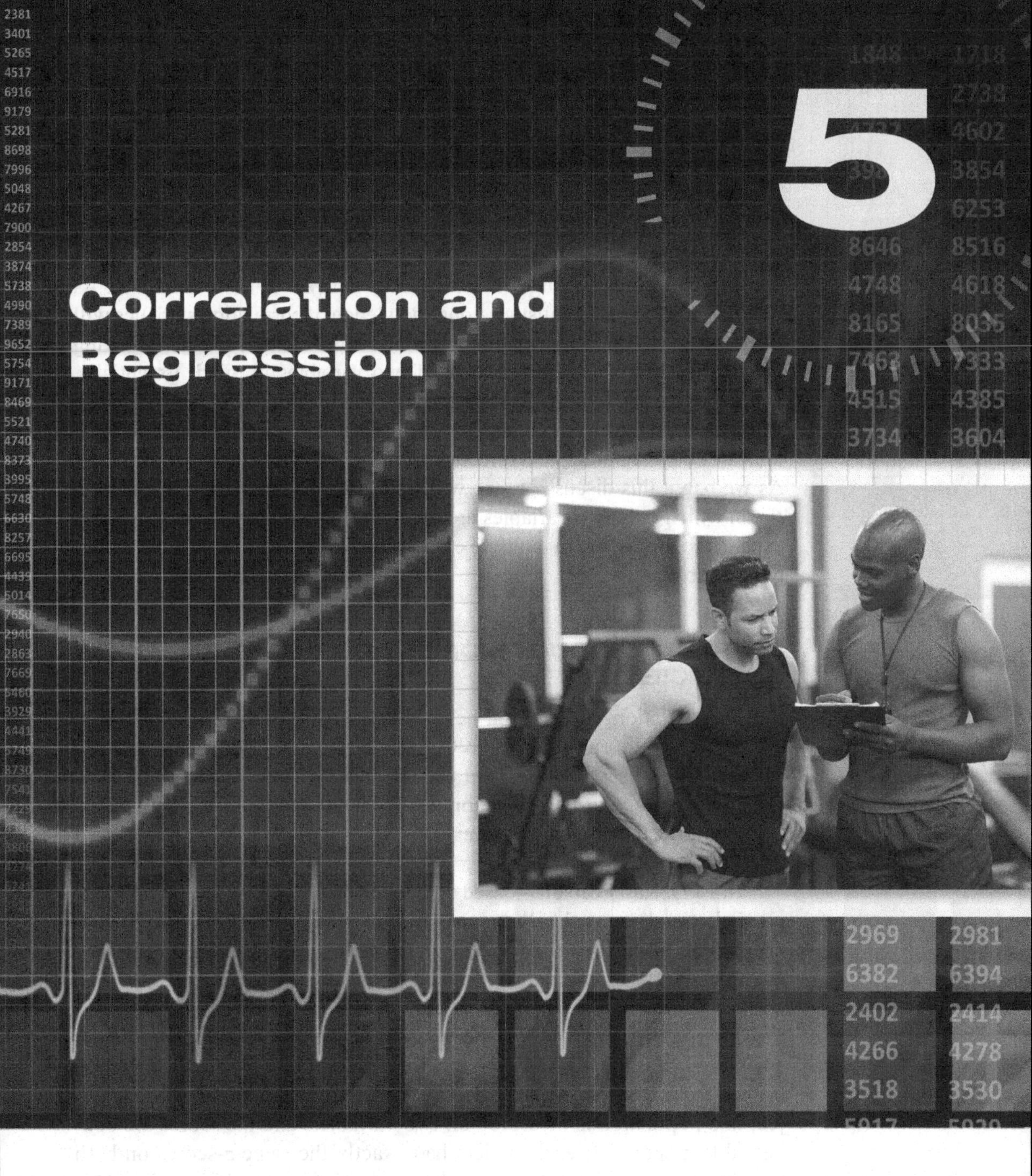

5 Correlation and Regression

INTRODUCTION

In the previous two chapters, we concentrated on descriptive statistics. These are tools that allow us to (1) describe a distribution of scores resulting from the measurement of some variable, (2) describe the location of any given score in a distribution, and (3) compare one distribution to other distributions. These descriptive statistics will also be critical to your understanding of the inferential statistics discussed in later chapters, as well.

Until now, we have presented and discussed a single variable at a time. In this chapter, we will begin to consider more than a single variable. We will begin to think of "pairs of variables"—for example, skinfold measurements and DXA-determined percentage of body fat. Height and weight is another example of paired variables. Undoubtedly, you can think of many examples of paired variables in your areas of interest.

We are concerned in this chapter with the concept called *correlation.* Correlation has a descriptive nature: it is a method for describing the relationship that might or might not exist between two or more variables. As with other descriptive statistics, your knowledge of correlation statistics will be necessary to your understanding of inferential statistics. When thinking about correlation, keep the concept of two variables in mind. Essentially, you are interested in how the two variables relate to one another.

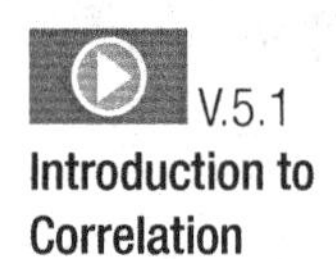
V.5.1
Introduction to Correlation

The **correlation coefficient** is a correlation statistic (i.e., a number) that indicates the magnitude and direction of the relationship between variables. As we begin this discussion, we will limit ourselves to simple correlation *(r)*, in which only two variables are involved. Later in this chapter, we will expand the discussion to multiple correlation *(R)*, which involves more than two variables.

PEARSON PRODUCT-MOMENT CORRELATION COEFFICIENT

Is there a relationship between height and weight? Is there a relationship between a 1-mile run time and VO_2max? Is there a relationship between amount of feedback provided and performance? The answer to all of these questions is yes. Our next question is, "Is there a way to describe numerically how much of a relationship exists?" The answer to this question is also yes.

In 1895, Karl Pearson—based on previous work by his mentor, Sir Francis Galton—devised the **Pearson product-moment (PPM) correlation coefficient,** also called simply Pearson's *r*. It is widely used in all sciences, including human performance, to quantify the magnitude and direction of a correlation between any two variables.

From a mathematical perspective, the value of *r*, which ranges from +1.0 to −1.0, reflects the extent to which the *z*-scores in one distribution are related to the *z*-scores in another distribution. Assume that each subject in a population is measured on two variables, and the scores are converted to *z*-scores. If each subject had exactly the same *z*-scores on both variables, the resulting PPM correlation coefficient would be +1.0. This would signify a perfect, *positive* correlation. For example, suppose a 600 yd run/walk test was administered to a group of subjects. After recording the times, the researcher finds the course to be slightly less than 600 yds and so decides to add five seconds to everyone's score. In this case, the PPM correlation coefficient between the original times and the "corrected" times would be +1.0. This is because adding five seconds to everyone's time would not change anyone's *z*-score. Remember the equation for the z-score is:

$$z\text{-}score = \frac{\bar{X} - X}{s}$$

For the corrected times, the numerator would remain the same as for the original times, because not only would X increase by 5, but so would $\bar{X}$. The denominator would not change, because adding five seconds to everyone's time does not change the variability of the scores.

Correlation Direction

The sign of the PPM correlation coefficient indicates the direction of the relationship between the two variables. If the coefficient is positive, this indicates that, generally, the high scores in the first distribution (Xs) are connected or related to the high scores in the second distribution (Ys), and the low scores in the first distribution are associated with the low scores in the second distribution. For example, if the r between height and weight is +.70, this signifies that taller people tend to weigh more than shorter people, and shorter people tend to weigh less than taller people. Obviously, this is not true for *all* people. Another example is the relationship between systolic blood pressure (SBP) and diastolic blood pressure (DBP). Figure 5.1 is a *scattergram* (we will elaborate on this type of graph later in this chapter) of these data, where the correlation is approximately +.70. Note that to create this plot, we must have a value for each variable (X and Y—systolic and diastolic blood pressure, in this case) for each subject.

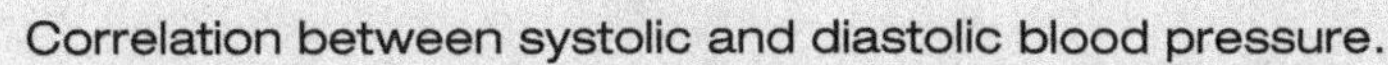

FIGURE 5.1 Correlation between systolic and diastolic blood pressure.

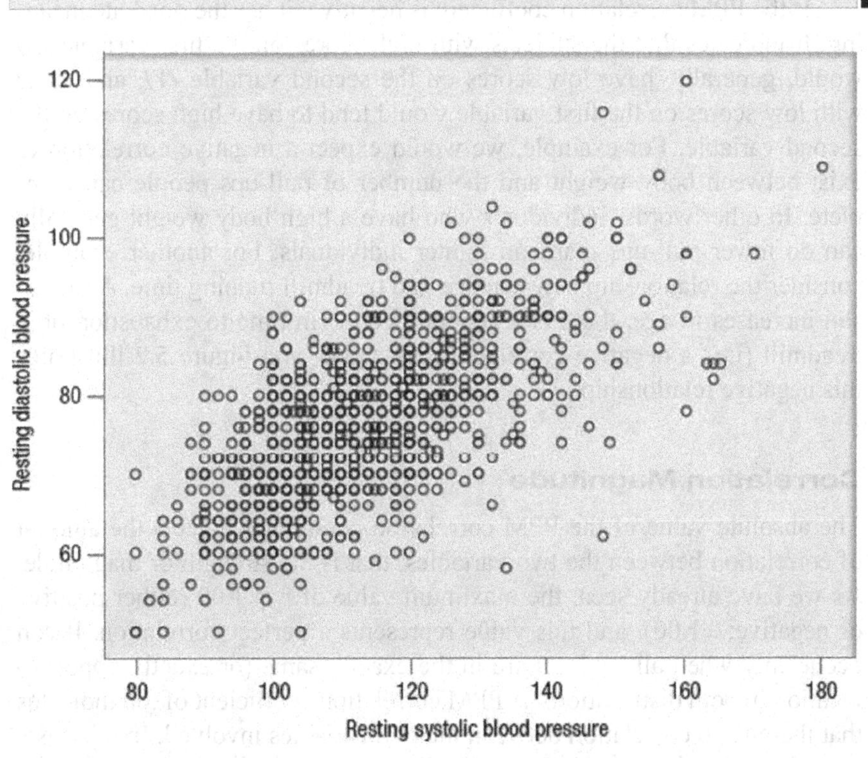

FIGURE 5.2 Negative correlation between age and treadmill time to exhaustion.

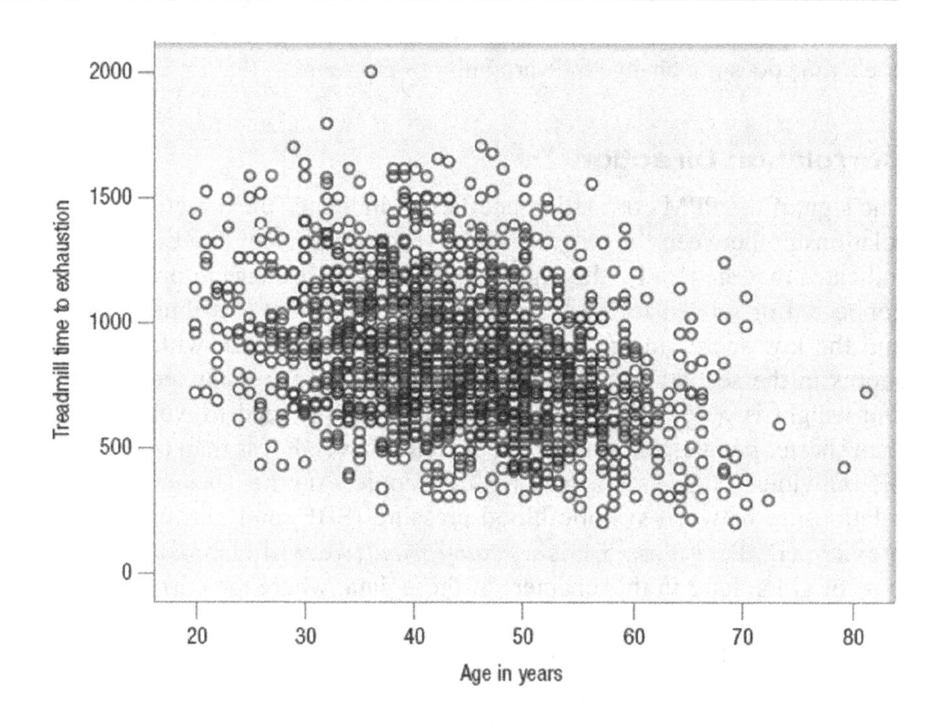

If the PPM correlation coefficient is negative, it has the opposite meaning. It indicates that the subjects with high scores on the first variable *(X)* would, generally, have low scores on the second variable *(Y)*, and those with low scores on the first variable would tend to have high scores on the second variable. For example, we would expect a negative correlation to exist between body weight and the number of pull-ups people can complete. In other words, individuals who have a high body weight generally can do fewer pull-ups than can lighter individuals. For another example, consider the relationship between age and treadmill running time. As a person increases in age, there is a general decline in time to exhaustion on a treadmill (i.e., a negative correlation; $r = -.34$), and Figure 5.2 illustrates this negative relationship.

Correlation Magnitude

The absolute value of the PPM correlation coefficient reflects the amount of correlation between the two variables, that is, its strength or magnitude. As we have already seen, the maximum value of r is 1.00 (either positive or negative; ±1.00), and this value represents a perfect correlation. It can occur only when all subjects are in the exactly same (or exactly opposite) location in both distributions. A PPM correlation coefficient of .00 indicates that there is no correlation between the two variables involved. In this case, any particular score on the first variable could be associated with *any* score

Little to no correlation between weight and age in adults. FIGURE 5.3

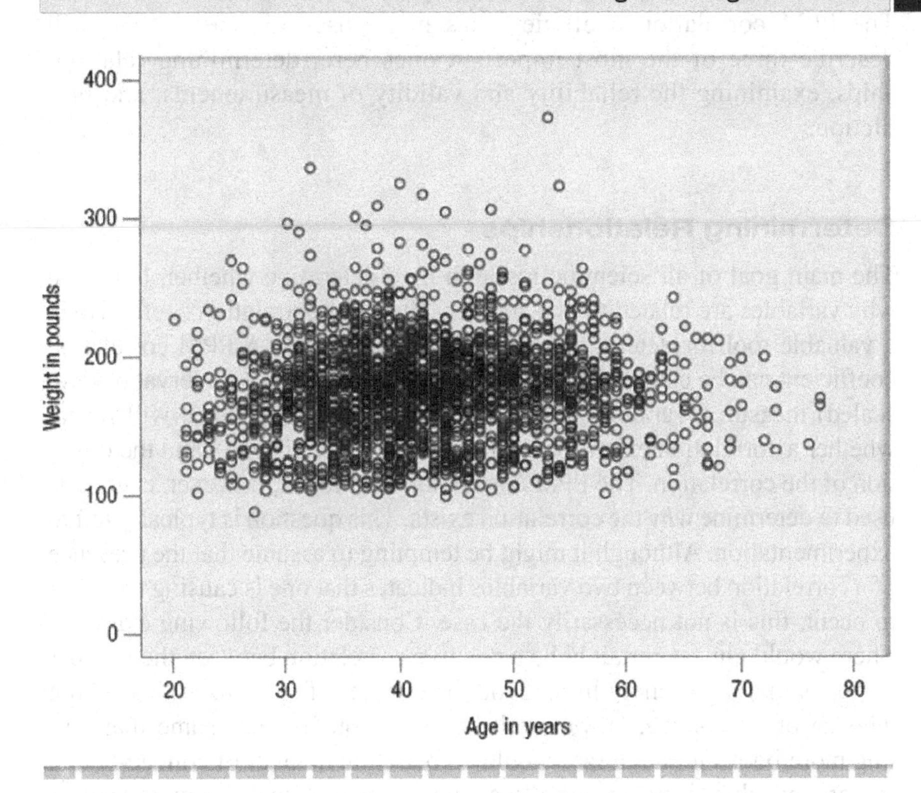

on the second variable. Figure 5.3 illustrates the relationship between age in years and weight for adults, where the correlation is nearly zero.

As the value of *r* increases from .00 and goes toward 1.00, in absolute value, the strength of the correlation increases, and the possible range of scores on the second variable *(Y)* for all subjects with a particular score on the first variable *(X)* decreases. This is shown in the scattergrams in Figure 5.4, illustrating correlations of approximately .30, .60, and .90.

Correlations of approximately .30, .60, and .90. FIGURE 5.4

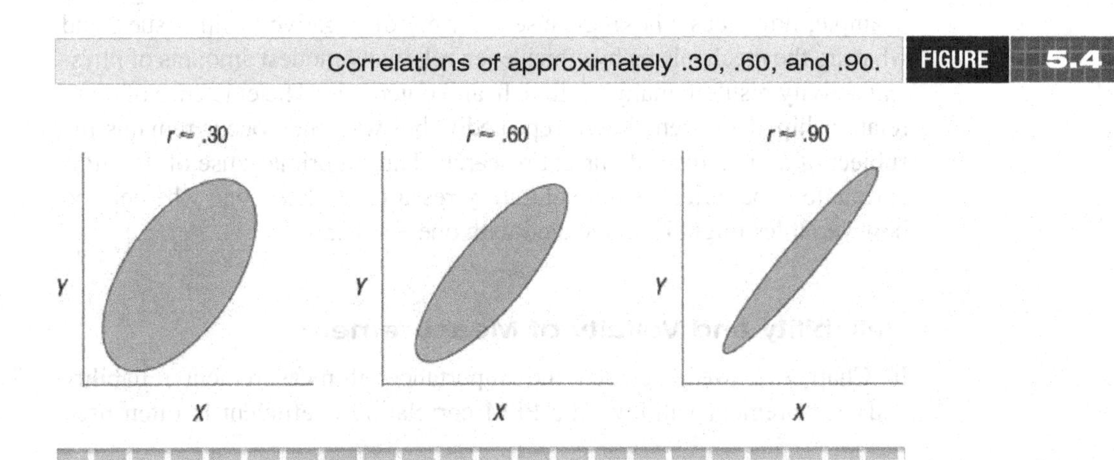

USES OF THE PPM CORRELATION COEFFICIENT

The PPM correlation coefficient has many uses in science. We will describe three of the most important ones here: determining relationships, examining the reliability and validity of measurements, and prediction.

Determining Relationships

The main goal of all scientific research is to determine whether, how, and why variables are related to one another. The PPM correlation coefficient is a valuable tool for determining "whether" and "how". A PPM correlation coefficient can be calculated for any two continuously (i.e., interval or ratio scaled) measured variables available for the same subjects. It will reveal whether a correlation exists, how strong it is (if there is one), and the direction of the correlation. The PPM correlation coefficient, however, cannot be used to determine *why* the correlation exists. This question is typically left to experimentation. Although it might be tempting to assume that the presence of a correlation between two variables indicates that one is causing the other to occur, this is not necessarily the case. Consider the following example: There would almost certainly be a positive correlation between the number of books in a high school library and the number of students at that school who go on to college. However, it is not reasonable to assume that putting more books in the library would increase the number of college-bound seniors, nor does it make sense to suppose that if, somehow, more seniors were convinced to go to college, more books would magically appear in the library. Neither of these variables is *causing* the other to occur. The positive correlation probably has something to do with the comparative wealth of various communities. Both the number of books in the school library and the number of students who go to college are probably higher in wealthier communities. Lower values for the two variables probably occur in less affluent communities, and this may be the reason for the positive correlation.

Even when empirical evidence seems to confirm a cause-and-effect relationship, it may take years of research to determine the mechanisms by which a change in one variable causes the other variable to change. For example, how does smoking cause a variety of negative health issues, and what are the mechanisms by which even relatively modest amounts of physical activity result in many positive health outcomes? The existence of these relationships has been shown repeatedly, but why they occur remains the subject of a great deal of current research. Thus, a primary use of the PPM correlation coefficient is in exploratory research to determine whether and how variables might be associated with one another.

Reliability and Validity of Measurement

In Chapter 2, we discussed the importance of measurement reliability and measurement validity. The PPM correlation coefficient is often used (sometimes incorrectly) to assess the amount of these two important characteristics of the measurement processes.

Reliability of Measurement

Recall that reliability of measurement is concerned with the consistency of the measurement process. Thus, if a measurement is made twice under the same conditions, a correlation coefficient calculated for the two sets of scores should be high if the measurement process is reliable (i.e., consistent). Suppose, for example, you want to determine the reliability of using a particular scale to measure body weight. If subjects were weighed and then immediately re-weighed, the correlation between the two sets of scores would reflect the reliability of the scale. Would the correlation show a perfect reliability? Probably not, because there are several potential sources of error. Perhaps each subject did not stand on the scale in the same place both times; the scale may have been moved to a slightly different spot between the weighings; the person recording the weight may have moved and viewed the scale dial from a different perspective; or subjects may have added or subtracted a piece of clothing or something from a pocket between the two measurements.

Use of the PPM correlation coefficient to estimate reliability is termed an **interclass correlation** procedure. Technically, the use of the PPM correlation coefficient in this type of test–retest protocol is not correct, because this statistic cannot account for mean differences in the two sets of scores, For example, if the spring in the scale was weakened in some way between weighings, and this caused everyone's weight measurement to increase by exactly 5 pounds (and no other errors occurred), the PPM correlation coefficient between the two distributions would be +1.00, indicating perfect reliability—even though the weight for each subject would not be consistent from the first to the second weighing. There would be a 5-pound difference between the means of the two weighings. We will suggest a solution to this problem later in the textbook. Nevertheless, in a test–retest situation, the PPM correlation coefficient can provide a reasonable estimate of reliability—so long as there is no significant mean difference between the two measurements, as in the weakened spring example.

Validity of Measurement

The PPM correlation coefficient can be used in the assessment of both concurrent and predictive validity of measurement (see pg., 37). Recall that validity of measurement is concerned with the degree to which a measurement process actually results in an assessment of the characteristic being investigated. A method to determine validity is to administer a criterion measurement process known to measure the characteristic accurately (although the criterion process might be unduly costly or very time consuming) and also administer a new measurement process thought to assess the same characteristic. The PPM correlation coefficient for the two resulting sets of scores provide evidence about the validity of the second process.

As discussed in Chapter 2, in the case of concurrent validity, the criterion and the new measurement process are both administered in the same timeframe. In the case of predictive validity, the criterion measure is obtained sometime in the future for which predictive validity evidence is being sought.

The following example illustrates the use of the PPM correlation coefficient for determining concurrent validity. Using a treadmill, expired air collection equipment, and various exercise protocols, we can calculate a subject's VO_2max. Values on this test are considered to be excellent indicators of cardiorespiratory fitness level.

In the 1960s, Dr. Kenneth H. Cooper (1968) was seeking a field test for estimating VO_2max that would be practical and would not require much equipment. He developed what became known as the Cooper 12-minute run test for determining the maximum distance a person can run (or walk) in 12 minutes. Although the PPM correlation coefficient differed for various subsets of subjects, Cooper found a high correlation ($r = .90$) between scores on the VO_2max treadmill test and scores on the Cooper 12-minute run for some categories of subjects. This correlation represents evidence for the concurrent validity of the Cooper 12-minute run test for some subjects.

Prediction

The third use of the PPM correlation coefficient is for prediction. When two variables are correlated to some degree, we can predict a subject's score on one of the variables if we know the score on the other variable, better than if no correlation exists between the two variables. The stronger the correlation, the more accurate the prediction will be. With a perfect correlation (±1.00), perfect prediction is possible. With less than a perfect correlation, some amount of error will exist in the prediction, but in this case, we can calculate a statistic (the **standard error of estimate**) to assess the accuracy of the prediction.

The use of correlation for prediction is quite common in many aspects of life. Insurance companies (e.g., predicting age at death based on percentage of fat at age 20), college admissions offices, physicians, coaches, weather forecasters, and gamblers are just a few examples where individuals rely heavily on the ability to make accurate predictions. Specific human performance examples include the use of skinfold measurements to estimate body fat and distance runs to estimate VO_2max. In addition, lifestyle behaviors (e.g., smoking, physical activity, and blood chemistry measurements) are frequently correlated with quality of life variables, including all-cause mortality. The mechanics of calculating predictions will be explained later in this chapter. Let's first look at pictorial representations of the correlation between two variables.

PICTORIAL REPRESENTATIONS OF THE CORRELATION BETWEEN TWO VARIABLES

Scattergrams

Constructing a **scattergram** (also called a **scatterplot**) is often the next step after you have gathered the scores for the two variables for which you wish to determine the relationship. Usually, a statistical computer program is used to produce the scattergram. Using the program, we plot each subject's two scores on one graph, locating one score on the *X*-axis and the other score on the *Y*-axis. The result is a visual display of the relationship between the two variables. We have already presented examples of scattergrams. The scattergram is an important tool, because the PPM correlation coefficient is accurate only for *linear* relationships between variables. If a linear relationship exists between the variables, when you inspect the scattergram you will be able to draw a line through the middle of all the data points, and the line will be straight, not curving.

V.5.2 SPSS Correlation Scatterplots and Matrices

V.5.2R RStudio Correlation Scatterplots and Matrices

Again, the PPM correlation coefficient estimates only linear correlations between variables. Variables can be related in a curvilinear, rather than linear, fashion. For example, Keeley, Zayac, and Correia (2008) found a curvilinear relationship between statistics anxiety and performance in a statistics course. Students with moderate statistics anxiety performed better than did students with either high or low anxiety. Methods exist for calculating the level of correlation in nonlinear relationships, but we will not present them in this textbook. Figure 5.5 illustrates a nonlinear relationship. Here, subjects who scored low or high on *X* tended to score lower on *Y*, but subjects with middling scores on *X* tended to score higher on *Y*.

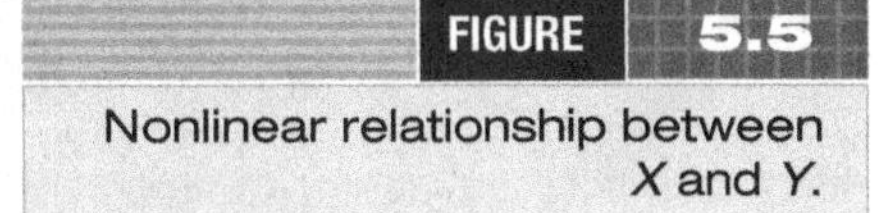
FIGURE 5.5 Nonlinear relationship between *X* and *Y*.

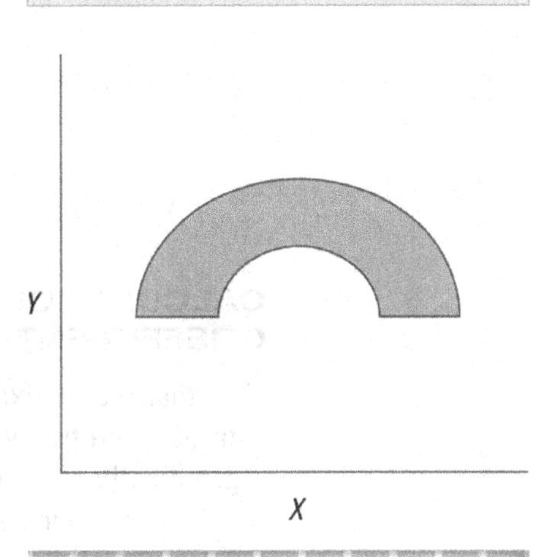

Line of Best Fit

Once we have determined, through visual examination of a scattergram, that the relationship between the two variables of interest is linear, we often wish to calculate the equation of the straight line that best fits the data. Various criteria could be used to define "best fit", but most commonly researchers use the *least squares criterion.* This definition of best fit produces the line that minimizes the result when the vertical distance between the line and every data point is determined, squared, and summed. The line of best fit is also called the *regression line* or *prediction line*. Its method of calculation according to the least squares criterion and its use will be illustrated later in this chapter. The distance between the line of best fit and any given data point is called a residual (or error); thus, the least squares criterion minimizes the sum of the squared residuals. We will return to the topic of residuals later in this chapter. Figure 5.6 illustrates the line of best fit for the systolic and diastolic blood pressure scattergram.

FIGURE 5.6 Line of best fit.

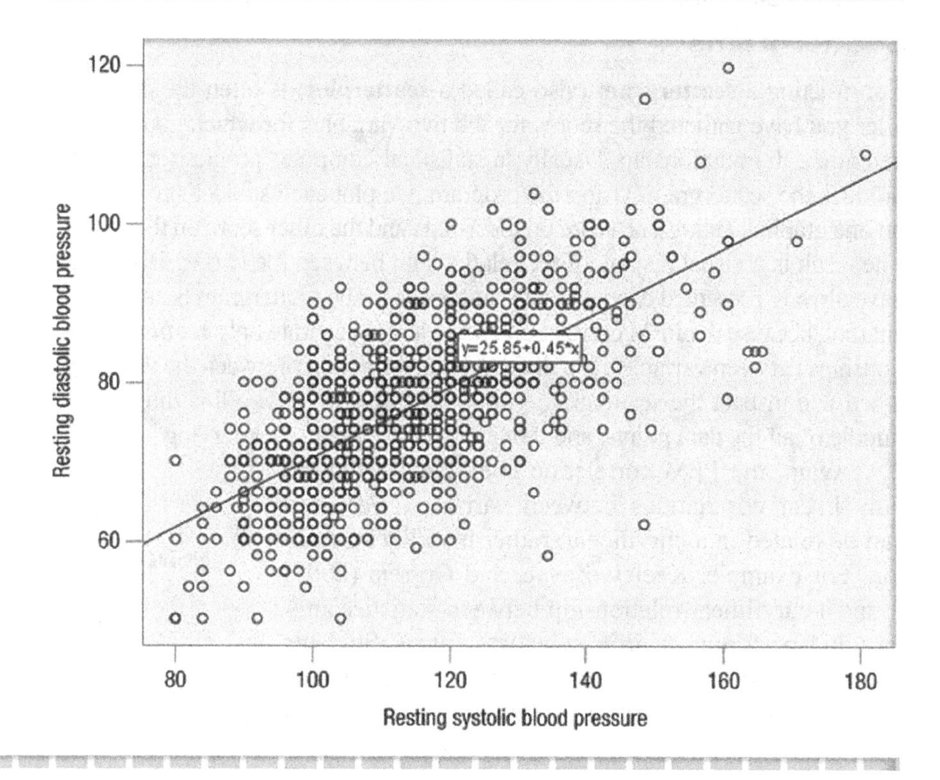

CALCULATION OF THE PPM CORRELATION COEFFICIENT

Tow that we've examined pictorial representations depicting the relationship between two variables, we turn to the calculation of r. Any of several algebraically equivalent equations can be used. Your choice will depend on the form of your data (i.e., raw scores, standard scores, or descriptive statistics), as will be discussed below. In any case, because a large number of calculations are required, the most efficient and accurate way to calculate r is to enlist a statistical computer program.

Definitional Equation

The definitional equation for the PPM correlation coefficient is

$$r = \frac{\sigma_i^N \left(Z_{X_i} * Z_{Y_i}\right)}{N}$$

Eq. 5.1

where

r = the PPM correlation coefficient,
Z_{Xi} = the z-score for a particular subject on variable X,
Z_{Yi} = the z-score for the same subject on variable Y, and
N = the number of pairs of scores.

This equation should be unsurprising to you, given our discussion of the connection between r and z-scores earlier in the chapter. The following illustration will help you see how this equation works to determine the value of r. Consider the small data set of scores 1, 2, 3, 4, and 5 to be a population of scores. Assume each subject has the same value for both X and Y. If we convert each score in the two distributions to a z-score (by using the population equation for calculating the standard deviation), the results are −1.41, −0.7, 0.0, 0.7, and 1.41, for both distributions. If we multiply the z-scores for each observation and sum the resulting products, we find a numerator of 5 (i.e., 2 + 0.5 + 0.0 + 0.5 + 2 = 5). Notice that this answer is equal to N. Therefore, if each subject's z-scores for variable X and variable Y have the exact same location, the numerator of the definitional equation for r is equivalent to squaring the z-score for either of the distributions. The result is N/N, or 1.00—a perfect correlation. If Z_{Xi} and Z_{Yi} are not identical, the numerator will necessarily be smaller than N, and the absolute value of r will be less than 1.00.

Calculating Formula for PPM Correlation Coefficient Using Means and Standard Deviations

Although the definitional equation for calculating the PPM correlation coefficient helps us see how the value of r is derived, it is not very convenient for calculating r. Ironically, some algebraically equivalent equations exist that appear more cumbersome but actually are simpler to apply. For example, if the two variables' means and standard deviations are available, you can use the following equation to calculate r:

$$r = \frac{\frac{\sigma(X * Y)}{N} - \bar{X} * \bar{Y}}{\sigma_X * \sigma_Y} \quad \text{Eq. 5.2}$$

where

r = the PPM correlation coefficient,
$\bar{X} * \bar{Y}$ = the product of a subject's X and Y scores (called the *cross product*),
N = the number of pairs of scores,
$\bar{X}$ = the mean of variable X,
$\bar{Y}$ = the mean of variable Y,
σ_X = the standard deviation for variable X, and
σ_Y = the standard deviation for variable Y.

Consider the data shown in Table 5.1. Using Equation 5.2, we calculate the correlation as follows:

$$r = \frac{\frac{55}{5} - 3 * 3.2}{1.41 * 1.17}$$

$$= \frac{11 - 9.6}{1.65} = \frac{1.4}{1.65} = .85$$

TABLE 5.1

Data set.

	X	Y	XY
	1	2	2
	2	3	6
	3	2	6
	4	4	16
	5	5	25
Sum	15	16	55
Mean	3	3.2	
σ	1.41	1.17	

Calculating Formula for Correlation Coefficient Using Raw Scores

The PPM correlation coefficient can also be calculated from raw scores. Although the equation looks imposing, it requires only N, sums of the raw scores, the sum of the cross products, and sums of squares of the raw data. The equation is:

$$r = \frac{N(\sigma XY) - (\sigma X)(\sigma Y)}{\sqrt{[N(\sigma X^2) - (\sigma X)^2][N(\sigma Y^2) - (\sigma Y)^2]}} \quad \text{Eq. 5.3}$$

where

r = the PPM correlation coefficient,
N = the number of pairs of scores,
σXY = the sum of the cross products,
σX = the sum of the X variable scores,
σY = the sum of the Y variable scores,
σX^2 = the sum of the squares of the X scores, and
σY^2 = the sum of the squares of the Y scores.

TABLE 5.2

Data set.

	X	Y	X^2	Y^2	XY
	1	2	1	4	2
	2	3	4	9	6
	3	2	9	4	6
	4	4	16	16	16
	5	5	25	25	25
Sum	15	16	55	58	55

When we use the same data as in the previous calculation, we find the same value for the PPM correlation coefficient. See Table 5.2.

Thus,

$$r = \frac{5(55) - (15)(16)}{\sqrt{[5(55) - (15)^2][5(58) - (16)^2]}}$$

$$= \frac{275 - 240}{\sqrt{[275 - 225][290 - 256]}}$$

$$= \frac{35}{\sqrt{[50][34]}} = \frac{35}{\sqrt{1700}} = \frac{35}{41.23} = .85$$

INTERPRETATION OF THE PPM CORRELATION COEFFICIENT

The PPM correlation coefficient allows us to describe the magnitude and direction of the relationship between two variables and to compare one correlation to another. However, understanding what the actual value of the correlation coefficient (say, .70) means is not intuitive. It is relatively easy to grasp the meaning of a correlation of 1.00 or .00, but how do we compare an r of .30 to an r of .70 or −.40? Unfortunately, the PPM correlation coefficient is not a linear scale, so we cannot say that an r of .80 is twice as strong as an r of .40. The notion that a correlation of .70 implies 70% of a possible perfect correlation is also an inaccurate interpretation. Following are three ways to approach the interpretation of a particular value of r: verbal description, the coefficient of determination, and statistical significance.

Verbal Description of r

It is tempting to try to attach verbal descriptions to various values of r, and this is sometimes seen in the research literature. For example, the verbal descriptions given in Table 5.3 might seem reasonable. Note that only positive values are presented here; recall that we interpret the magnitude of the PPM correlation coefficient by looking at the absolute value of r.

Verbal descriptions such as those given in Table 5.3 are inappropriate because the meaning of the value of r is situation specific. In other words, in some circumstances a correlation coefficient of .70 might be considered low, and in other situations this same value would be considered high. For example, the creators of a nationally administered test to be used to determine high school scholastic achievement for the purpose of assessing future college performance would consider a reliability coefficient of .70 to be unsatisfactory. On the other hand, when the characteristics of interest are highly variable and only imprecise measuring instruments are available, such as in the measurement of self-esteem following an exercise treatment, a correlation coefficient of the same value (.70) might be considered acceptable.

TABLE 5.3

Verbal descriptors of values of r.

VALUE OF r	DESCRIPTION
1.00	perfect correlation
.80–.99	very high correlation
.70–.79	high correlation
.40–.69	moderate correlation
.30–.39	fairly low correlation
.20–.29	low correlation
<.19	very low correlation
0.00	no correlation

Thus, for verbal descriptions to be helpful, we must know and consider the details of the situation, including how the correlation was calculated, the nature of the variables involved, the precision of the measurement process, and the use to be made of the correlation coefficient. Perhaps the only verbal "rule of thumb" that is generally useful is that values of r greater than .70 are required for meaningful predictions. Another method to help us grasp the meaning of a particular value of r is to calculate the coefficient of determination.

Coefficient of Determination (r^2)

The **coefficient of determination** is obtained by squaring the value of r. The value of r^2 reflects the amount of variance in the Y variable (the criterion variable) that is accounted for by the variation in the X variable (the predictor variable). For example, if the correlation between height and weight is .70, then 49% of the variance in the weight scores can be attributed to the variance in the height scores (see Table 5.4). The remaining 51% of the variance in the weight scores is attributable to factors other than the differences in height. The unaccounted-for variation is often called *residual*, or error. Notice in Table 5.4 how much variance in Y is accounted for by X with different values of r.

Statistical Significance of r

If two variables are not correlated (r = .00) in a population, we would expect that if we drew a random sample of N subjects from that population and

TABLE 5.4 The coefficient of determination (r^2).

r	r^2	% OF VARIANCE ACCOUNTED FOR	% OF VARIANCE NOT ACCOUNTED FOR (RESIDUAL OR ERROR)
0.10	0.01	1	99
0.20	0.04	4	96
0.30	0.09	9	91
0.40	0.16	16	84
0.50	0.25	25	75
0.60	0.36	36	64
0.70	0.49	49	51
0.80	0.64	64	36
0.90	0.81	81	19
0.95	0.90	90	10
0.99	0.98	98	2
1.00	1.00	100	0

calculated the correlation coefficient for the sample, the value of r should be .00, or close to .00. However, some low value of r (either positive or negative, but close to zero) could result by chance. We might ask, how large would the sample r (the correlation coefficient for the sample) have to be before it would become untenable to believe that the sample could have come from a population with no correlation (i.e., $r = .00$) between X and Y? In other words, how large does r need to be, to be considered statistically significantly different from zero?

Determination of the statistical significance of r is a function of only three factors:

1. the value of r,
2. the size of the sample (N), and
3. the **level of confidence** that you want to have that your decision regarding the statistical significance of r is correct.

A table has been developed to give the statistical significance of r for various r values, sample sizes, and desired confidence levels. This table (Statistical Significance of the Correlation Coefficient) can be found in Appendix A.2.

The first column of the table is labeled *df*, which stands for degrees of freedom (see Chapter 3). In the table, the value of *df* is equal to the number of pairs of scores used to calculate r, minus 2 (i.e., $N - 2$). For example, if a sample contains 22 subjects, $df = 20$.

The next three columns have to do with the level of confidence you can have that you are making a correct decision about the statistical significance or true value of r. The headings of .10, .05, and .01 denote the probability that the value of r could have occurred by chance when, in fact, the true value of r in the population is actually .00. In other words, these values indicate how often you would be wrong (as a percentage) if you were to declare *r* to be significantly different from zero.

Let's look at an example. Assume that your variables are BMI and the sum of skinfolds. The calculated value of *r* for your sample is .50, and the degrees of freedom are 20. By referring to Appendix A.2, you can determine that, if the correlation coefficient for the population is actually .00, the probability that a correlation as high as .360 could occur by chance is 10%—and your *r* value is higher than that. Hence, you could declare your correlation coefficient of .50 to be statistically significant (i.e., different from zero) and be 90% confident in making this statement. In fact, a correlation as high as .423 could occur by chance only 5% of the time, so your correlation of .50 could be declared significantly different from zero with 95% confidence. Could you be 99% confident in declaring an *r* value of .50 to be statistically significant? No. To be able to do that, you would need an *r* value greater than .537. Based on the evidence at hand, the most reasonable conclusion is that the correlation coefficient for the population is *not* .00. In fact, the most reasonable estimate is that the correlation coefficient for the population is 0.50.

Notice the relationship that exists between sample sizes and *r* values in the table. Can you see how as the sample size *(N)* increases, the value of the correlation decreases in each row? This suggests that with a large sample, you can claim that the correlation coefficient is significantly different from zero even though it might be quite small and account for relatively little variance. With large *N*s, the value of *r* required for a determination of statistical significance can be small; conversely, smaller *N*s require larger values of *r*. This is because a small sample is more likely to produce a spuriously high value of *r*, when the correlation coefficient for the population is actually zero, than is a larger sample.

Remember that these three methods—verbal description, coefficient of determination, and statistical significance—are methods that can help you interpret correlation coefficients. They are not definitive: You still must consider the details of each situation. For example, let's say that with a sample size of 95, you find an *r* value of .22. Use the table in Appendix A.2 to verify that this value of *r* is statistically significant different from zero with 95% confidence. Now, calculate the coefficient of determination (r^2). Only 4.8% of the variability of the *Y* variable is accounted for by the variability of the *X* variable. Despite the information given in the Statistical Significance of the Correlation Coefficient table, the ability to predict *Y* from *X* is actually quite poor.

PREDICTION

As we pointed out in Chapter 1, two important purposes for conducting science are to predict and to explain relationships. Predicting is the easier of the two. Explanation is more difficult. Consider that there is a relationship between cigarette smoking and the occurrence of lung cancer. This relationship has been verified repeatedly. However, what is the exact mechanism that causes cigarette smoking to increase the risk of lung cancer? It involves behaviors, habits, environment, genetics, physiology, and a host of other

FIGURE 5.7 Positive correlation between systolic and diastolic blood pressures.

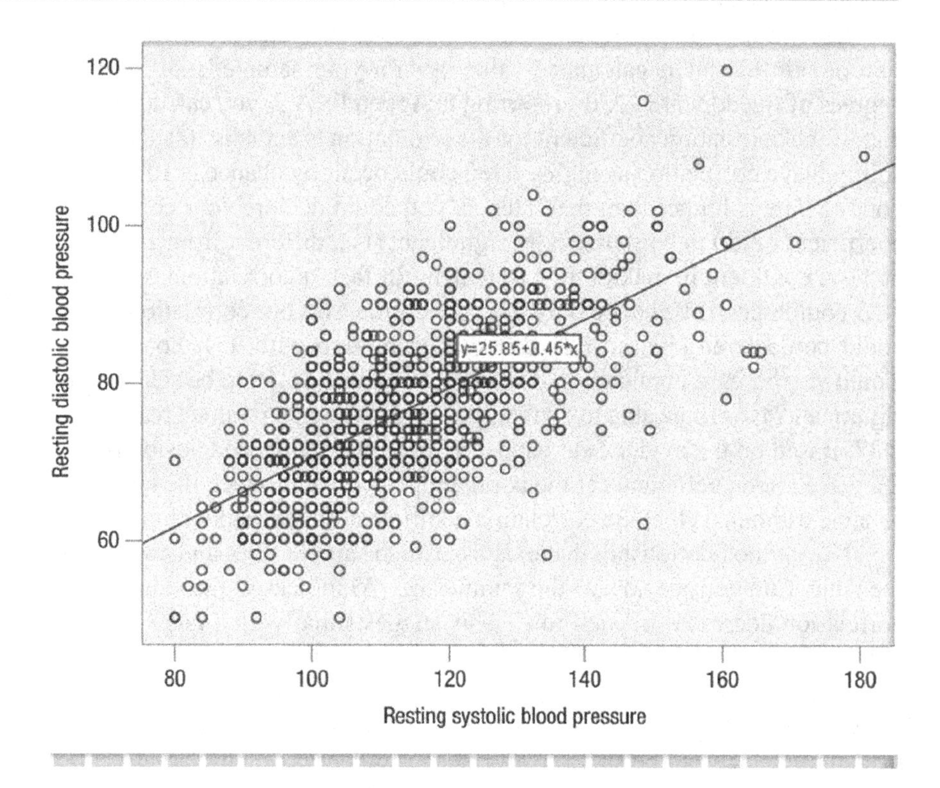

intervening factors. For the time being, it is important that you realize that if two variables are related, one can be predicted from the other. However, predictions are virtually never perfect in real life. While some smokers do not develop lung cancer, some individuals who have never smoked do develop lung cancer.

Let's return to the example of blood pressure to illustrate how the existence of a relationship between two variables permits prediction (see Figure 5.7). Notice that as systolic blood pressure increases, so does diastolic blood pressure. However, since the relationship is not perfect, individuals with generally lower systolic values have a range of diastolic values. This is true in the extremes but perhaps is most noticeable in the middle portions of the distributions.

Calculating the Line of Best Fit

The straight line in Figure 5.7 illustrates the line of best fit or regression line previously described. Another name for it is *prediction line*, reflecting the fact that if you know a value of the X variable, you can predict the corresponding value of Y by starting at the known value on the X axis, moving vertically up to the line, and then moving to the left, parallel with the X axis, to the Y axis, to find the predicted Y value, which is designated as $\hat{Y}$. If the relationship is perfect (±1.00), the predictions would have no error.

Perfect correlation between length in inches and length in centimeters. FIGURE 5.8

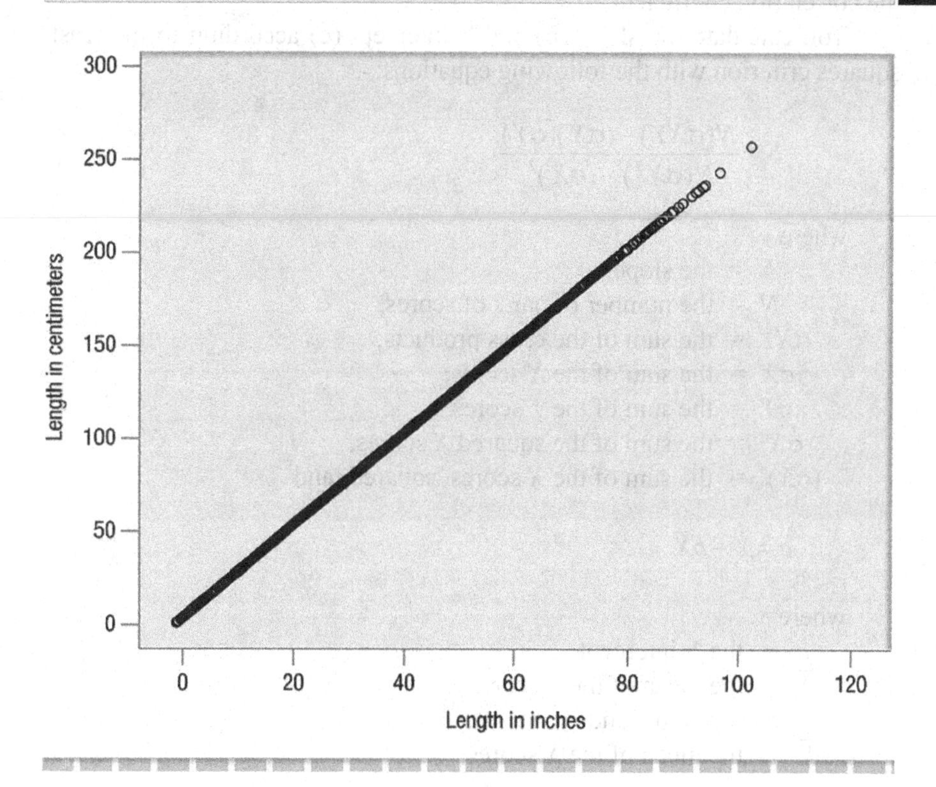

Figure 5.8 illustrates this. The equation for predicting length in centimeters from length in inches is:

length in centimeters = 2.54 * length in inches

You may recall from studying geometry that the equation for a straight line is

$$\hat{Y} = bX + c \quad \text{Eq. 5.4}$$

where
$\hat{Y}$ = the predicted value on Y (predicted from a value on X),
b = the slope of the line, and
c = the Y-intercept.

For any value of X, you can enter it into the equation and, by performing the calculation, estimate (i.e., predict) the value for Y based on the relationship between X and Y.

Note that the slope of the line of best fit is directly related to the Pearson product-moment correlation coefficient presented earlier. The slope will have the same sign as the correlation coefficient. In fact, if the values for variables X and Y are in the form of z-scores, the value of the slope *(b)* will be equal to the correlation coefficient. This further illustrates the point we discussed earlier in the chapter about the connection between *z*-scores and the correlation coefficient.

You calculate the slope (b) and Y-intercept (c) according to the least squares criterion with the following equations:

$$b = \frac{N(\sigma XY) - (\sigma X)(\sigma Y)}{N(\sigma X^2) - (\sigma X)^2}$$ **Eq. 5.5**

where

b = the slope,
N = the number of pairs of scores,
σXY = the sum of the cross products,
σX = the sum of the X scores,
σY = the sum of the Y scores,
σX^2 = the sum of the squared X scores,
$(\sigma X)^2$ = the sum of the X scores, squared, and

$$c = \bar{Y} - b\bar{X}$$ **Eq. 5.6**

where

c = the Y-intercept,
$\bar{Y}$ = the mean of the Y scores,
b = the slope, and
$\bar{X}$ = the mean of the X scores.

V.5.3
SPSS Line of Best Fit and Example

V.5.3R
RStudio Line of Best Fit and Example

Notice that b is calculated from some of the values that were used to calculate the PPM correlation coefficient (see Eq. 5.3). We then use b and the mean values of X and Y to calculate c, the Y-intercept. In the equation for predicting length in centimeters, the value of c is zero. (Review Figure 5.8 to verify this.)

Note the differences in the three graphs and lines presented in Figure 5.9. Can you identify the one with the highest correlation

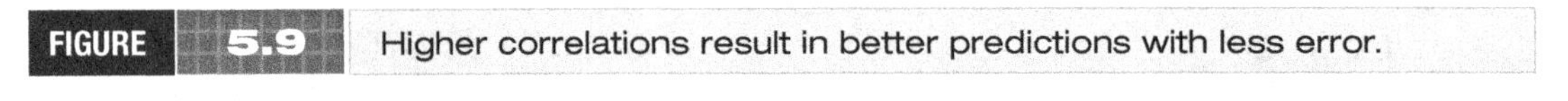
FIGURE 5.9 Higher correlations result in better predictions with less error.

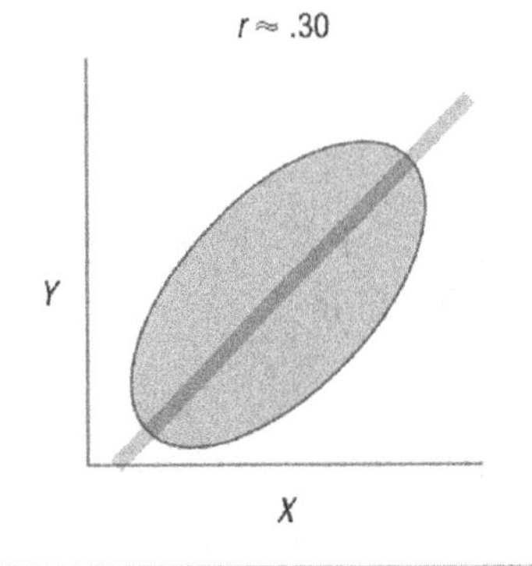

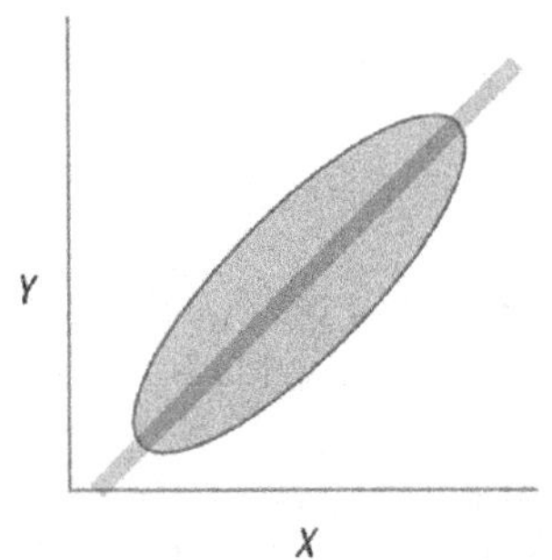

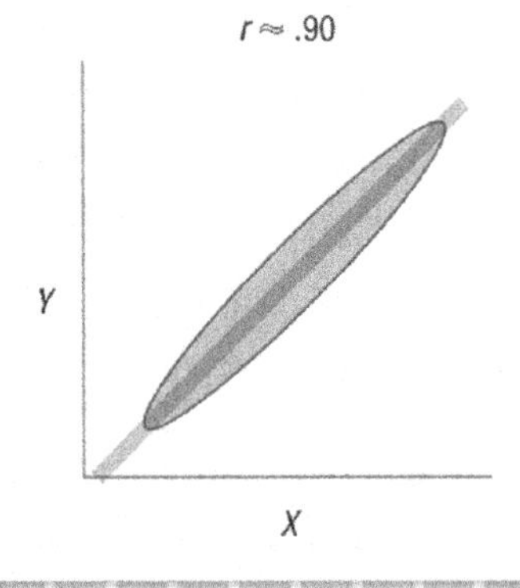

coefficient, which, therefore, would result in the best predictions? Recognize, however, that none of these examples is a perfect correlation, so all three will produce some amount of prediction error. We will examine this concept next.

PREDICTION ERROR

The amount of prediction error resulting from a line of best fit is a function of the correlation coefficient and the variability in the X and Y variables. Generally, as the correlation coefficient approaches 1.00 in absolute value, the predictions increase in accuracy. However, unless the correlation is perfect, there will always be some amount of error in the prediction. This error, introduced previously, is called by a number of different terms, including standard error of estimation (SEE), standard error of prediction (SEP), and standard error (SE).

You should be aware that the term *standard error* can refer to many different statistics. We suggest that, when reporting research, you use the term standard error carefully and fully explain what you mean by it. We recommend, for now, that you use SEE to refer to the amount of error in a prediction. The equation for the standard error of estimation is

$$s_{y \cdot x} = s_Y \sqrt{1 - r^2} \quad \text{Eq. 5.7}$$

where

$s_{y \cdot x}$ = the standard error of estimating Y from X,
s_Y = the standard deviation of the Y scores (those you are predicting), and
r = the correlation coefficient between the X and Y variables.

The SEE is actually the standard deviation of the errors of the predicted Y scores for a given value of X. Notice that the SEE depends on the s_Y (standard deviation of the Y scores) and the correlation *(r)* between X and Y. The higher the absolute value of the correlation coefficient, the closer the SEE is to zero. Recall that r^2 is the coefficient of determination (the amount of variation accounted for), so $1 - r^2$ is the residual or error; thus, SEE is the standard error of estimating Y from X. You should take a moment to confirm for yourself that regardless of the standard deviation of Y, when the correlation between X and Y is perfect, the SEE is zero. That is, when there is a perfect correlation between two variables, there is no error in predicting one from the other. The value of is $s_Y \sqrt{1 - r^2}$ algebraically equivalent to the standard deviation of the residuals. Let's illustrate this concept with the simple data set introduced earlier. See Table 5.5.

The equation for predicting Y from X is:

$$\hat{Y} = bX + c$$

TABLE 5.5 Data set.

	X	X^2	Y	XY	$\hat{Y}$	RESIDUAL ($Y - \hat{Y}$)
	1	1	2	2	1.8	0.2
	2	4	3	6	2.5	0.5
	3	9	2	6	3.2	−1.2
	4	16	4	16	3.9	0.1
	5	25	5	25	4.6	0.4
Sum	15	55	16	55	16	0.00
Mean	3		3.2		3.2	0.00
sd	1.41		1.17			.616

Let's first calculate b and c:

$$b = \frac{N(\sigma XY) - (\sigma X)(\sigma Y)}{N(\sigma X^2) - (\sigma X)^2}$$

$$b = \frac{5(55) - (15)(16)}{5(55) - (15)^2}$$

$$b = \frac{275 - 240}{275 - 225} = \frac{35}{50} = .70$$

$$c = \overline{Y} - b\overline{X}$$

$$c = 3.2 - .7 * 3 = 3.2 - 2.1 = 1.1$$

Thus, the equation for predicting Y from X for these data is:

$$\hat{Y} = .7X + 1.1$$

Let's calculate the predicted and residual value for Y from a score of 1 on X.

$$\hat{Y} = .7 * 1 + 1.1 = .7 + 1.1 = 1.8$$

Thus, the residual is 2 − 1.8 = 0.2. Take a moment to calculate the predicted Y and residual for the remaining four X values.

Now that you have the residuals, calculate the standard deviation of the residual scores by using Equation 3.7 from Chapter 3. Confirm that the value is 0.616.

Now, let's use Equation 5.7 and confirm that we find the same value:

$$\begin{aligned} s_{y \cdot x} &= s_Y \sqrt{1 - r^2} \\ &= 1.17\sqrt{1 - (.85)^2} \\ &= 1.17\sqrt{1 - .7225} \\ &= 1.17\sqrt{.2775} \\ &= 1.17 * .5267 = .616 \end{aligned}$$

As previously suggested, there will virtually always be some error in prediction. Figure 5.10 illustrates this in graphic form. Notice that for each value of *X*, there is a distribution of observed values of *Y*. The prediction equation gives a best estimate for *Y* for the given *X* (provided from the equation for the line) that is the mean of the scores on *Y* among those subjects who scored the given value on *X*. Confirm this concept with Figures 5.10 and 5.11. Figure 5.11 illustrates this concept with the blood pressure scattergram presented earlier in Figure 5.1.

Correlation, line of best fit, and standard error of estimate. FIGURE 5.10

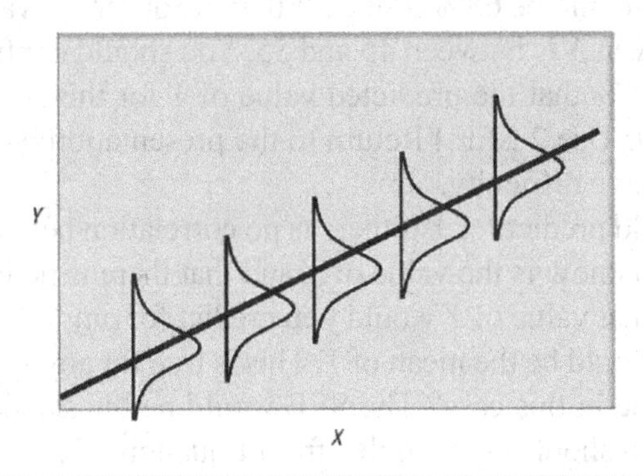

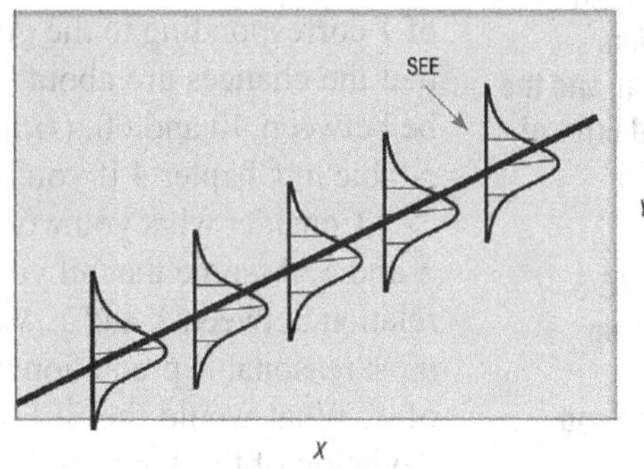

Blood pressure scattergram with line of best fit and standard error of estimate illustrated. FIGURE 5.11

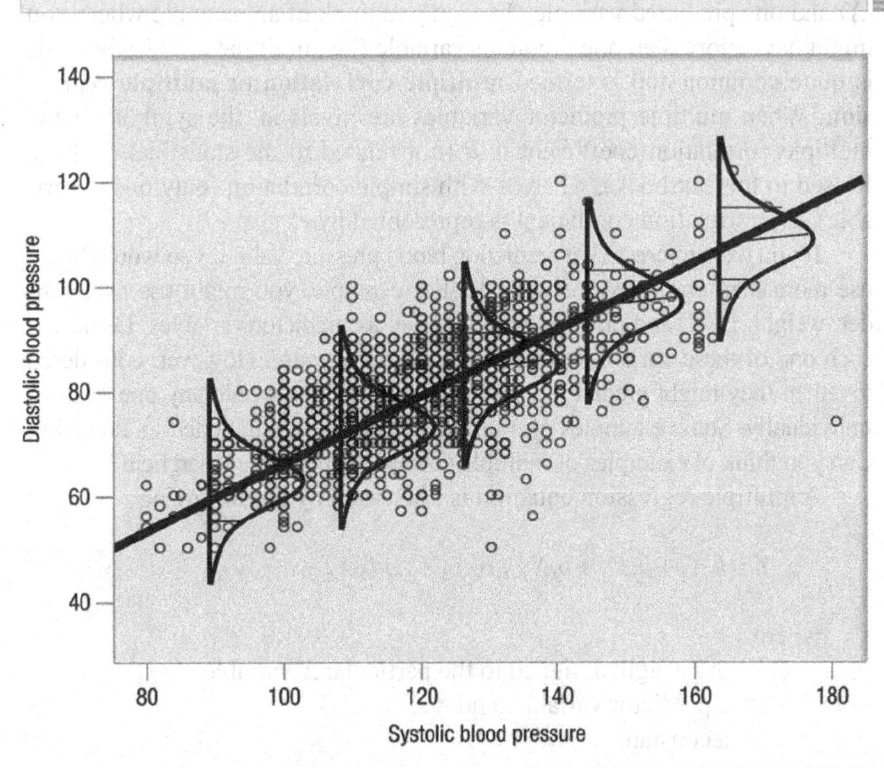

It is assumed that the distribution of values for the *Y* variable for any given *X* is normal (as illustrated in the figures). Since these are normal distributions, each has a mean (which is the predicted value for *Y*) and some variability around the predicted value. The variability around the predicted value is reported as a standard deviation (as previously presented in Chapter 3). In this case, the standard deviation is the SEE, the standard deviation of the errors of prediction when predicting *Y* from *X*.

At this point, the same logic that was presented in Chapter 4 regarding use of the standard deviation and the *z*-table applies to the SEE. For example, if the SEE is 5 units and the predicted value of *Y* is 50, approximately 68% of the actual values of *Y* will be between the mean predicted value and ±1 SEE. In other words, you could be 68% confident that the observed value of *Y* corresponding to the given *X* is between 45 and 55. You should confirm that the chances are about 95% that the predicted value of *Y* for this *X* will be between 40 and 60. (*Hint:* Use 2 SEE.) Return to the presentation of the *z*-table in Chapter 4 if you have difficulty.

V.5.4 **Correlation, Prediction, and the Standard Error of Prediction**

V.5.5 **Relationship Between Prediction and Error**

Consider what you would predict for *Y* if there is no correlation between *X* and *Y*. Assume that all you know is the value of *X* and that there is no correlation between *X* and *Y*. What value of *Y* would you predict for *any X*? The most reasonable prediction would be the mean of *Y*. This is true for any value of *X*. What would the SEE be in this case? The SEE would be the standard deviation of the *Y* scores. You should confirm this from Equation 5.7.

MULTIPLE CORRELATION

So far, our discussion of correlation has involved just one predictor variable (*X*) and one predicted variable (*Y*). Can you think of an example where you might have more than one predictor variable (i.e., multiple *X*s)? In fact, this is quite common and is termed **multiple correlation** or **multiple regression**. When multiple predictor variables are involved, the symbol for the multiple correlation coefficient is *R* (not related to the statistical package R used in this textbook), whereas with simple correlation (only one *X* variable), the correlation coefficient is represented by *r*.

If you were interested in predicting blood pressure values, you would likely use more than one predictor variable. For example, you might use age, gender, weight, BMI, and physical activity level as predictor variables. Logically, each one of these variables is related to blood pressure. However, considered together, they might predict blood pressure better than would any one of them individually. Some examples of multiple correlation are included in Table 5.6. Can you think of examples of multiple correlation in your chosen field?

A multiple regression equation is illustrated by the following:

$$\hat{Y} = b_1X_1 + b_2X_2 + b_3X_3 + b_4X_4 + \ldots\ b_kX_k + c \quad \text{Eq. 5.8}$$

where

b = the weight assigned to the particular *X* variable,
X = a predictor variable, and
c = a constant.

TABLE 5.6 Examples of multiple correlation.

Y (OUTCOME VARIABLE)	*X*s (PREDICTOR VARIABLES)
College success	High school GPA, high school rank, ACT scores, SAT scores
Graduate school success	Undergraduate GPA, GRE scores
Risk of death	Family history, current health status, physical activity level, gender
VO_2max	Age, gender, weight, physical activity level, treadmill test time

FIGURE 5.12 Illustration of reduction in SEE with additional predictors.

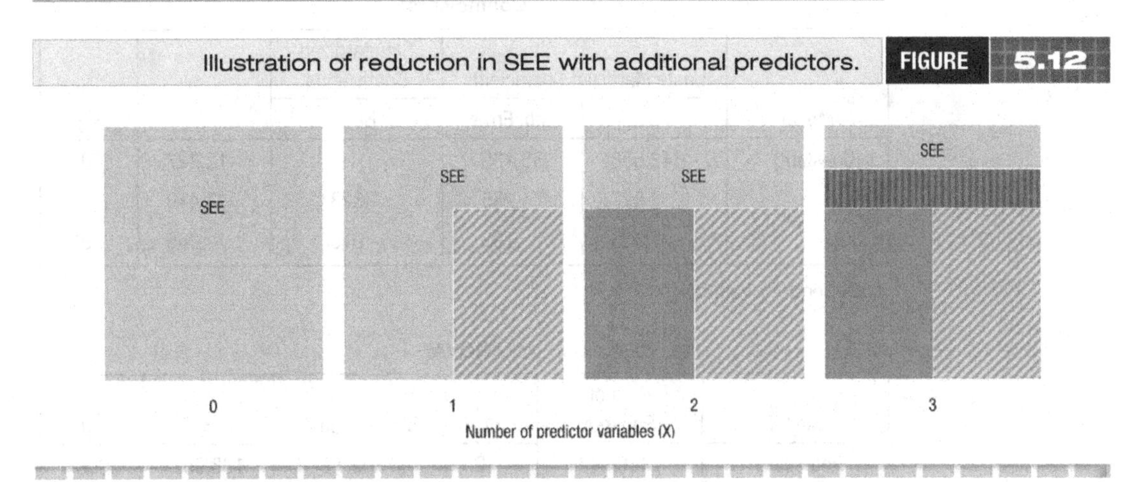

Here, we are adding variables to the regression equation to predict *Y* from multiple *X*s rather than just one *X*. What do you think happens to the SEE when we do this? Generally, you would expect the error in prediction (i.e., the SEE) to decrease as you add variables to the equation, because the variation in *Y* that is accounted for usually increases with additional predictor variables. You would generally *not* want to add variables that are unrelated to *Y*. Also, you would want to consider including variables that provide "new" information about *Y* and not information already provided by other *X* variables. That is, you want the new variables that you include in the equation to relate to *Y* but to be unrelated to the other *X* variables. This is illustrated in Figure 5.12, where each of the three variables reduces the SEE, and there is little or no overlap among the predictor variables. Each square represents variance in the *Y* variable, and each differently shaded section indicates the effect of adding a predictor variable *(X)* to the equation.

In general, the *Y* variable is what you want to predict, and the *X*s are the predictor variables. The *Y* variable may also be termed the *dependent variable*, the *outcome variable*, and the *criterion*. Among other terms for the *X* variables are the *independent variables* and the *predictor variables*.

Interpreting the Results From Multiple Correlation

Many computer programs are available to help you develop the regression equation and determine the associated weights. In general, the output of these programs resembles that presented in Figure 5.13.

FIGURE 5.13 SPSS output for multiple regression.

Model Summary

Model	R	R Square	Adjusted R Square	Std. Error of the Estimate
1	.811[a]	.657	.654	14.051

a. Predictors: (Constant), X2,X1

Coefficients[3]

Model	Unstandardized Coefficients		Standardized Coefficients	t	Sig.
	B	Std. Error	Beta		
1 (Constant)	−649.658	53.135		−12.227	.000
X1	3.827	.285	.823	13.419	.000
X2	−.035	.132	−.016	−.265	.791

a. Dependent Variable: Y

ANOVA[b]

Model		Sum of Squares	df	Mean Square	F	Sig.
1	Regression	74605.441	2	37302.721	188.952	.000[a]
	Residual	38891.621	197	197.419		
	Total	113497.062	199			

a. Predictors: (Constant), X2, X1
b. Dependent Variable: Y

For the example presented in Figure 5.13, the regression equation for predicting Y is:

$$\hat{Y} = 3.827X_1 - 0.035X_2 - 649.658$$

In this figure, the b coefficients (appearing as B in SPSS and Estimate in R) and the constant (c) for this equation are found in the Coefficients table under Unstandardized Coefficients. In the Model Summary table, we find that the multiple correlation coefficient *(R)* is .81, and the SEE is 14.05. The coefficient of determination (r^2), presented earlier, is also appropriate here. Here, $R^2 = (0.8112)^2 = 0.657$. Thus, 65.7% of the variation in Y can be accounted for by the values of X_1 and X_2.

In Figure 5.13, in the Coefficients table, the numbers marked "Beta" are the regression coefficients (weights), assuming the variables are all in z-score form. The numbers marked "t" and "Sig". are tests of whether the constant, X_1 coefficient (b_1), and X_2 coefficient (b_2) are significantly different from zero.

Essentially, if a coefficient is zero, then this variable doesn't really help in predicting Y. Notice that b_2 (−0.35, associated with X2) is not significantly different from zero. You will learn more about this later, when we discuss tests of significance. The RStudio output for this example is presented here.

Coefficient	Estimate	Std. Error	t value	Pr(>|t|)		
(Intercept)	−649.6579	53.1350	−12.227	<2e-16***		
X1	3.8270	0.2852	13.419	<2e-16***		
X2	−0.0350	0.1319	−0.265	0.791		
Signif. codes:	0***	0.001**	0.01*	0.05	0.1	1
Residual standard error: 14.05 on 197 degrees of freedom						
Multiple R-squared: 0.6573, Adjusted R-squared: 0.6539						

The ANOVA (analysis of variance) portion of Figure 5.13 illustrates two concepts that we have already discussed: the coefficient of determination (r^2) and the standard error of estimate (SEE). When you use the sum the squares due to regression (74605.441) and divide the result by the total sum of squares (113497.062), the result is r^2:

$$r^2 = \frac{74605.441}{1134947.062} = .656$$

Note, also, that when you take the square root of the mean square residual (197.419), the result is equal to the SEE (14.05) calculated with Equation 5.7. We will go into much more detail about ANOVA in Chapter 9.

V.5.6 SPSS Multiple Regression and Example

Creating the Multiple Regression Equation

V.5.6R and 5.7R RStudio Multiple Regression and Calculating Example

To create a multiple regression equation, you will be selecting the variables to include and calculating a weighting coefficient for each variable. Several methods have been developed to help you do this. Perhaps the most important point is that, in the final step, all of them provide the same general information. However, different amounts of information and different interpretations result from the various methods for deriving the regression equation. The methods are presented in Table 5.7.

Regardless of the method chosen, the ultimate goal is to increase the accuracy of the prediction equation and consequently, reduce the amount of error (i.e., the size of the SEE) when predicting Y from the various Xs.

V.5.7 SPSS Calculating Multiple Regression

Issues in the Development of Multiple Regression Equations

As scientists learn more and more about a specific variable, regression equations involving that variable might be updated, adjusted, and modified in researchers' attempts to create better regression equations. "Better" means that the correlation increases and the SEE decreases. The box on p. 124 describes additional issues for you to consider when creating a regression equation.

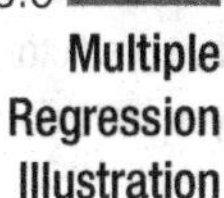

V.5.8 Multiple Regression Illustration

TABLE 5.7 Methods for creating a multiple regression equation.

METHOD	DESCRIPTION
Enter	All ***X*** variables are entered into the equation at once, and the researcher then interprets the unstandardized weights (***b***s) and the standardized weights (Betas).
Hierarchical	This is a theory-driven method. The researcher decides on the variables to be entered into the equation and their order. The order might be based on previous research, pilot work, or some theoretical reason for entering a particular variable. For example, one might enter gender into the equation first because gender is not a variable that can be changed. The researcher can then test after the addition of each variable to determine whether the amount of additional variation accounted for in the ***Y*** variable has been significantly increased.
Stepwise	This method is similar to the hierarchical method, except that instead of the researcher determining the choice and order of the variables, these determinations are made solely by the computer. As in the hierarchical method, the variables are chosen one at a time, and tests can be conducted to see whether a significant improvement to the prediction equation occurs with the addition of each chosen variable. The first variable selected for entry is the one with highest correlation with ***Y***. The next variable selected is the ***X*** with the highest correlation with ***Y***, ***after*** the variation in the ***Y*** variable that is accounted for by the first variable is removed. This is called a ***partial correlation***. Each added variable has a diminishing impact on the variation accounted for in ***Y***. You can generally set a lower limit so that no more variables are added after their impact is less than the designated limit.
Forward or backward	In the methods described above, variables are added "forward", meaning included in the equation one at a time. Methods also exist for entering the variables "backward". In these methods, all of the variables are included in the equation at first, and then they are "backed out" (removed) one at a time while the researcher observes the impact of each removal on the prediction equation.

Considerations When Using Multiple Regression

Number of X Variables

You want to include as few as you can, so that your model is as parsimonious as possible. On the other hand, it might be difficult to measure certain variables, so you should consider feasibility of measuring the *X* variables you have chosen. Also consider how the *X* variables might relate individually to the *Y* variable, how the various *X* variables relate to one another, and how reliably each *X* can be measured.

Sample Size

A number of "rules of thumb" can help you determine an appropriate sample size. Generally, the more subjects you have, the better. The "stability" of the weightings for the variables (the *b*s) is a function of the sample size. Small samples result in regression equations that are not very stable and cannot be trusted. You should strive to have at least 10 to 20 subjects for each *X* variable. Thus, if you have five predictor variables, you want at least 50 subjects where you have data on all subjects for all variables (*X*s and *Y*).

Cross Validation

Once you have created a regression equation on a sample, you should test it with another, similar sample to see whether it cross validates. That is, do you achieve the same multiple correlation *(R)* and SEE? Often, you will find a reduced correlation (i.e., *shrinkage* or *adjusted*) when you apply the equation to another sample. Most computer programs can estimate the shrinkage for you, but you should go through the actual process of cross validation. Some researchers will split their sample into two groups: the validation group (consisting of 60% to 70% of the sample) and the cross-validation group (the remaining sample). Some computer programs use a "bootstrapping" procedure in which different subsamples are repeatedly chosen and analyzed in an attempt to validate the regression equation without having to gather additional data.

Multicollinearity

Recall that you want to choose predictor variables that are as unrelated to one another as possible. That is, you want each *X* variable to add something "new" to your ability to predict *Y*. If any of the *X* variables are highly correlated, this results in *multicollinearity*. Essentially, some of the *X* variables (at least two of them) are so highly correlated that it is difficult (both factually and with a computer program) to determine which variable should be entered into the equation. The result is that the equation cannot be developed. Careful selection of the *X* variables can help you eliminate multicollinearity. An example of multicollinearity would occur in the use of a health history questionnaire with yes-or-no answers. If you treated the number of items answered "yes" as one *X* variable and the number answered "no" as another *X* variable, you would create multicollinearity. These two variables are perfectly inversely correlated. Thus, knowing one of them tells you exactly what the other one is and using both does not provide more information than using just one.

Range Restriction

In general, if a correlation exists between *X* and *Y* variables, the more variable *X* and *Y* are, the more likely it is that you will be able to observe the correlation. If for any reason a researcher restricts the range of one of the variables (either *Y* or X), this will reduce the ability of the *X* variable to predict (and will increase the SEE). An example is the use of the GRE to predict graduate school success. Some people criticize the use of GRE scores and undergraduate GPA to predict graduate school success, because the resulting multiple correlation coefficient is less than .50. But consider a few things: How many and what types of people go to graduate school? Generally, academically better students apply for graduate school. Thus, generally, those with higher undergraduate GPAs apply, and those with lower GPAs do not. Thus, the range for GPA is "restricted". Also consider the "restricted range" that occurs in graduate school grading. Course grades and, therefore, overall GPAs in graduate school are generally higher than undergraduate grades and GPAs. The "typical" grade in a graduate course might be a B (or even an A), and grades of C occur less frequently than in undergraduate courses. Thus, the graduate GPA is restricted. Thus, when you correlate undergraduate GPA and GRE scores with graduate school GPA, the ranges on all of the variables are restricted, and this results in a lower multiple correlation coefficient and a greater error of prediction. Equations exist for estimating what the correlation coefficient would be if the range were not artificially restricted. These equations are based on the variability (i.e., variance) in the total population. Removing the range restriction will increase the magnitude of the correlation.

OTHER MEASURES OF CORRELATION

This chapter has described two common types of correlation: simple regression and multiple regression. Both typically involve the use of variables measured on a continuous (i.e., interval or ratio) scale. In both, the *Y* variable measurement *must* be continuous, but it is possible for the predictor variables to be categorical (nominal) variables. Gender and group membership are good examples. Table 5.8 describes other types of correlation that you might encounter. In all cases, the essential purpose remains the same: you are determining the association (i.e., relationship) between variables. However, in some types of correlation you will have multiple *Y* variables, and in others you will have multiple *X* variables. The general principle is referred to as the **canonical model.** That is, regardless of the procedure you are using, you have a set of *Y* variables (from one to several) and a set of *X* variables (from one to several), and you are interested in the correlation (relationship) between the *Y* and *X* sets. This canonical model approach will serve you well when you begin to study more advanced statistical procedures (as you will see in later chapters).

TABLE 5.8 Other measures of correlation.

CORRELATION TYPE	DESCRIPTION/COMMENTS
Partial correlation	This is the correlation between two variables when the effect of one or more other variables has been removed from the original variables. For example, after removing the effect of body weight, you might be interested in the relationship between strength and power. Effectively, a residualized strength score is correlated with a residualized power score. Thus, you would be looking at the relationship between strength and power, with the impact of body weight removed from both of them.
Spearman rank order correlation (rho)	This correlation is used when the variables are ranks (i.e., not continuously measured). It gives an estimate of the PPM correlation coefficient between ranks. This correlation is presented in Chapter 14.
Phi	Phi gives the correlation between variables when both are dichotomously scored (0 or 1). This correlation is also presented in Chapter 14.
Point biserial	This correlation is used when one of the variables is a dichotomy (e.g., gender) and the other is a continuous variable.
Biserial	Akin to the point biserial correlation, the biserial correlation is also used when one of the variables is a dichotomy and the other is continuous. However, in this case the dichotomous variable is assumed to have an underlying normal distribution.
Curvilinear	The PPM correlation coefficient describes a linear relationship between variables. However, variables might be related in a curvilinear manner. If so, the PPM correlation coefficient will underestimate the true relationship between the variables. Methods exist to modify the variables to identify any curvilinear relationship in the data.
Logistic regression	When the Y variable is a dichotomy (0 or 1), we can use logistic regression to estimate the change in odds of Y being classified in one category or the other. Epidemiologists often use this correlation in situations where the outcome variable is dead/alive or diseased/not diseased. See Chapter 13.
Discriminant function analysis	Discriminant function analysis is similar to logistic regression. The intent is to determine which continuously measured X variables help most to classify subjects on a categorically measured Y variable. The ultimate intent is to be able to observe the X variables and then predict into which category of the Y variable the observed unit would most likely be classified. The Y variable might have more than two categories. This correlation might be useful to a track coach who needs sprinters, distance runners, and field event participants. She might use multiple anthropometric and performance variables in a discriminant function analysis to estimate the most likely candidates for each of the categories. See Chapter 13.
Canonical correlation	The canonical correlation is the generalized model. Any number of Y and X variables can be included in this model. The intent is to relate the set of Y variables to the set of X variables to determine the number of dimensions developed, in order to see, in turn, which variables in the Y set are related to which particular variables in the X set. In the process, a number of "discriminant functions" are created that help identify these relationships. If there is only one Y and one X variable, this constitutes the simplest case of the canonical correlation, which is the PPM correlation.

SUMMARY

The concept of relationships underlies all inferential statistics. Researchers study relationships with the intent of identifying how changes in an outcome variable (Y) are a function of changes in other variables (Xs). Essentially, researchers seek to identify relationships between and among variables. Researchers use a variety of statistical procedures to identify these relationships. The specific type of procedure that is used depends on the number and levels of measurement of the Y and X variables and on the questions that the researcher is attempting to answer. However, the underlying concept of studying relationships among variables is common to all of these procedures.

Review Questions and Practice Problems

PART A (ANSWERS PROVIDED IN APPENDIX C)

1. Which correlation coefficient must have been calculated incorrectly?

   A. 0.1

   B. –0.5

   C. +1.00

   D. –1.03

2. The correlation between miles driven and tread left on a tire would most likely be what?

   A. Positive

   B. Negative

   C. Zero

   D. No way to know

3. What would happen to the correlation coefficient between X and Y if we subtract 10 from every X score and add 3 to every Y score?

   A. It would increase in value.

   B. The sign would change from + to –.

   C. Nothing.

   D. Depends on the original value of r.

4. If the 68% confidence level for predicting Y from X is 44–56, and the standard deviation of the Y scores is 10, what is r?

   A. .40

   B. .60

   C. .80

   D. Cannot be determined.

5. The correlation coefficient between X and Y is .50. What is the equation of the regression line to predict Y from X, if both variables are converted to T-scores?

6. The correlations of four predictor variables with the criterion variable are .41, –.62, .01, and .34. In a stepwise regression, which predictor variable would be entered first?

   A. .42

   B. –.62

   C. .01

   D. .34

PART B

7. Use the following data and three equations (definitional, descriptive statistics, and raw score) given in this chapter to determine the r between X and Y. Round answers to nearest the nearest tenth. Assume you are working with population values.

X	Y
1	2
2	1
3	4
4	3

8. For the data in question 7, how much of the variance in the Y variable is attributable to the variance in the X variable?

   A. 0.04

   B. 0.36

   C. 0.60

   D. 0.80

9. If the four subjects in question 7 are considered to be a sample, is the r obtained statistically significant?

   A. Yes, if the level of confidence is .10

   B. Yes, if the level of confidence is .05

   C. Yes, if the level of confidence is .01

   D. No

10. What percent fat (PF) would you predict for Maureen, given the following information?

    PF mean = 25%

    PF standard deviation = 4%

    The *standardized* regression equation is:

    predicted percent fat = $.4(X_1) + .25(X_2) - .8(X_3)$

    Maureen's standardized measurements on X_1, X_2, and X_3 are 2, 6, and 3.5, respectively.

11. In question 10, what are the values of .4, .25, and −.8 called?

    A. Standard error of estimates

    B. Residuals

    C. *b* coefficients

    D. Beta weights

COMPUTER ACTIVITIES

Correlation

ACTIVITY 1 Bivariate Correlation

The data in Table 5.9 represent the scores on two youth fitness measurements, PACERLAPS and treadmill-determined VO2MAX, for 25 subjects. Enter or import the data and then perform the indicated operations.

Using SPSS

A. On the toolbar, go to Graphs, Legacy Dialogs, select Scatter/Dot. . ., and then select Simple Scatter. Click Define. Move VO2MAX to the *Y*-axis and PACERLAPS to the *X*-axis. Select **Titles . . .** and label the graph appropriately. Click **Continue.** Click OK.

When the graph is displayed, double-click on the graph to open the Chart Editor. Locate and click on the normal-curve icon. This brings up a Properties screen (click on the Linear circle). This adds a straight line of best fit to your scattergram.

1. What are the following statistics for each of the two tests?
    a. mean
    b. standard deviation
    c. range

B. Go to **Analyze** on the toolbar, select **Correlate** and then select **Bivariate.** Move PACERLAPS and VO2MAX into the Variables box, make sure Pearson is checked, and click OK.

2. What is the correlation between PACERLAPS and VO2MAX? Explain briefly what this means. (Hint: notice the sign and magnitude.)

C. Suppose the researcher decides to give each subject an additional five PACERLAPS because of a timing error. Go back to your data set and add a new column to achieve the researcher's goal. (*Hint:* you can do this manually, or you could use the Compute command, under Transform on the toolbar.) Now, go back to **Analyze** and select **Correlate** and **Bivariate** again. Put all three tests in the variables box. Click OK.

3. What is the correlation between VO2MAX and the "corrected" PACERLAPS? What does this tell you?

D. To help answer the next question, change PACERLAPS and the "corrected" PACERLAPS to *z*-scores. To do this, go to **Analyze → Descriptive Statistics → Descriptives** and move the variables to the right window. Be certain to click on **Save standardized values as variables.**

4. What do you notice about the two sets of *z*-scores? Hint: RStudio users should look in the Environment pane.
5. What is the correlation between PACERLAPS and the "corrected" PACERLAPS (PACERLAPS5)? What does this and your *z*-scores tell you?

TABLE 5.9

Data set.

STUDENT	PACERLAPS	VO2MAX
1	47	33
2	54	49
3	48	40
4	47	44
5	50	48
6	45	36
7	50	35
8	56	50
9	54	46
10	48	37
11	53	40
12	47	39
13	38	32
14	47	42
15	48	39
16	49	37
17	53	42
18	53	40
19	49	40
20	52	47
21	58	51
22	39	30
23	50	40
24	44	34
25	55	47

Using R

Use the following directions to answer items 1 through 5.

1. Open RStudio.
2. Install any necessary R packages.

   For Chapter 5, we do not require any additional packages.

3. Set up your RStudio workspace.

   In the RStudio menu bar, click on **File → New File → R Script.**

   On the first line of the R Script (upper left pane of RStudio), type #Chapter 5.

RStudio Instructions

1. Enter the variable names (STUDENT, PACERLAPS, VO2MAX) and corresponding values found in Table 5.9 into a spreadsheet and create a csv file named "pacer" to be imported into R. (Alternatively you could import Table_5_9_Data from the textbook website).
2. If you created the pacer file, now import it into R (see Chapter 3).

   You should see the pacer data frame listed under the "Environment" tab (upper right pane of RStudio).

3. If you created the pacer file, generate a scatterplot of the data using the plot function and labeling the x and y axes:

- plot(pacer$PACERLAPS, pacer$VO2MAX, xlab="PACER Laps", ylab="VO2MAX")

Note: If you imported Table_5_9_Data, you need to change pacer to Table_5_9_Data here and in the steps below.

When you've finished typing, highlight the entire function you've just typed in and click on **Run**.

You should now see the scatterplot displayed in the lower right pane of RStudio under the "Plots" tab.

4. We can also add a fitted line (generated by the lm function) to the plot using the abline function:

    - abline(lm(pacer$VO2MAX ~ pacer$PACERLAPS), col="red")

    When you've finished typing, highlight the entire abline function and click on **Run**.

    You should now see a red fitted line overlaid on the scatterplot.

5. Enter the following RScript commands to obtain descriptive statistics to answer question 1.

    - mean(pacer$PACERLAPS)
    - mean(pacer$VO2MAX)
    - sd(pacer$PACERLAPS)
    - sd(pacer$VO2MAX)
    - range(pacer$PACERLAPS)
    - range(pacer$VO2MAX)

    Highlight the above 6 lines and click on **Run**.

Use the following RScript commands to obtain the data to answer questions 2 through 5.

6. Next, run a Pearson correlation between PACERLAPS and VO2MAX.

    - cor(pacer$PACERLAPS, pacer$VO2MAX, method="pearson")

    The Pearson correlation coefficient will be displayed.

7. To add an additional 5 PACERLAPS to each subject, you could do the following:

    - pacer$PACERLAPS5 <- pacer$PACERLAPS + 5

    When you've finished typing, highlight the entire function you've just typed in and click on **Run**.

    You should now have a new variable created in your pacer data set.

8. Run another Pearson correlation, this time using the new variable PACERLAPS5 and VO2MAX.

    - cor(pacer$PACERLAPS5, pacer$VO2MAX, method="pearson")

9. To calculate *z*-scores for PACERLAPS and corrected PACER laps (PACERLAPS5) using scale function.
   - pacer$PACERLAPSZ <- scale(pacer$PACERLAPS)
   - pacer$PACERLAPS5Z <- scale(pacer$PACERLAPS5)

   Highlight what you've typed and click **Run** to execute.

ACTIVITY 2 Correlation Matrix

Using SPSS

The data in Table 5.10 represent the scores of 10 students on four different sports rules written exams. Use SPSS to obtain a correlation matrix involving all four exams (**Analyze → Correlate → Bivariate** and move the four exam scores to the variables box). Use the options button to obtain descriptive statistics to answer the following questions:

TABLE 5.10 Data set.

EXAM1	EXAM2	EXAM3	EXAM4
49	85	61	43
53	74	54	22
51	78	56	13
49	58	53	30
48	83	62	24
49	70	58	10
49	76	65	39
51	90	63	33
50	84	57	31
52	63	63	22

6. What is the correlation between each examination and each of the other three examinations?
7. (Hint: use Descriptives for questions 7 and 8.) If each of the examinations contained 100 questions, which test was the easiest and which was most difficult?
8. On which exam were the scores most scattered?
9. Which two of the exams have the highest correlation with each other?
10. Which two exams show the least amount of correlation with each other?
11. If a new student took Exam 1 and scored above 50.1, would you predict he would score above or below 59.2 on Exam 3? Why?
12. Construct a scattergram for Exams 1 and 4 by considering Exam 1 as the *X* variable and Exam 4 as the *Y* variable. Which person falls closest to the line of best fit on the graph? (SPSS Hint: Use Scatter/Dot under **Graphs** on the top toolbar, and add a regression line as done above.)

Using R

Use the following directions to answer items 6 through 12.

1. Enter the variable names (EXAM1, EXAM2, EXAM3, EXAM4) and corresponding values found in Table 5.10 into a spreadsheet and create a csv file named "exams" to be imported into RStudio. (Alternatively, you could import Table_5_10_Data from the textbook website).
2. Import "exams" into RStudio (see Chapter 3).

You should see the exams data frame listed under the "Environment" tab (upper right pane of RStudio).

3. Generate correlation matrix for the exam data.
   - cor(exams,method="pearson")

     The correlation matrix (Pearson) will be displayed.
4. Generate descriptive statistics for the exam data.
   - mean(exams$EXAM1)
   - mean(exams$EXAM2)
   - mean(exams$EXAM3)
   - mean(exams$EXAM4)
   - sd(exams$EXAM1)
   - sd(exams$EXAM2)
   - sd(exams$EXAM3)
   - sd(exams$EXAM4)
5. Create a scattergram with a Line of Best Fit.
   - plot(Table_5_10_Data$EXAM1, Table_5_10_Data$EXAM4, xlab= "EXAM1", ylab="EXAM4")
   - abline(lm(Table_5_10_Data$EXAM4~Table_5_10_Data$EXAM1), col="red")

REFERENCES

Cooper, K. H. (1968). A means of assessing maximum oxygen intake. *Journal of the American Medical Association, 203*(3), 201–204.

Keeley, J., Zayac, R., & Correia, C. (2008, May). Curvilinear relationships between statistics anxiety and performance among undergraduate students: Evidence for optimal anxiety. *Statistics Education Research Journal, 7*(1), 4–15.

6

Introduction to Inferential Statistics

INTRODUCTION

The main purpose of inferential statistics is to estimate, as accurately as possible, information about a population of subjects by examining the characteristics of a subset of that population. The subset is called a *sample*. A simple example might be to estimate the mean BMI score for all of the middle school students in a particular school district (the population). To do this, we would randomly select some number of the middle school students in the district (the sample), measure their BMIs, calculate the mean BMI for the sample, and project the results to the entire school district population.

In this example, we would need to consider the following issues:

1. How we would define who is a member of the population,
2. How we would select the specific students to be in the sample,
3. How we would decide how many students are needed to comprise the sample, and
4. How we would determine the accuracy of our projection to the population.

Once we learn to perform this statistical procedure and address the issues involved, we will be able to apply this technique to increasingly complex problems.

An obvious question is, "Why not just measure the BMIs of *all* the middle school students in the population?" Then there would be no need to estimate the mean BMI—we could calculate it precisely. That is absolutely true. However, if we can estimate the population mean BMI quite accurately with a relatively small sample, we can save a lot of time and effort. Also, consider another example: a company that manufactures light bulbs would like to be able to tell customers how long the bulbs should last on average. If the company measured the length of time the entire population of bulbs lasted, they wouldn't have any left to sell. Measuring the life of a sample of bulbs and using this information to make an estimate for the entire population, assuming this can be done accurately, seems more sensible.

SAMPLING

Once a population has been defined, the next step is to select a subset of subjects to be included in the sample. The most important feature of the sample is that it is representative of the population. A larger sample is generally better than a smaller sample, but this is true only when the selected subjects accurately represent the population.

Random Sampling

The best way to be assured that a sample accurately represents a population is to use a procedure called **random sampling.** Two conditions must be met to obtain a random sample. First, every subject in the population must have an *equal* chance of being selected for the sample. Second, every subject in the population must have an *independent* chance of being selected. That is, the selection of any given subject has no influence on the selection of any other subject. If either or both of these conditions are not met, the resulting sample may be *biased* and, thus, not representative of the population.

As the definition of a population increases in breadth, it becomes more difficult to meet the random sample conditions. A very narrowly defined population, on the other hand, makes it easier to meet these conditions. However, any resulting findings apply to a smaller set of subjects. For example, it would be difficult to select randomly a set of people from the state of Texas, even though the researchers might desire to generalize to that population.

Definitions Overview

Before we begin to learn how inferential statistical techniques work, we need to review and expand some definitions of terms that will be involved.

Population

A *population* is any set of subjects that have at least one attribute in common. We use the term *population* to refer to the group to which we wish to project our sample information. It is important to define clearly who is a member of a particular population and to make sure that all members of a population can be identified. Of course, for a given study, the population could be people, animals, or objects.

Parameter

A *parameter* is a fact about a population. The facts about the population are usually numerical, such as the mean value of some variable. Parameters are generally identified by symbols taken from the Greek alphabet. For example, the mean of a population is identified by the Greek letter μ.

Sample

A *sample* is a subset of a population. The sample is the group of subjects we select from the population (by using methods described in the next section) for the purpose of estimating parameters.

Statistic (or Estimate)

A *statistic* is a fact about a sample. As with population parameters, these facts are generally expressed numerically, such as the mean value of some variable. Statistics are generally identified by symbols taken from the Roman alphabet. For example, the mean of a sample is identified by $\bar{X}$.

Point Estimate

A **point estimate** is the value of a sample statistic used to estimate a parameter. For example, if the mean BMI value of our middle school sample was 20, this value is our point estimate of μ. In other words, our best guess of the mean BMI value of the population from which the sample was selected is 20.

Interval Estimate

The **interval estimate** is a range of values around a sample statistic, which includes the parameter with a given level of confidence. Assume the population mean (μ) is 20. If you were to obtain hundreds of samples and determine 95% confidence intervals around the sample means (the $\bar{X}$s), 95% of the confidence intervals would capture the population mean (i.e., $\mu = 20$). How the range of the interval estimate and the level of confidence are determined will be discussed later.

Sampling Error

Sampling error is the difference between a statistic and a parameter. It is recognized that every sample statistic will not be exactly equal to the parameter it represents, because every sample cannot be precisely representative of the population. The difference between the statistic and the parameter that results from sampling is termed sampling error. Since the true value of the parameter is not known, sampling error is estimated. How we arrive at this estimation will be explained following the information on sampling.

Conversely, it would be relatively easy to select randomly a sample from the students in a particular human performance class, but the findings from that sample would then apply to a rather limited population.

Because random sampling is often difficult to achieve, you may encounter other methods of obtaining samples. Some of these methods and their advantages and disadvantages are identified below.

Systematic Sampling

To obtain a **systematic sample,** one would select every *n*th subject from a list of every subject in the population. For example, if a population contains 1,000 subjects and a sample of 100 subjects is desired, one would select every tenth subject on the list. From a table of random numbers, a number between 1 and 10 would be obtained (say, 7). The sample would then consist of subject 7, subject 17, subject 27, and so forth, on the list. The advantages of systematic sampling are that it is usually easier to perform than random sampling and that it usually can be considered to be, and is treated as though it were, a random sample. The disadvantage is that if there happened to be some cyclic pattern in the population list, and the sampling pattern happened to coincide with the cycle, the resulting sample could be biased. For example, imagine a list that is organized in a male, female manner, say, a list of married couples that gives the husband's name first for each couple. Selecting every second name from such a list would result in a sample that includes only females.

Stratified Sampling

The **stratified sampling** method of deriving a sample from a population ensures that the sample is representative of the population on selected variables. For example, suppose that the population contains 1,000 college-aged subjects (60% females and 40% males) and that the 600 females are distributed across class levels such that 240 are freshmen, 180 are sophomores, 120 are juniors, and 60 are seniors. The 400 males are distributed as 140 freshmen, 100 sophomores, 80 juniors, and 80 seniors. Then, the numbers and percentages of the two strata (gender and class) are as follows:

240 freshmen females = 24%	140 freshmen males = 14%
180 sophomore females = 18%	100 sophomore males = 10%
120 junior females = 12%	80 junior males = 8%
60 senior females = 6%	80 senior males = 8%

If the sample is to consist of 100 subjects, then 24 freshmen females would be required for the sample, and they would be selected randomly from the 240 freshmen females in the population. The required 18 sophomore females would be randomly selected from the 180 available. This process would continue until the 100 subjects were selected. The resulting sample would contain 60 females (24 freshmen, 18 sophomores, 12 juniors, and 6 seniors) and 40 males (14 freshmen, 10 sophomores, 8 juniors, and 8 seniors). The resulting sample would be exactly proportional to the population in regard to the strata selected. The disadvantage of stratified sampling is that the sample is not necessarily representative of the population for characteristics other than those selected for stratification (e.g., exercisers and non-exercisers). Thus, it would be important to stratify on all variables that could influence the dependent variable(s) involved in the research.

Cluster Sampling

The **cluster sampling** technique is sometimes used when the sampling unit is composed of natural groups of subjects. If the population of interest is fifth-grade physical education students in Colorado, it would be difficult to obtain a random sample of this population. Cluster sampling in this case might involve randomly selecting some number, say five, of the 64 counties in Colorado and then randomly selecting some number of school districts, say two, in each of the five counties. This might be followed by random selection of some number, say three, of the elementary schools from each of the 10 selected school districts and then randomly selecting some number, say two, of the fifth-grade classes from each of the 30 selected schools. The resulting sample of 60 fifth-grade physical education classes would be assumed to represent the population. Although the advantage of cluster sampling is that it does achieve some randomization, the very important disadvantage is the violation of the independent selection condition of a random sample. The students from the counties not selected, the school districts not selected, the schools not selected, and the fifth-grade classes not selected have no chance to be selected for the sample.

Convenience/Accidental Sampling

Although it might be convenient to select all of the patients in a particular rehabilitation facility as a sample of the population of all of the patients in rehabilitation facilities in a particular community, it is not appropriate to do so. Sadly, the convenience/accidental sampling technique is more common than it should be. Although it is relatively quick and simple to obtain a sample in this way, to do so violates both the equal likelihood requirement and the independent selection requirement of random sampling. You would be very wise to be skeptical of the results of any experiment in which the researchers use this technique for obtaining sample subjects.

V.6.1 **Overview: Sampling Procedures**

THE SCIENTIFIC METHOD AND HYPOTHESIS TESTING

As you learned in Chapter 1, researchers use the scientific method to move science forward by looking at aspects of our world and studying the relationships between variables. For example, is there a relationship between diet (or exercise) and health outcomes? Decisions about this relationship could be arrived at capriciously or could be based on sources of evidence, such as personal experience, that do not allow for valid conclusions. (Most individuals' experiences are limited and do not generalize well.) Rather, making decisions based on data is at the heart of the scientific method. There are many ways to make decisions, but the merit of the scientific method is that its decisions are reproducible. You can never be absolutely certain of the relationships that are identified in a single scientific inquiry, but the scientific method allows for repeated experiments that accumulate evidence for a conclusion. The preponderance of evidence can provide a basis for making a good judgment about the relationship between variables.

The statistical parts of the scientific method consist of:

1. Generating a research hypothesis about the relationship between variables,
2. Developing a statistical **null hypothesis** (H_0),
3. Developing an **alternative hypothesis** (H_1),
4. Gathering data, and
5. Making a decision based on the probability (likelihood) that the null hypothesis is true.

People often use the scientific method without really thinking about it. Consider the decision of whether or not to take an umbrella on your way to work. You know there is a relationship between the appearance of the sky and "weather" it will rain. You look out the window and see a clear sky. You hypothesize that it will not rain. The alternative hypothesis is that it will rain. Based on the data you have gathered (by looking out the window), you make a decision about the likelihood of rain and then choose whether or not to take an umbrella with you. Let's expand on each of these points a bit.

Think back to Chapter 5 and our discussion about correlation. We presented methods of illustrating graphically, with symbols, and with numbers, the relationship (i.e., correlation) between two variables, *X* and *Y*. In our presentation of correlation, we wanted to know whether there was a relationship between *X* and *Y* so that we could predict *Y* from knowledge about *X*. The scientific method builds on the same logic. Science examines how we can predict one variable from another (or several others) so that, ultimately, humans can make decisions with deeper understanding of the world.

Let's illustrate these points by considering whether there is a relationship between gender and VO_2max. Consider the illustration in Figure 6.1.

On the *X*-axis we have a variable (Gender) with two levels (Females and Males), and on the *Y*-axis we have a measure of VO_2max. As stated earlier, the outcome variable, *Y*, is termed the dependent variable, and the *X* variable is termed the independent variable. Scientists wish to learn whether the relationship between the independent variable and the dependent variable is sufficient to allow us to predict *Y* from *X*. The logic is the same as that presented in Chapter 5. In Figure 6.1, the mean value of VO_2max for females and for males is the same. In fact, the difference between the mean for females and for males is zero $(\bar{Y}_{\text{females}} - \bar{Y}_{\text{males}} = 0)$. In other words, the average value for females and for males is the same $(\bar{Y})$.

Consider, now, that you gathered this information from a sample of females and males (perhaps $N = 20$ each, but it could be any number). You found no difference between your sample means. Can you conclude that, for the population to which you wish to generalize, there is no difference in the population means for females and males? This example helps to illustrate the fact that larger samples yield better estimates of population parameters.

The relationship between gender and VO_2max: no relationship. FIGURE 6.1

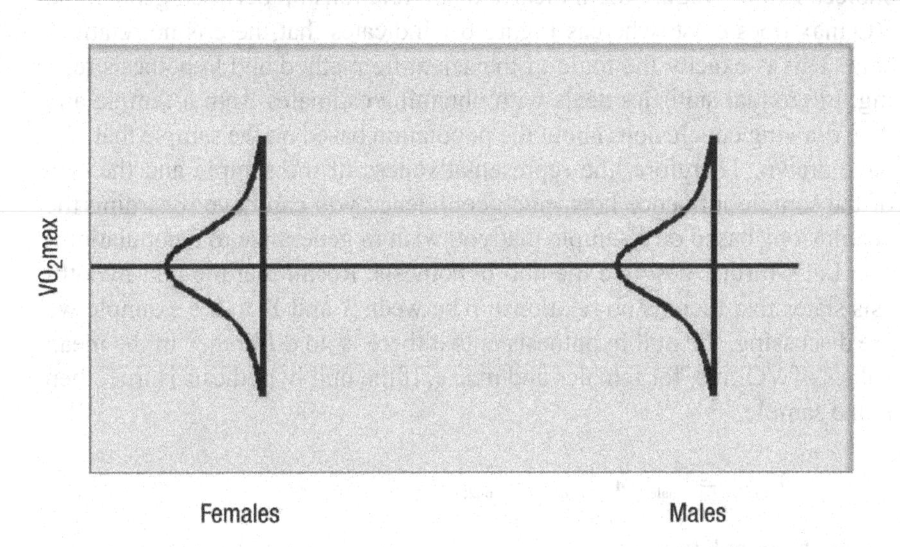

(Can you see why?) However, as was discussed previously, although large samples are generally better than small ones for estimating population values (in this case μ, the population mean for females and males), more important than sample size is the representativeness of the sample when compared to the population.

Now look at Figure 6.2, which has the same dependent and independent variables as Figure 6.1. Notice that here, the means for the genders are different. Based on the information in Figure 6.2, you would predict that a

The relationship between gender and VO_2max: a relationship exists. FIGURE 6.2

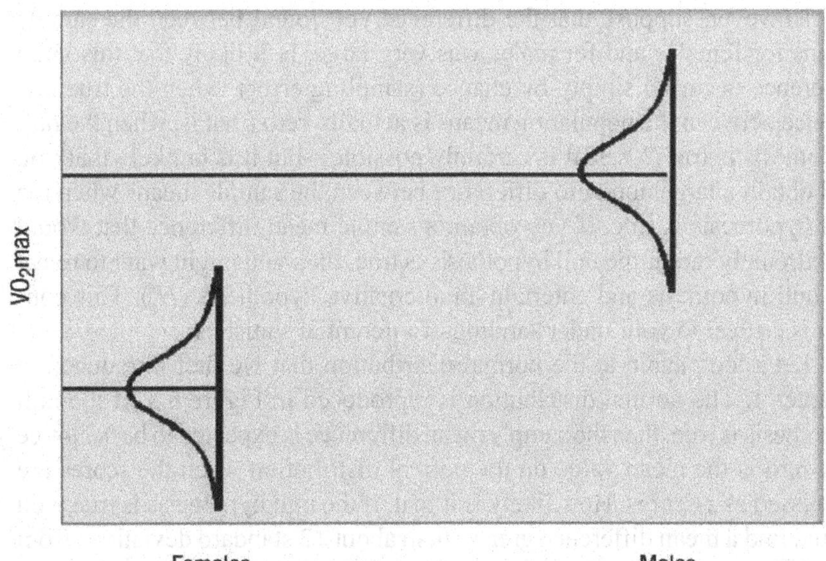

male subject's value for VO_2max would be a good bit higher than a female subject's value. Figure 6.2 indicates that a relationship between gender and VO_2max does exist, whereas Figure 6.1 indicates that there is no relationship. This is exactly the logic of the scientific method and hypothesis testing. Inferential statistics deals with obtaining estimates from a sample and then drawing conclusions about the population based on the sample that you have drawn. Therefore, the representativeness of the sample and the size of the sample influence how much confidence you can have regarding the conclusions based on a sample that you wish to generalize to a population.

Let's further describe the null hypothesis. Recall that the null hypothesis states that there is no relationship between X and Y. In the example we are discussing, the null hypothesis is that there is no difference in the mean values of VO_2max for females and males. If the null hypothesis is true, then in the sample

$$\bar{Y}_{\text{females}} = \bar{Y}_{\text{males}} \text{ or } \bar{Y}_{\text{females}} - \bar{Y}_{\text{males}} = 0$$

and in the population

$$\mu_{\text{females}} = \mu_{\text{males}} \text{ or } \mu_{\text{females}} - \mu_{\text{males}} = 0$$

Consider the value of zero that we find when the null hypothesis is true. If the null hypothesis is true, is it possible that you might find a non-zero value for the difference between the means in your samples of females and males? Certainly, it is possible to find values that are non-zero but that are relatively close to zero, even if the null hypothesis is true. This outcome is possible because of what we earlier defined as sampling error. That is, the difference between the means actually *is* zero in the population, but sampling error resulted in slight differences in the sample means, such that the difference between the sample means is not exactly zero. Logically, this could happen quite easily.

However, suppose that the difference you found between the sample means for females and for males was very large. Is it likely that this large difference occurred simply by chance (sampling error), when the true difference between the population means is actually zero (that is, when the null hypothesis is true)? Yes, it is certainly possible—but it is unlikely that you will obtain a large non-zero difference between the sample means when the null hypothesis is true. If you obtain a sample mean difference that would be extremely rare if the null hypothesis is true, then you might want to reject the null hypothesis and entertain an alternative hypothesis (H_1). This concept is critical to your understanding of inferential statistics.

Let's look again at the normal distribution that we first introduced in Chapter 4. The normal distribution is reproduced in Figure 6.3. If the null hypothesis is true, then the sample mean difference is expected to be 0. Notice that zero is the mean value on the normal distribution when the scores are expressed as z-scores. How likely is it that, if the null hypothesis is true, you would find a mean difference greater than about ±2 standard deviations from zero? *Not very likely*—in fact, only about 5% of the time would we expect

The normal distribution. FIGURE 6.3

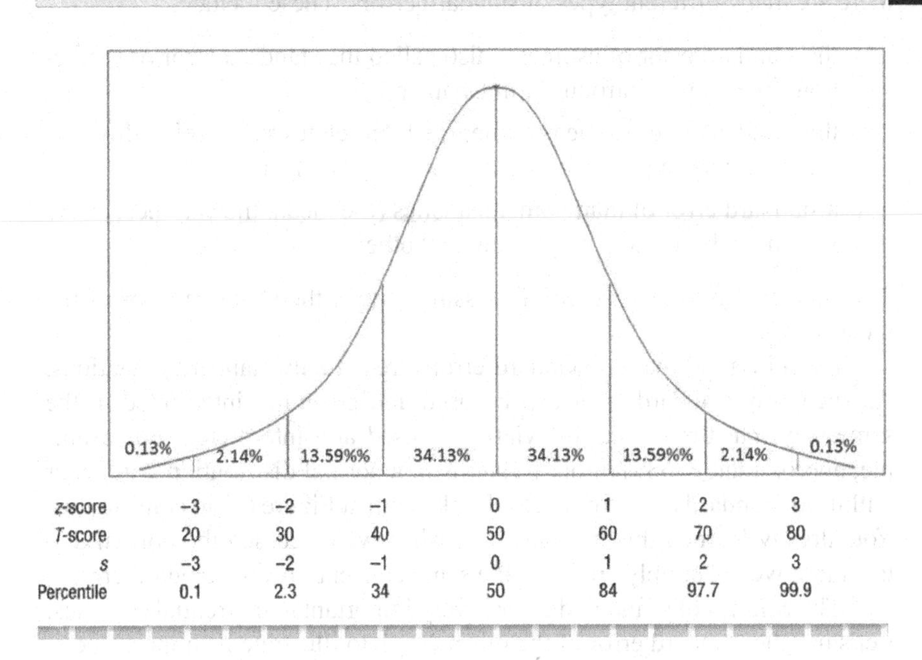

to find a value greater than ±2 standard deviations between the means, when the null hypothesis is true. Thus, if you obtain a value in the extremes of the distribution, you would consider rejecting the null hypothesis and entertaining an alternative hypothesis: that there *is* a relationship (or, in the example we've been discussing, a mean difference). You could, of course, be wrong to reject the null hypothesis, because approximately 5% of observations *will* be greater than ±2 standard deviations from zero even when the null is true. You can confirm this statement by examining Figure 6.3.

The next question we'll consider is: how large would the difference between the sample means have to be, for it to occur more than ±2 standard deviations away from zero in the normal distribution? To determine this, we need to examine another standard error, the **standard error of the mean.** The standard error of the mean is denoted by $\sigma_{\bar{X}}$.

$$\sigma_{\bar{X}} = \frac{\sigma}{\sqrt{N}}$$ Eq. 6.1

where

$\sigma_{\bar{X}}$ = the standard error of the mean,
σ = the population standard deviation, and
N = the sample size.

Standard Error of the Mean

The estimation of a standard error is an important concept related to all statistical hypothesis testing. We'll repeat a caveat that we mentioned in an earlier chapter: one must be careful with use of the term "standard error", as there are many different types of standard errors. These include:

- the standard error of estimate (also called the standard error of prediction), which was introduced in Chapter 5;
- the standard error of measurement, which relates to the reliability of a measurement (and was introduced in Chapter 2); and
- a standard error of many other statistics (the mean, the variance, skewness, kurtosis, a proportion, and others).

The standard error that we are discussing now is the standard error of the mean $\left(\sigma_{\bar{X}}\right)$.

Keep in mind that *all* standard errors are actually standard deviations. As such, any standard error can be used and generally interpreted in the same way that the standard deviation is used and interpreted. For example, approximately 68% of observations in a normal distribution will occur within ±1 standard error from zero. Look again at Figure 6.3 to confirm this. You already learned this in Chapter 4, when we discussed the normal distribution; we are simply applying the same concept now to standard errors.

The concept of a standard error is very important to inferential statistics. Let's use the standard error of the mean $\left(\sigma_{\bar{X}}\right)$ to illustrate its importance.

1. First, consider a population of interest to you. Choose a population that is quite large, so that it would be unlikely that you would ever attempt to measure a variable on all subjects in the population.
2. The population has a mean of μ. Remember that this value in the population is termed a parameter.
3. Now, suppose you would like to estimate this parameter. You would start by obtaining a random sample from the population.
4. Pick a sample size of any number. We will use a sample size of 10. You will repeatedly draw different samples of size 10 from your population, and for each sample you will measure the variable and calculate the mean of the measured variable. You might do this 1,000 or even 10,000 times. Each time you calculate a mean, you add it to a frequency of means distribution (see Figure 3.3), so that you can generate a histogram of frequencies for all of the sample means. This frequency distribution, which is based on the means that you calculated for many, many samples, is called the **sampling distribution** of the mean. The mean of the sampling distribution will be very close to the mean of the population.
5. Now, consider increasing your sample size, say from 10 to 20 or 100 or even more. Each sample mean, because it will be based on a larger sample size, will be a somewhat better estimate of the population mean than the estimation from a smaller sample size. What do you suppose the shape of the sampling distribution of the means will look like when it is based on larger samples? It will still center around the mean of the population; however, because of the larger samples and the **central limit theorem,** there will be less variability in all of the estimated

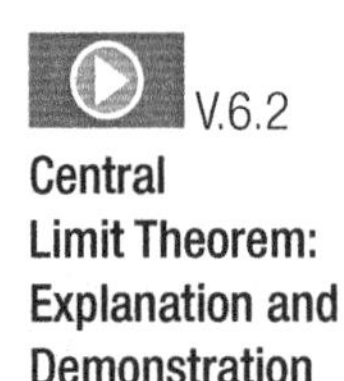

Central Limit Theorem: Explanation and Demonstration

means. This is illustrated in Figure 6.4, where the sample size increases from 2 to 5 to 30. Note that the mean of the sampling distribution, regardless of the sample size, is very close to the mean of the population from which the samples were drawn. Importantly, the heterogeneity of the curves is greater for smaller sample sizes.

6. Now we come to the key point of this discussion: *The standard deviation of the sampling distribution is the standard error of the mean.* Recall that the standard error is a standard deviation. Thus, in each of the panels in Figure 6.4, there is a standard deviation. The larger the sample size, the smaller the standard error of the mean. That is, the standard deviation of the errors in estimating the mean will be smaller as the sample size increases. As an aside, this is one of the reasons why researchers generally want to include large numbers of subjects in their research.
7. Every statistic has a standard error associated with it. For example, let's consider the median. How would you obtain the standard error of the median? You would take the steps as above: select samples, measure them, calculate the median for each sample, and develop a histogram of the median values. The standard deviation of that distribution would be the standard error of the median. You could obtain the standard error of the variance in a similar fashion: draw repeated samples, calculate the variance of each sample, and plot them on a histogram; the standard deviation of the variance estimates would be the standard error of the variance.

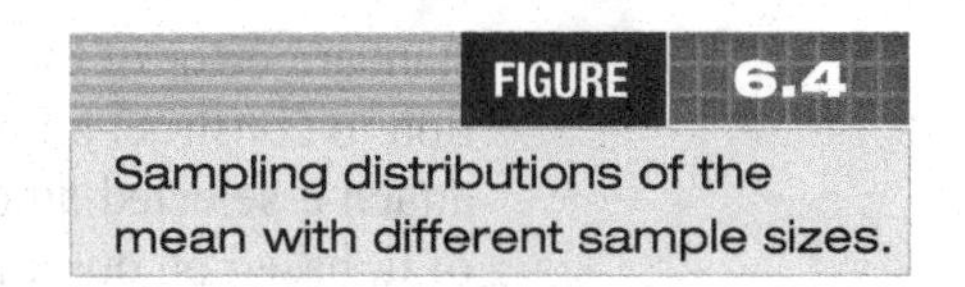

FIGURE 6.4

Sampling distributions of the mean with different sample sizes.

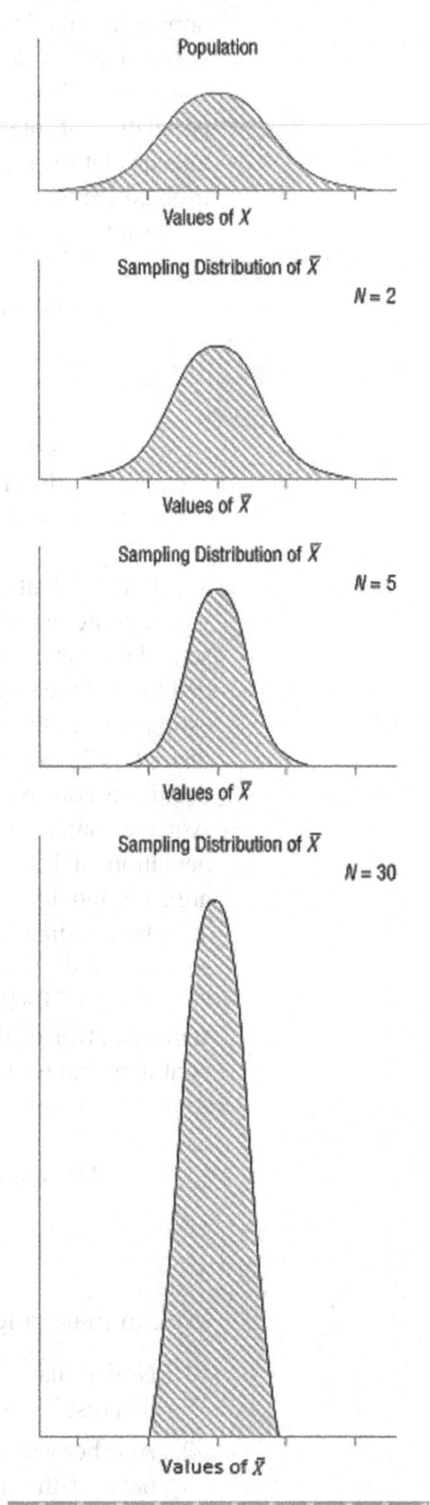

You have learned about proper sampling, sampling distributions, and the standard error of the mean. With this knowledge, you are now ready to enter the world of inferential statistics. We will begin this journey by introducing the concept of **confidence intervals.**

Confidence Intervals

When a sample is correctly drawn from a population, the best point estimate of the population mean, μ, is the sample mean, $\bar{X}$. However, because of sampling error, we know that every sample mean will not be exactly equal to μ. We can use the value of the standard error of the mean, $\Sigma_{\bar{X}}$, and our knowledge of the normal distribution to construct an interval around the point estimate of μ. Further, we can attach a selected amount of confidence that the interval contains the value of μ, based on our knowledge of the values associated with *z*-scores and the normal distribution. As just pointed out above, the standard error of the mean is the standard deviation of the sampling distribution that is *theoretically* constructed from a very large number of samples of a given *N* taken from the population of interest. Further, the central limit theorem, mentioned above, tells us that the means of this large number of samples will be normally distributed around the population mean. By combining this information, we can arrive at the following equation to construct a confidence interval:

$$\text{confidence interval} = \bar{X} - z * \sigma_{\bar{X}} \text{ to } \bar{X} + z * \sigma_{\bar{X}} \quad \text{Eq. 6.2}$$

where
$\bar{X}$ = the mean of the sample,
z = a value from the normal-curve table (Appendix A.1) that will differ depending on the amount of confidence we desire, and
$\sigma_{\bar{X}}$ = the standard error of the mean.

Let's look at an example to illustrate the concept of a confidence interval. Assume that we properly sampled 100 middle school students from the school district mentioned earlier in the chapter and measured their BMI. We determined the mean BMI for our sample of 100 students to be 20. Our best estimate of μ, the mean BMI of all the middle school students in the school district, is 20 (a point estimate). Because we know about sampling error, we decide to construct a 95% confidence interval around our point estimate. Assume that we know from national surveys that the mean and standard deviation of BMI scores for middle school students in the United States are 21.5 and 4.9, respectively. We calculate the standard error of the mean, $\sigma_{\bar{X}}$, by dividing the known standard deviation for this measurement by the square root of *N*. (See Equation 6.1)

The σ of BMI in the general population is 4.9; hence, the value of the standard error of the mean $\left(\sigma_{\bar{X}}\right)$ is $\frac{4.9}{\sqrt{100}} = 0.9$). Since we wish to be 95% confident that we will capture μ in our interval, we will use the value of 1.96 from the normal-curve table (Appendix A.1). Thus,

$$\begin{aligned} 95\% \text{ confidence interval} &= 20 - (1.96 * .49) \text{ to } 20 + (1.96 * .49) \\ &= 20 - .96 \text{ to } 20 + .96 \\ &= 19.04 \text{ to } 20.96 \end{aligned}$$

We can make a few interesting observations about this procedure:

1. Notice that we did not take a very large number of samples to obtain this result. We only took one.
2. Another way to think about this is that because the value of $\sigma_{\bar{X}}$ is 0.49, 68% of the means in the sampling distribution would differ by this amount or less from μ.

3. If the mean BMI of the general population is 21.5, we can be at least 95% confident that the mean BMI of the middle school students in this school district is significantly lower than the general population, because the value of 21.5 is *not* within our confidence interval.
4. All of this is well and good, except for one issue. It is not common—in fact, it is rare—that we know σ, the standard deviation of the variable of interest in the general population. The value of σ is required to obtain the value of $\sigma_{\bar{X}}$. We will present a solution to this problem in the next chapter.

Confidence Levels

The amount of confidence in a statistical decision is a function of the *z*-value (from Appendix A.1) that is used to obtain the value we add and subtract from the statistic to estimate the parameter. In the above example, the 95% **confidence level** is associated with a *z*-value of 1.96 because the percentage of the scores in a normal distribution that lie between −1.96 and +1.96 standard deviations from the mean is 95%. The higher the level of confidence we desire, the wider the range of values containing the parameter will become.

Although it is possible to construct a confidence interval with any level of confidence desired (for instance, we could use a *z*-value of .84 for 60% confidence), researchers commonly use *z*-values of 1.65 (90%), 1.96 (95%), and 2.58 (99%).

V.6.3 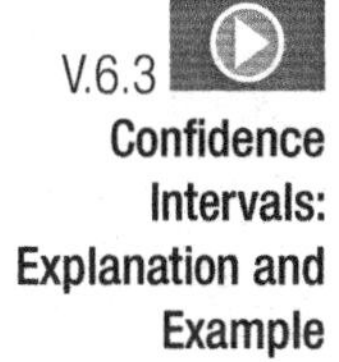
Confidence Intervals: Explanation and Example

PROPERTIES OF ESTIMATORS

We use a sample statistic to estimate a parameter of the population from which the sample was obtained. As you know, statistics (also termed *estimators*) have some amount of error attached to them, because not every sample will precisely mirror the population it represents. We will examine three properties of estimators that affect how well they serve to represent a population: **bias, consistency,** and **efficiency.**

Bias

A statistic is a biased statistic if the mean of its sampling distribution is *not* equal to the population parameter. The sample mean, $\bar{X}$, is an example of an unbiased statistic. If all possible samples of a given size were obtained, for each sample calculated, and the frequency of each $\bar{X}$ thus obtained plotted in a frequency distribution, the resulting sampling distribution would be normal in shape, and its mean would be the population mean, μ. Thus, $\bar{X}$ meets the definition of an unbiased statistic.

Not all statistics are unbiased. For example, if all possible samples of a given size were obtained, the range determined for each sample, and the frequency of each range thus obtained plotted in a frequency distribution, the mean of the resulting distribution would not be equal to the range of the population. None of the samples would overestimate the population range, few would result in the actual range, and most would underestimate the range.

It is interesting to note that the variance of a sample when calculated with $N - 1$ in the denominator is an unbiased estimate of the population variance, but the square root of this sample variance (standard deviation) is a *biased* estimate of the population standard deviation. This is true because taking the square root is a nonlinear transformation.

Consistency

Consistency refers to the fact that as the size of the sample that is used to estimate a parameter increases, the estimate comes closer to the value of the parameter. This is true for both biased and unbiased estimators. Generally, sample estimates are consistent. That is, as the sample size increases, the statistic is a more accurate estimate of the population parameter.

Efficiency

Although we could use the *median* of a large number of samples to construct a sampling distribution to estimate the population *mean*, μ, this would not be an efficient method. Using the sample means would be more efficient. This is because the sampling distribution of the medians would have a larger standard error than the sampling distribution of the means.

Ideally, the best estimators are unbiased, consistent, and efficient.

ONE-SAMPLE CASE

With the knowledge that we've accumulated to this point, we are ready to use statistics to test a hypothesis. Our eventual goal will be to determine experimental outcomes derived from sample data and then, from this information, to make inferences about the population of interest with a specific amount of confidence that what we have learned applies to the population. We will begin with the simplest situation possible—the "one-sample case". We will illustrate this with an example.

Let's assume that Professor Kline has been teaching the exercise physiology class in her department for many years and has kept data on her final examination for all these years. She decides to alter only one thing about her class this semester: she intends to introduce a technological component—a polling method—and wants to see whether this teaching aid changes student learning. Instructors can choose from various methods of polling students during a lecture, asking students to respond using clickers or apps (via cell phone or tablet) to a question (typically a multiple-choice question) presented during the lecture (e.g., on a PowerPoint® slide). This provides instant feedback to both the instructor and the students, reflecting student comprehension of the topic being presented. When we refer to Professor Kline's teaching modification here and in subsequent chapters we will use the term *polling application*. She has no idea if the polling application will improve the student's learning because they will like the use of technology or if it will prove to be a distraction and reduce the students' learning. Although this is not a very good experimental design (for example, maybe this particular class of students is unusually bright—or dull—and that might influence the results more than the application), we will not worry about that here. Our goal is solely to illustrate the statistics Professor Kline would use.

Based on all the data Professor Kline has collected over the years, she expects the mean and standard deviation of the final examination scores to be 75 and 10, respectively. (Real data seldom produce nice round numbers, but why make this example any more complex than it has to be?) On the final examination, the "polling class" of 36 students achieved a mean score of 80. This class did better than Professor Kline's previous population of students, whose mean was 75. However, we might ask some questions about these means: Is the difference of 5 points in the means a relatively rare event, or could this difference be reasonably explained as occurring by chance? In other words, how likely is it that this difference of 5 points is a result of sampling error?

The null hypothesis for this experiment is that the sample mean is equal to the population mean of 75, that is, $H_0: \bar{X} - \mu = 0$. In other words, we hypothesize that there will be no difference between the sample mean from the class that used the new technology and the population mean (i.e., the original teaching method). The research hypothesis is $H_1: \bar{X} \neq \mu$. That is, the two means will differ, but we do not stipulate which mean will be higher.

Professor Kline will test the null hypothesis to see whether it is tenable or not. To do this, she needs to determine how reasonable it is to obtain a sample mean of 80 from a population with a mean of 75. Professor Kline calculates the standard error of the mean, $\sigma_{\bar{X}}$, from the equation $\sigma_{\bar{X}} \cdot = \frac{\Sigma}{\sqrt{N}}$ to be $\frac{\sigma}{\sqrt{N}} = \frac{10}{\sqrt{36}} = \frac{10}{6} = 1.67$. She multiplies 1.67 by 1.96 (from the normal-curve table in Appendix A.1) and adds and subtracts the resulting product (3.27) to and from 80. In this way, Professor Kline constructs a 95% confidence interval of 76.73 to 83.27. This range is obtained from 80 – (1.96 * 1.67) to 80 + (1.96 * 1.67) = 80 – 3.27 to 80 + 3.27 = 76.73 to 83.27.

Notice that the population mean of 75 is not contained in this confidence interval. Thus, Professor Kline's result indicates that the population from which the sample, (with its mean of 80), is taken is unlikely to have come from a population with a mean of 75. The event of obtaining a sample mean of 80 from a population with a mean of 75 would happen less than 5 times out of 100 by chance. Thus, this result gives Professor Kline evidence that will allow her to reject the null hypothesis and accept the alternative hypothesis. That is, she concludes that the "polled" class did significantly better on the examination than previous classes.

Let's look at this example from a different perspective, called hypothesis testing. Let's determine where this event (obtaining a sample mean of 80) would be on a sampling distribution constructed from Professor Kline's population data (a mean of 75 and a standard deviation of 10, resulting in a standard error of 1.67). To do this, we convert the score of 80 to a *z*-score:

$$\text{z-score} = (80 - 75)/1.67 = 2.99$$

Therefore, this event is located nearly 3 standard deviation units above the mean (75), and, thus, is extremely rare. In fact, comparing 2.99 to 1.96 suggests that this event would occur less than 5 times out of 100 by chance,

One-Sample Case Hypothesis Test

if the null hypothesis is true. This line of reasoning leads to the same conclusion—rejection of the null hypothesis—as did examination of the confidence interval.

Final Thought

The concepts involved in this one-sample case, where we were able to make a decision about the null hypothesis (we rejected it in this instance), form the basis for most of the inferential statistics described in the rest of the book. Be sure to understand them completely.

SUMMARY

In this chapter, we defined several terms used in inferential statistics, described sampling techniques used by researchers, and outlined the procedures of hypothesis testing. Hypothesis testing involves use of the standard error, confidence intervals, and probability. As you will see in later chapters, the information presented in this chapter is fundamental to all inference testing.

Review Questions and Practice Problems

PART A (ANSWERS PROVIDED IN APPENDIX C)

A nationally standardized test measuring positive attitude toward statistics has a mean of 75 and a standard deviation of 15. Twenty-five randomly selected human performance majors took the test, and their mean score was 83. You wonder if these students are significantly different from the nationally tested group.

1. What is the null hypothesis for your research?
2. What is your research hypothesis?
3. What is the standard error of the mean?
4. Are these students like the population, or do they have a statistically significantly better attitude toward statistics? Use 95% confidence level.

Use the diagrams below to answer questions 5, 6, and 7.

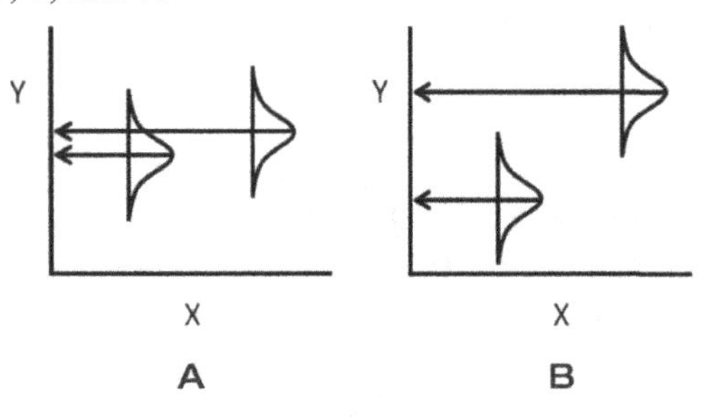

5. With which figure is a researcher more likely to reject the null hypothesis?
   A. A
   B. B
   C. They are equally likely.
   D. That is what the null hypothesis is all about, no relationship between variables.
   E. Impossible to determine from the figures
6. With which figure is a researcher most likely to retain the null hypothesis?
   A. A
   B. B
   C. They are equally likely.
   D. That is what the null hypothesis is all about, no relationship between variables.
   E. Impossible to determine from the figures
7. In which figure does there appear to be a relationship between variable *X* and variable *Y*?

A. A

B. B

C. They are equally likely.

D. That is what the null hypothesis is all about, no relationship between variables.

E. Impossible to determine from the figures

8. What is sampling error?

A. Difference between a parameter and its estimate

B. Expected value of the statistic being sampled

C. Imprecision in the measurement of a sample

D. Bias that results from small samples

PART B

Use the following information from four independent samples to answer questions 9 and 10.

Group	$\overline{X}$	σ	N
1	45	5	25
2	50	5	36
3	47	6	36
4	53	7	25

9. Which group has the smallest standard error of the mean?

A. 1

B. 2

C. 3

D. 4

10. Which group has the most variable scores?

A. 1

B. 2

C. 3

D. 4

11. In a sampling distribution based on random samples where $N = 25$, the standard error of the mean in a particular situation is 4. What size sample would be required to reduce the standard error of the mean to 2?

A. 50

B. 100

C. 1600

D. Cannot determine unless standard deviation is known

Use the following information to answer questions 12 and 13.

A population has a variance of 64. A random sample ($N = 400$) has a sample mean of 50.

12. What is the standard error of the mean?

13. Which of the following is the 95% confidence interval for the sample mean?

A. 42.0–58.0

B. 46.8–53.2

C. 49.2–50.8

D. 49.7–50.3

14. How would you draw a random sample of students from your university?

15. On what variables might you stratify if you were conducting stratified sampling at your university?

COMPUTER ACTIVITIES

Inference

TABLE 6.1

MCT scores for 25 students.

21	18	21	15	23
23	21	16	14	20
21	33	35	29	32
21	22	19	27	21
14	21	20	35	23

For a national Medical Certification Test (MCT), the mean is 25 and the standard deviation is 6. Suppose the scores in Table 6.1 were obtained from the 25 majors in your department who took the MCT. Your interest lies in whether the students' average score is significantly different from the national average. Enter the data for your students and answer the questions below.

1. What is the mean for the 25 students?
2. What is the standard deviation for the 25 students?
3. Calculate the standard error of the mean by hand, by using Equation 6.1. Use the national standard deviation and the sample size.

Using SPSS

Now use SPSS to test whether your students' average was significantly different from 25. Go to **Analyze→Compare Means→One-Sample T-test.** Move the MCT scores into the Test Variables window. Enter 25 in the Test Value box. Click OK. This is the mean value (greek symbol μ) to which you are comparing the students' mean.

Using R

1. Create a data frame of MCT scores from Table 6.1:
   - MCT <- c(21,23,21,21,14,18,21,33,22,21,21,16,35,19,20,15,14,29,27,35,23,20,32,21,23)

   Highlight what you've typed and click **Run** to create the vector of MCT scores.
2. Enter and run the following RScript commands to obtain descriptive statistics:
   - mean(MCT)
   - sd(MCT)
3. Conduct a one-sample T-test using the t.test function to test whether the average MCT scores are significantly different than 25 (mu = 25):
   - t.test(MCT, mu=25)

   Highlight what you've typed and click **Run**

t-test output will be displayed including test statistic (t), degrees of freedom (df) and p-value

Alternatively, you could read the file from Excel or csv. The format then would be:

t.test(filename$VARIABLE, mu = 25) where "filename" is the name of the saved file, followed by $ and "VARIABLE" is the name of the variable in the Excel file that is the dependent variable in the analysis (MCT from Table 6.1).

4. What is the mean difference between the national test and the students' results?
5. What is the value of the "t" that appears in the results?
6. What is the value of the "Sig". (p-value) that appears in the results?
7. Is the Sig. (p-value) value less than .05?
8. Is the student group significantly different from the national average?

7 Two-Sample *t*-test

INTRODUCTION

The one-sample case we examined at the end of Chapter 6 contains many of the important concepts of inferential statistics, but it has rather limited application in human performance research and in science in general. As we saw, the primary use of the one-sample case is to compare the mean of a sample ($\overline{X}$) to a hypothesized mean of a population (μ). We learned that we can compare the two means and either accept the hypothesis that they are not different, except for sampling error (this is called the null hypothesis and symbolized as H_0), or reject this hypothesis. We also saw that we can even attach various levels of confidence to our decision.

We learned two ways to reach our decision. First, we learned how to construct a confidence interval around the sample mean and then to check to see whether the hypothesized population mean is located within this interval. If it is, we decide the null hypothesis is true and conclude that the difference between the two means is simply the result of sampling error. If the hypothesized population mean is not located in the interval, we reject the null hypothesis (that the means are not different) and accept the alternative hypothesis (symbolized as H_1) that they are, in fact, different. Recall that the width of the confidence interval is a function of the level of confidence that is desired, the size of the sample, and the variability of the values in the population (represented by Σ).

The second way to reach the same conclusion regarding the equality of the means, called hypothesis testing, is to determine where the sample mean is located on a hypothetical sampling distribution. This sampling distribution would, theoretically, be constructed by taking many, many different samples of the same size *(N)* from the population for which we know μ and σ, calculating the mean of each sample, and plotting them in a frequency distribution. The result would be a normal curve with a mean of μ and a standard deviation (here called the standard error of the mean, $\sigma_{\overline{X}}$) equal to $\frac{\sigma}{\sqrt{N}}$. It is important to understand that the standard error of the mean is actually the standard deviation of the sampling distribution of the means that were calculated across many, many samples and then plotted. Although we never actually construct this sampling distribution, we do use our knowledge of the normal curve and the value of $\sigma_{\overline{X}}$ to locate where our sample mean would reside in the hypothetical sampling distribution. We can create a 95% confidence interval by multiplying the value of $\sigma_{\overline{X}}$ by 1.96 and adding and subtracting the resulting product to and from the sample mean.

The hypothesis testing way to decide whether to accept or reject the null hypothesis is to conduct what is called a *z*-test.

z-Test

To conduct a *z*-test, we convert the distance between the sample mean and the population mean into a *z*-score by using the following equation:

$$z = \frac{\overline{X} - \mu}{\sigma_{\overline{X}}}$$ Eq.7.1

where

z = the *z*-score,
$\overline{X}$ = the sample mean,
μ = the population mean, and
$\sigma_{\overline{X}}$ = the standard error of the mean.

The numerator of this equation represents the difference between the two means, and we convert this difference into standard error units by dividing it by $\sigma_{\overline{X}}$ Note that $\sigma_{\overline{X}}$ is a standard deviation, so Equation 7.1 is equivalent to the z-score formula presented in Equation 4.2. We can then look up the resulting z-score in the z-table to decide whether to accept or reject the null hypothesis. If the absolute value of the z-score (Appendix A.1) is less than 1.96, we conclude the null hypothesis is true, but if it is 1.96 or greater we reject the null hypothesis. In this case, we are 95% confident that the sample mean and the population mean are significantly different (i.e., the difference between the means is unlikely to be the result of sampling error).

Technically, we are testing whether the difference between the hypothesized mean and the observed mean is zero. If the null hypothesis is true, then the expected difference between the observed mean and hypothesized mean is zero. However, the exact difference will typically not be precisely zero, due to sampling error. However, if the null hypothesis is true, then in the long run the expected value of the difference between the hypothesized mean and observed mean will indeed be zero. Think back to the normal distribution presented in Chapter 4. Recall that the mean value for z-scores was zero and that values beyond ± 1.96 would occur, by chance, only 5% of the time. Recalling the normal distribution and the general range of values will help you understand the decision-making that occurs with hypothesis testing. Regardless of which method you use, you will reach the same conclusion. Some recommend that researchers report both the probability and the confidence interval.

WHAT IF WE DON'T KNOW Σ

Before we can move on to the two-sample case, which has a bit more experimental applicability than the one-sample case, we need to resolve the issue, identified earlier, that we seldom know the actual value of the population standard deviation, σ. Fortunately, this problem was solved by an English statistician, William Gosset, in 1908. Gosset, a friend of Karl Pearson's, worked for the Guinness brewery. The brewery, because another of its employees had published some trade secrets, forbade any further publication by its employees. Gosset discovered how to make an adjustment so that the standard deviation of a *sample* could be used to represent the standard deviation of the population from which the sample was taken. He convinced the Guinness company that this discovery would not benefit other breweries and was allowed to publish it under a pseudonym. He chose to publish under the name "Student" and, thus, what probably would have been known as Gosset's t-distribution came to be called the Student's t-distribution. The adjustment that Gosset devised is to use the Student's t-distribution instead of the z-distribution (the normal distribution) when we do not know the standard deviation of the population and must estimate it by using the sample standard deviation.

To use the Student's t-distribution, instead of using the z-test described previously, we use what is called the t-test. Its equation is:

$$t\text{-}ratio = \frac{\bar{X} - \mu}{s_{\bar{X}}}$$

Eq.7.2

where

t-ratio = the difference between the sample mean and population mean expressed in $s_{\bar{X}}$ units,
$\bar{X}$ = the mean of the sample,
μ = the hypothesized mean of the population, and
$s_{\bar{X}}$ = the standard deviation of the sample divided by $\sqrt{N}$ (i.e., $\frac{s}{\sqrt{N}}$).

Notice that the symbol for the standard error is the Roman letter s rather than the Greek letter σ (a population fact), indicating that it is a fact about the sample.

Unlike the z-distribution, there is not just one t-distribution but rather a series or family of t-distributions. A different t-distribution exists for each different value of the degrees of freedom. As N (the sample size) increases, the standard deviation of a sample, s, becomes an increasingly more accurate estimate of the population standard deviation, σ. Because of this, the shape of the t-distributions approaches the shape of the z-distribution as N increases, and at $N = \infty$ they are identical. When N is small, the t-distributions are symmetrical, like the z-distribution, but are not fully normal in shape. The t-distributions with small Ns are a bit platykurtic, but they become more normally distributed as N increases. Fortunately, the t-distributions approach the z-distribution rather quickly. Figure 7.1 illustrates the changes one sees in the shape of the t-distribution depending upon the degrees of freedom. Note that as the sample size increases, the curve becomes more normal. The values for the Student's t-distributions for various degrees of freedom and levels of confidence have been tabulated. This table appears in Appendix A.3.

Let's look at an illustration to learn how to use Appendix A.3. Assume that in the example at the end of Chapter 6, Professor Kline knew the population mean of her exercise physiology final examination to be 75, but she did not know the value of the population standard deviation. She can calculate the standard deviation of her current sample of 36 students, which had a mean of 80. She did so and determined the standard deviation to be 12. Now, instead of using a z-test to determine whether 75 and 80 are significantly different, she can use a t-test. To do this, she first needs to calculate $s_{\bar{X}}$, which is $\frac{12}{\sqrt{36}} = 2$. The resulting t-ratio is $\frac{80-75}{2} = 2.5$. If Professor Kline had calculated a z-ratio of 2.5, she would have known immediately that the means were significantly different at the 95% confidence level, because 2.5 is larger than 1.96. Does a t-ratio of 2.5 lead to the same conclusion? To determine this, she would go to the t-distribution table (Appendix A.3) and go down the df column to 35 ($df = N - 1$ in the one-sample case). Because 35 doesn't appear in Appendix A.3, however, she would have to use 30 df (to be conservative). She would go across to the third column in the table (two-tailed test, .05; we will explain about two-tailed tests later), where she would find the value of 2.042. This value is often referred to as the "critical" value, because to reject the null hypothesis, you must obtain a calculated t-value that exceeds this "critical" number. Because her t-ratio is 2.5, she would

Changes in the shape of the *t*-distribution as a function of sample size. FIGURE 7.1

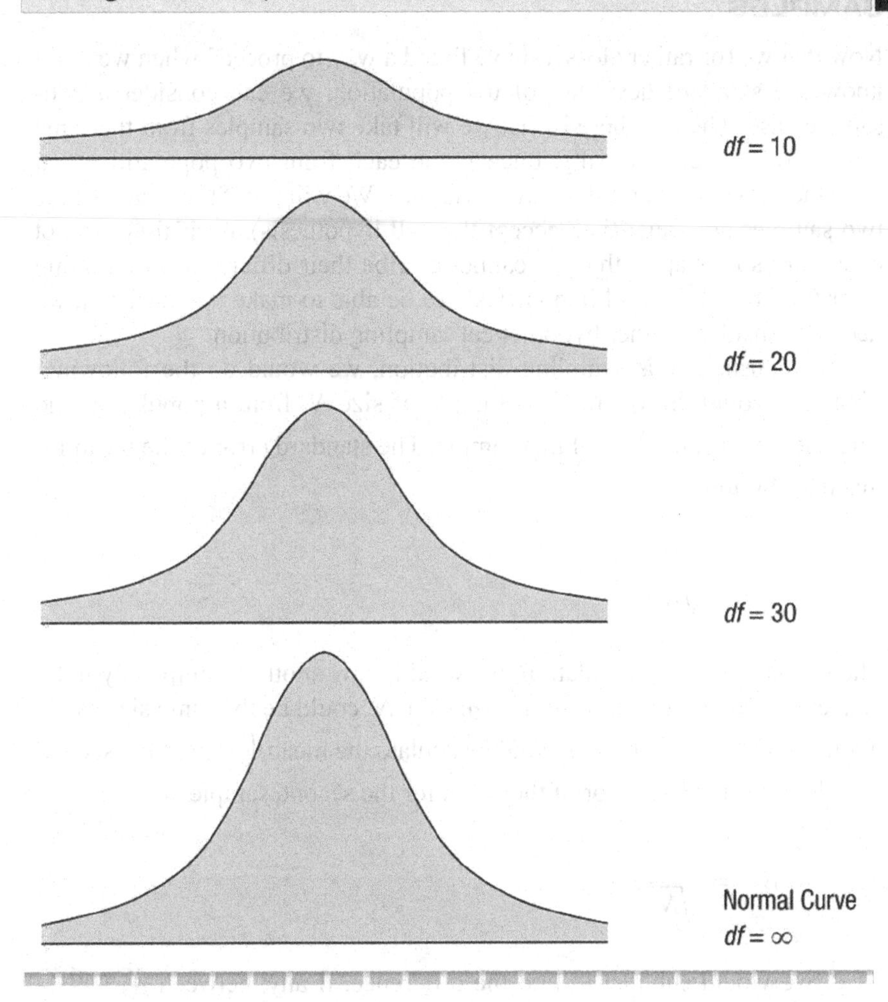

reject the null hypothesis, as before. In other words, the event of obtaining a mean difference of 5 points would happen less than 5% of the time if the two means were actually equal. In this case, Professor Kline rejects the null hypothesis.

Because the Student's *t*-distributions are symmetrical but platykurtic in comparison to the normal curve, the *t*-ratio required to determine a significant difference between the sample mean and population mean is greater than the corresponding *z*-ratio. (Notice, however, that 2.042 is not very much different from the 1.96 that we would have used in the *z*-test.) Familiarize yourself with the *t*-distribution table. Confirm that if Professor Kline desired to be 99% confident that the two means were significantly different, the *t*-ratio of 2.5 would *not* have allowed her to reject the null hypothesis because the critical *t*-ratio value in this case is 2.75. Confirm, also, that if she had had only five degrees of freedom, she would have to retain the null hypothesis at the 95% confidence level, because the obtained value does not exceed the critical value (2.571) listed in the *t*-table.

TWO-SAMPLE *T*-TEST CASE—INDEPENDENT SAMPLES

Now that we (or rather Gosset) have found a way to proceed when we don't know the standard deviation of the population, we can consider a two-sample case. The idea here is that we will take two samples from the same population (or, conceptually, one-sample each from two populations that have the same mean and standard deviation). We will see if the means of the two samples are equal (i.e., accept the null hypothesis), or, if they are not equal, are so far apart that we cannot ascribe their difference to sampling error (i.e., reject the null hypothesis). To be able to make this decision, we need to consider another hypothetical sampling distribution.

To construct this sampling distribution, we would do the following: First, we would draw a random sample of size N_1 from a population and calculate the mean $(\bar{X}_1)$ of this sample. The standard error of the mean for that distribution is:

$$S_{\bar{X}_1} = \frac{s_1}{\sqrt{N_1}}$$

Then, from the same population, we would draw another, completely independent random sample, of some size (N_2). N_2 could be the same size as N_1, but it doesn't have to be. We would calculate the mean $(\bar{X}_2)$ of the second sample. The standard error of the mean for the second sample is:

$$S_{\bar{X}_2} = \frac{s_2}{\sqrt{N_2}}$$

Next, we would subtract to find the difference, if any, between $\bar{X}_1$ and $\bar{X}_2$ and record the answer. Hypothetically, we would repeat this procedure many thousands of times and then construct a frequency distribution of all the mean differences. The result would, theoretically, be a normal curve with a mean of zero if the null hypothesis is true. Its standard deviation—here called the standard error of the difference between two means—is symbolized as $S_{\bar{X}_1 - \bar{X}_2}$. Note that $S_{\bar{X}_1 - \bar{X}_2}$ is the standard deviation of the distribution of *differences* between the means. We can calculate it by using the following equation:

$$S_{\bar{X}_1 - \bar{X}_2} = \sqrt{\left(s_{\bar{X}_1}\right)^2 + \left(s_{\bar{X}_2}\right)^2} \quad \text{Eq.7.3}$$

where

$S_{\bar{X}_1 - \bar{X}_2}$ = standard error of the difference between two independent-sample means,

$s_{\bar{X}_1}$ = standard error of the mean of the first sample, and

$s_{\bar{X}_2}$ = standard error of the mean of the second sample.

This equation may be used when the *N*s for the two samples are equal or nearly so. Later, we will present another equation for the standard error of the difference between two means that can be used with equal or unequal sample sizes.

Let's examine this theoretical sampling distribution so we can understand how to use it. If we take two independent random samples from the same population, in theory the means of the two samples should both be equal to the population mean and, thus, to each other. This would be expressed $\bar{X}_1 - \bar{X}_2 = 0$ for the sample and a null hypothesis as $\mu_1 - \mu_2 = 0$ for the population. However, because of sampling error, this would seldom actually be true. We would expect small differences between the sample means and the population mean, resulting from chance. Thus, we would expect to find some differences when we subtract one-sample mean from the other; we would not expect always to obtain the theoretical value of zero. Most of the time, these differences would be relatively small, and they could be either positive or negative, depending on the order in which the means are subtracted. Rarely would we expect to obtain large differences between the two-sample means if the null hypothesis (H_0) is true. Since it is chance that is causing any differences that do occur, and since we have learned that the normal curve is "the curve of chance", we can use our knowledge of the normal distribution to determine the probability that differences of various magnitudes will occur. For example, we would expect approximately 68% of the differences to lie within $\pm 1 * S_{\bar{X}_1 -}$ of zero and 95% of the differences to lie within $\pm 1.96 * S_{\bar{X}_1 -}$ of zero. Said another way, we would expect only 5% of the differences to be larger than $\pm 1.96 * S_{\bar{X}_1 - \bar{X}_2}$

Therefore, because we can calculate $S_{\bar{X}_1 - \bar{X}_2}$ from just two samples, we can (without actually constructing the hypothetical sampling distribution) determine, at chosen levels of confidence, whether the means of the two samples are different by chance or are too different for us to attribute their difference to sampling error. Let's look at an example to see how to perform a two-sample case *t*-test.

Earlier in this chapter, we reviewed how Professor Kline used the one-sample case *t*-test to compare a sample mean from scores on her exercise physiology final to a population mean score on the final, computed from several previous semesters when a polling application was not used. Let's change the experiment for illustrative purposes and have Professor Kline teach two sections of exercise physiology this semester, with the only difference between the two sections being the use of polling in one of them and not in the other. We pointed out that the one-sample case was not a very sound experimental design, because many other variables (besides the introduction of the polling application) could have caused differences to occur between the current class and those taught in the past. Similarly, our two-sample case example is not a very strong experimental design, because the students probably are not selected and placed randomly into the two sections. If some reason exists that would cause students to elect or be placed into one section rather than the other, and if this reason is related in any way to performance in class, any differences that are found could be the result of this confounding variable rather than an effect of using the polling application. However, as our task here is to demonstrate the use of a statistical procedure and not to analyze the experimental design, we will proceed with the scenario presented.

At the end of the semester, Professor Kline collected the following data:

	N	$\overline{X}$	s
Polling section	49	79.2	9.6
Regular section	64	74.5	9.5

The question of interest is: "Is the polling section's higher mean on the final examination reasonably attributable to chance (sampling error), or are the means significantly different?" To answer this question, Professor Kline would conduct an **independent *t*-test.** The word *independent* here means that the two samples are made up of entirely different subjects. In the next section of this chapter we will examine the dependent *t*-test.

The equation for the one-sample case *t*-test was presented earlier:

$$t\text{-}ratio = \frac{\overline{X} - \mu}{s_{\overline{X}}}$$

We will need to change this slightly for the two-sample case and an independent *t*-test, to:

$$t\text{-}ratio = \frac{\overline{X}_1 - \overline{X}_2}{s_{\overline{X}_1 - \overline{X}_2}} \qquad \text{Eq.7.4}$$

where

$\overline{X}_1$ = the mean of the polling section,
$\overline{X}_2$ = the mean of the regular section, and
$s_{\overline{X}_1 - \overline{X}_2}$ = the standard error of the difference between two independent means.

In the one-sample case, the numerator is the difference between the one-sample mean and the hypothesized population mean. In the two-sample case, the numerator is the difference between the two-sample means. In the one-sample case, the denominator is the standard error of the mean. In the two-sample case, the denominator is the standard error of the difference between two means. As you might recall, the standard error of the difference between two means is actually a standard deviation.

When we presented Equation 7.3, we mentioned that that equation may be used when the two samples have equal or nearly equal *N*s. In Professor Kline's study, she had a different number of students in the two sections, so we will use the equation for unequal *N*s. It is:

$$s_{\overline{X}_1 - \overline{X}_2} = \sqrt{\left[\frac{(N_1 - 1)(s_1)^2 + (N_2 - 1)(s_2)^2}{N_1 + N_2 - 2}\right]\left[\frac{1}{N_1} + \frac{1}{N_2}\right]} \qquad \text{Eq.7.5}$$

It is especially important to use this equation when the unequal *N*s are relatively small. One final consideration before we calculate the *t*-ratio for this example is to note that the degrees of freedom for a two-sample independent *t*-test are calculated as $(N_1 + N_2) - 2$. Equation 7.6 gives the full equation for calculating a *t*-ratio in the two-sample case with an independent *t*-test and unequal *N*s:

$$t - ratio = \frac{\bar{X}_1 - \bar{X}_2}{\sqrt{\left[\frac{(N_1 - 1)(s_1)^2 + (N_2 - 1)(s_2)^2}{N_1 + N_2 - 2}\right]\left[\frac{1}{N_1} + \frac{1}{N_2}\right]}} \quad \text{Eq.7.6}$$

Hence, in Professor Kline's study:

$$t\text{-}ratio = \frac{79.2 - 74.5}{\sqrt{\left[\frac{(49-1)(9.6)^2 + (64-1)(9.5)^2}{49 + 64 - 2}\right]\left[\frac{1}{49} + \frac{1}{64}\right]}}$$

$$= \frac{4.7}{\sqrt{\left[\frac{(48)(92.16) + (63)(90.25)}{111}\right][.020 + .016]}}$$

$$= \frac{4.7}{\sqrt{\left[\frac{(4423.68) + (5685.75)}{111}\right][.036]}}$$

$$= \frac{4.7}{\sqrt{[91.08][.036]}}$$

$$= \frac{4.7}{\sqrt{3.28}}$$

$$= \frac{4.7}{1.81}$$

$$= 2.60$$

Professor Kline would now go to Appendix A.3. Although she has 111 degrees of freedom (49 + 64 − 2), she would have to go to the row for 60 *df* (because the degrees of freedom column provides values only for 60 and 120) and locate the critical value in the two-tailed .05 column. The critical value is 2.000. Since the calculated *t*-ratio is 2.60, Professor Kline rejects the null hypothesis and concludes that the two means are significantly different. If the conclusion had resulted in retaining the H_0, Professor Kline might have sought out a more complete *t*-distribution table. Although this experiment does not constitute proof that polling improves performance (remember the weakness of the experimental design, and notice that, even though this event could happen by chance less than 5% of the time, it still could happen), the professor does have fairly strong evidence that polling may be beneficial.

V.7.1
SPSS *t*-Test for Independent Samples

V.7.1R
RStudio *t*-Test for Independent Samples

Another important statistic, which we will describe more fully later in Chapter 8, is *effect size* (ES). The ES helps researchers interpret results for *practical significance* in addition to the statistical significance resulting from the *t*-ratio. The ES is essentially another standard score. It is obtained by subtracting the two means (here = 79.2–74.5 = 4.7; note this is the numerator in the *t*-ratio). This difference is then divided by the standard deviation of the "control group". In Professor Kline's case, she uses the "traditional" course delivery as the "control group"; thus, ES = 4.7/9.5 = .49. Can you see how this looks much like a *z*-score? The .49 ES is interpreted to mean the typical student in the polling sample is about one-half of a standard deviation above the typical student in the traditional sample. An observation that the typical "treated" student increased by nearly one-half of a standard deviation is quite important. Not only are Professor Kline's results *statistically* important; they are also *practically* quite important. You will learn more about effect sizes, statistical significance, and practical significance in Chapter 8.

TWO-SAMPLE *T*-TEST CASE—DEPENDENT SAMPLES

As you have seen, the *t*-test is designed to determine whether or not the presence of a difference between two means can be explained reasonably by chance (sampling error). For the one-sample case, we examined the difference between a sample mean and a hypothesized population mean (μ), and for the two-sample independent *t*-test, we examined the difference between two independent-sample means. We will now look at the two-sample **dependent *t*-test.** Some refer to the dependent *t*-test as a *correlated t-test* or a *paired t-test*. All of these terms can be used interchangeably.

The use of the dependent *t*-test is appropriate when the samples are not independent but represent repeated measures of the same subjects, who have been measured twice. It is also appropriate when each subject is somehow matched with another subject, and the resulting two scores are paired with each other. In both of these cases, the subjects are not independent but rather are connected (i.e., correlated or paired). This assignment of subjects is called a **repeated measures design** or a **within-groups design.** A common application of this type of experiment is the *pre—post design*. Here, the researcher randomly selects subjects for the sample, measures them on some variable, administers some form of treatment, and then re-measures the subjects on the same variable. The dependent *t*-test is then used to determine whether there is a significant difference between the pre-treatment mean and the post-treatment mean. If the means are significantly different, then the administered treatment is implicated in causing the difference. If there is no effect from the treatment, the difference between the pre and post means is expected to be zero. As with the independent-samples *t*-test case (Professor Kline's experiment discussed above), other research design issues should be considered, but they do not pertain to our discussion here.

Although the equation for the dependent *t*-test looks exactly like the equation for the independent *t*-test, there is one major difference, and that is how the denominator (the standard error of the difference between two means) is calculated. The reason for this is that smaller differences would be expected to occur with a treatment administered to the same group of subjects than with treatments administered to two different groups of subjects. A subject who scores high on the pretest on the variable of interest—even if the treatment does increase scores on this variable—would probably not change very much, and this subject's two scores would probably not differ as much as would the scores from two independent subjects. This argument would also hold for subjects who score low on the pretest. Thus, the denominator of the dependent t-test must be reduced to result in the correct t-ratio, and this reduction is related to the amount of correlation that exists between the two measurements. The equation for the standard error of the difference between two dependent means is:

$$s_{\bar{X}_1-\bar{X}_2} = \sqrt{\left(s_{\bar{X}_1}\right)^2 + \left(s_{\bar{X}_2}\right)^2 - 2r\left(s_{\bar{X}_1}\right)\left(s_{\bar{X}_2}\right)} \quad \text{Eq.7.7}$$

where

$s_{\bar{X}_1-\bar{X}_2}$ = the standard error of the difference between two dependent means,
$s_{\bar{X}_1}$ = the standard error of the mean of the first mean,
$s_{\bar{X}_2}$ = the standard error of the mean of the second mean, and
r = the correlation between the two sets of scores.

Notice that if r is positive, it results in a *reduction* of the value of the $s_{\bar{X}_1-\bar{X}_2}$ Also, notice that if $r = 0$ (which would indicate no correlation between the sets of scores; they would be independent), the equation becomes identical to the denominator for the independent t-test (see Equation 7.3). Let's look at an example to illustrate the use of the dependent t-test.

Dr. Johnson, an exercise physiologist at a cardiac rehabilitation facility, wonders if allowing his clients to have more input in designing their exercise programs will have any effect on their attitudes toward the exercise programs. He happens to have access to a valid and reliable questionnaire, based on the theory of planned behavior, to measure this construct. He administers the test to a group of clients who are currently participating in a clinician-designed exercise program. He tests them again six weeks after they have participated in their clinician/patient-designed program. Dr. Johnson's data are as follows:

	N	$\bar{X}$	s
Pretest	49	80	15
Posttest	49	84	16

The correlation coefficient between the pre and post measures is .65 (see Chapter 5). Notice that the N in a dependent t-test will always be the same for the two means, because the same (or matched) subjects are being measured. Dr. Johnson's first step is to calculate the value for the standard error of the difference between two dependent means:

$$\begin{aligned} s_{\bar{X}_1-\bar{X}_2} &= \sqrt{\left(s_{\bar{X}_1}\right)^2 + \left(s_{\bar{X}_2}\right)^2 - 2r\left(s_{\bar{X}_1}\right)\left(s_{\bar{X}_2}\right)} \\ &= \sqrt{(2.14)^2 + (2.29)^2 - 2(.65)(2.14)(2.29)} \\ &= \sqrt{4.58 + 5.24 - 6.37} \\ &= \sqrt{3.45} = 1.86 \end{aligned}$$

Now he can calculate the t-ratio. Since there is an increase in positive attitude from the pretest to the posttest, Dr. Johnson decides to subtract the pretest mean from the posttest mean, so that the numerator will be a positive number. The order of subtraction doesn't really matter, because it is only the absolute value of the magnitude of the t-ratio that will determine whether the means are significantly different.

$$\begin{aligned} t - ratio &= \frac{\bar{X}_1 - \bar{X}_2}{\sqrt{\left(s_{\bar{X}_1}\right)^2 + \left(s_{\bar{X}_2}\right)^2 - 2r\left(s_{\bar{X}_1}\right)\left(s_{\bar{X}_2}\right)}} \\ &= \frac{84 - 80}{1.86} \\ &= \frac{4}{1.86} \\ &= 2.15 \end{aligned} \quad \text{Eq.7.8}$$

In the dependent-samples t-test, the degrees of freedom are $N - 1$, where N is the number of pairs of scores or the number of subjects who are measured twice. Thus, Dr. Johnson consults Appendix A.3, uses the row for 40 df (because it is the closest lower value to 48), and finds a critical value of 2.02. Because his t-ratio of 2.15 exceeds the critical value of 2.02, Dr. Johnson decides, with a confidence level of 95%, that the difference of 4 points between the pre and post means constitutes a significant difference. As with any statistical test, this finding is not proof that, on average, it was involvement in designing their exercise programs that increased his clients' positive attitude toward exercising, because several other potential variables could be involved. Also note the effect size in this case is (84 − 80)/15 = .27. One-fourth of a standard deviation is a relatively small ES.

V.7.2
SPSS *t*-Test for Dependent Samples

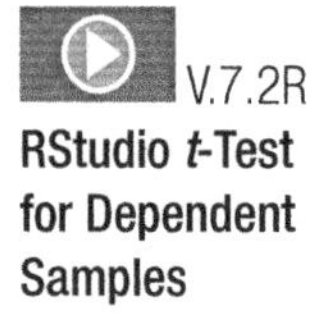
V.7.2R
RStudio *t*-Test for Dependent Samples

DIRECT DIFFERENCE METHOD FOR THE DEPENDENT *T*-TEST

An interesting feature of the pre–post or paired two-sample case is that we can test to see whether two means differ significantly by turning it into a one-sample case. This is called the *direct difference method*, and we will illustrate it with an example that uses paired subjects rather than the pre–post design.

Assume that Dr. Parker, a nutrition expert, is interested in determining whether a difference exists in the efficacy of two diet plans designed to reduce body weight. The researcher randomly selects 30 individuals from the population to which he is interested in being able to generalize his results. He is aware that there is a positive correlation between initial body weight and the potential to lose weight, and he would like to eliminate the influence of this possible confounding variable. Thus, he weighs all 30 subjects and then randomly assigns one of the two heaviest subjects to Diet 1 and the other to Diet 2. He follows this same procedure with the next two heaviest subjects and continues in this manner until he reaches the two lightest subjects, whom he again randomly assigns to the two diets. With the subjects paired in this manner, the mean body weights of the two groups should be nearly identical, and, thus, initial body weight should not be a factor in Dr. Parker's final results.

After the subjects completed the treatments (diets) for the length of time specified by the research design, Dr. Parker collected each subject's weight-loss data. The data for this study are presented in Table 7.1. The subjects are numbered 1a, 1b, and so forth because, although 1a and 1b (for example) are different individuals, they are matched on initial body weight. With this type of sample, the dependent *t*-test is the correct procedure to use.

We can summarize the data from Table 7.1 as shown in Table 7.2.

Notice that, as expected, the heavier subjects lost more weight than did the lighter subjects, some of whom actually gained weight while on the

TABLE 7.1 Dr. Parker's weight-loss data. Subjects are listed from heaviest to lightest.

DIET1		DIET2		
SUBJECTA	WEIGHT LOSS (LBS)	SUBJECTB	WEIGHT LOSS (LBS)	DIFFERENCE
1A	20	1B	18	2
2A	17	2B	18	–1
3A	15	3B	19	–4
4A	16	4B	17	–1
5A	12	5B	15	–3
6A	10	6B	8	2
7A	8	7B	6	2
8A	9	8B	10	–1
9A	6	9B	8	–2
10A	4	10B	4	0
11A	1	11B	3	–2
12A	2	12B	1	1
13A	5	13B	–3 (gained)	8
14A	2	14B	–2 (gained)	4
15A	0	15B	0	0

TABLE 7.2 Summary of data.

	DIET1	DIET2	DIFFERENCE
Sum of weight lost	127 lbs	122 lbs	5 lbs
N	15	15	
Mean	8.47 lbs	8.13 lbs	0.34 lbs
Standard deviation	6.38 lbs	7.72 lbs	3.02 lbs
Standard error of the mean	1.65 lbs	1.99 lbs	0.78 lbs

diet. Notice also that both diets were effective to some degree: on average, the subjects who followed Diet 1 lost 8.47 pounds, and those who followed Diet 2 lost an average of 8.13 pounds. The initial question was concerned with comparing the efficacy of the two diets: that is, is 8.47 significantly different from 8.13? If both diets were equally effective, then there will be no difference between the mean weight losses for the diets; the null hypothesis will be true. We can test this hypothesis by determining whether the mean difference for weight lost (.34 pounds) is significantly different from the hypothesized value of zero. This, in effect, turns the two-sample case into a one-sample case. An added bonus is that calculating r is not necessary, which would be required if we conducted the typical dependent t-test. The equation for the direct difference method is:

$$t - ratio = \frac{\bar{X}_D}{S_D}$$ **Eq.7.9**

where

$\bar{X}_D$ = the mean of the differences and
s_D = the standard error of the mean differences.

We calculate s_D, the standard error of the mean differences, by dividing the standard deviation of the differences by $\sqrt{N}$. That is, $s_D = \frac{3.02}{\sqrt{15}} = .78$ Hence,

$$\begin{aligned} t\text{-}ratio &= \frac{\bar{X}_D}{S_D} \\ &= \frac{.34}{.78} \\ &= .44 \end{aligned}$$

After consulting Appendix A.3, Dr. Parker concludes that the null hypothesis is true and that the two diets were equally effective (i.e., that the difference between 8.47 pounds and 8.13 pounds is not significant and is reasonably attributable to sampling error). Use Appendix A.3 to confirm that, to reject the null hypothesis with 95% confidence, Dr. Parker would have to obtain a *t*-ratio of 2.145. (Recall that, in this situation, the *df* are $N - 1$.)

V.7.3 SPSS Dependent *t*-Test as One-Sample Case

We will now demonstrate that the direct difference method is equivalent to the dependent *t*-test previously discussed. The correlation between the paired observations is .926 (see Chapter 5). When we use Equation 7.7 to calculate the standard error of the difference between two correlated measures, we find the following *t*-ratio:

V.7.3R RStudio Dependent *t*-Test as One-Sample Case

$$t\text{-}ratio = \frac{\bar{X}_1 - \bar{X}_2}{\sqrt{\left(s_{\bar{X}_1}\right)^2 + \left(s_{\bar{X}_2}\right)^2 - 2r\left(s_{\bar{X}_1}\right)\left(s_{\bar{X}_2}\right)}}$$

$$= \frac{8.47 - 8.13}{\sqrt{(1.65)^2 + (1.99)^2 - 2(.926)(1.65)(1.99)}}$$

$$= \frac{.34}{\sqrt{2.72 + 3.96 - 2(.926)(1.65)(1.99)}}$$

$$= \frac{.34}{\sqrt{2.72 + 3.96 - 6.08}}$$

$$= \frac{.34}{\sqrt{.60}} = \frac{.34}{.77} = .44$$

ASSUMPTIONS OF THE *T*-TEST

For the *t*-test to be an accurate comparison between two means, certain statistical assumptions must be met. Early statisticians assumed that the following conditions had to be present when the *t*-test was employed:

- *Normality:* The variable of interest is normally distributed in the population.
- *Random sampling:* Simple random sampling is used to draw the sample(s) from the population.
- *Level of measurement:* The variable of interest is measured on a continuous scale (either interval or ratio measurement).
- *Homogeneity of variance:* All random samples will have the same (**homogeneity**) variance (within the limits of sampling error). This assumption is necessary because we estimate the variability of the population by using the combined data from both samples. In the next chapter, you will see how SPSS and RStudio check for the validity of this assumption.

V.7.4 SPSS *t*-Test Assumptions

Although it was originally thought that these assumptions had to be met, this is no longer the case. Over time, the *t*-test has been found to be reasonably **robust** to these assumptions. The word *robust,* here, means that the *t*-test functions quite well even when some of these assumptions are not completely met. The *t*-test is particularly robust to these assumptions when the *N*s are equal and reasonably large (>30).

V.7.4R RStudio *t*-Test Assumptions

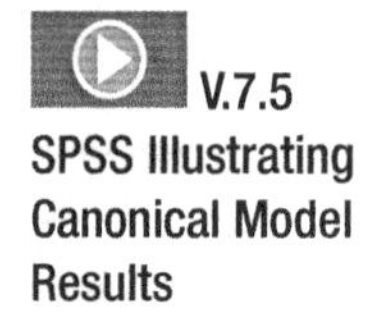
V.7.5
SPSS Illustrating Canonical Model Results

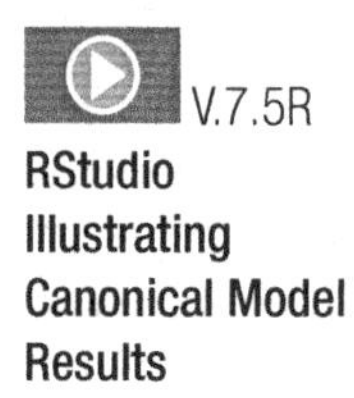
V.7.5R
RStudio Illustrating Canonical Model Results

ILLUSTRATION OF THE GENERAL CANONICAL MODEL

Recall from Chapter 5 that with the canonical model approach, you have a set of *Y* variables (from one to several) and a set of *X* variables (from one to several), and you are interested in the correlation or relationship between the *Y* and *X* sets. In a very real sense, when you use the independent *t*-test, you are testing whether a relationship exists between the group that a subject is in and the outcome variable. Essentially, you are testing whether there is a relationship between group membership (the *X* variable) and the resulting score (the dependent variable, *Y*). We will illustrate that you would reach the same decision about the existence of a significant relationship between group membership and the outcome variable if you simply used the Pearson product-moment correlation (presented in Chapter 5). This demonstrates the general nature of inferential statistics: that it is about relationships between variables.

For this illustration, we will return to Professor Kline's polling application experiment data. Assume that the 49 students in the polling section are coded 0 (zero), and the 64 students in the regular section are coded 1. With these codes, Professor Kline now can calculate the correlation between group membership and the dependent variable. This is the point biserial correlation described in Table 5.8. The point biserial correlation for this example is −.243. Now, the professor can test to determine whether −.243 is significantly different from zero by consulting Appendix A.2. She would use 90 *df* even though the actual value is 111 *df* because of the values available in Appendix A.2. In the appendix, Professor Kline finds that a correlation of .205 is significantly different from zero at the 95% confidence level. Because the absolute value of her correlation is .243, which is larger than the critical value of .205, she concludes that there is a significant relationship between group membership and the dependent variable. In this case, she concludes that a significant mean difference does exist between the two groups. The negative correlation coefficient indicates that the polling section (coded 0) scored higher on the test than did the regular section (coded 1). This is exactly the same conclusion Professor Kline reached with the independent *t*-test. As we will continue to illustrate for you, many inferential tests can be thought of as tests of relationships between one or more *X* variables and one or more *Y* variables.

SUMMARY

The one-sample *t*-test and the two versions (independent and dependent *t*-test) of the two-sample case represent one of the simplest levels of inferential statistics. In the one-sample case, the researcher defines a population, randomly samples subjects from the population, and measures the subjects on the variable of interest to infer something about the population mean. In the two-sample case, the researcher places the subjects into two groups to be used in an experiment. The researcher applies treatments and then compares the means of the two groups. The researcher tests the null hypothesis (H_0) and generalizes the results to the population.

We can summarize the steps for independent and dependent sample *t*-tests as follows:

1. Draw random samples,
2. Calculate the sample means,
3. Calculate the appropriate standard error,
4. Use the appropriate statistical test (independent or dependent *t*-test) and calculate the *t*-ratio,
5. Determine the appropriate critical value, depending on the degrees of freedom and level of confidence desired,
6. Compare the calculated *t*-ratio with the critical value from Appendix A.3, and
7. If the absolute value of the obtained *t*-ratio exceeds the critical value from Appendix A.3, reject the null hypothesis (that there is no difference), conclude that the difference is not the result of chance, and accept the alternative hypothesis. If the absolute value of the obtained *t*-ratio does not exceed the critical value, retain the null hypothesis.

In this process, many inferential statistics issues require consideration. We will address these in the next chapter, before we go on to examine more complex inferential statistics. These issues are relatively easy to address with the *t*-test, but they can be generalized to the more complex tests we will describe later.

Review Questions and Practice Problems

PART A (ANSWERS PROVIDED IN APPENDIX C)

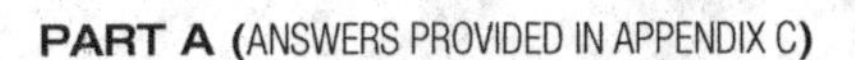

1. Which of the following *t*-ratios could not possibly be statistically significant at the 95% confidence level?

   A. 1.90

   B. 2.00

   C. 2.58

   D. Need to know degrees of freedom

2. Which of the following is *not* an assumption of the *t*-test?

   A. The dependent variable is normally distributed.

   B. The independent variable is normally distributed.

   C. The samples are randomly selected.

   D. The measurement level is continuous.

3. What are the degrees of freedom for a *dependent* *t*-test with 30 subjects measured on both the pretest and the posttest?

   A. 60

   B. 59

   C. 58

   D. 29

4. If the null hypothesis is true, what is the expected value of the sum of the differences in a direct difference dependent *t*-test?

   A. A small negative value

   B. A large positive value

   C. Zero

   D. Depends on *N*

5. Given the following values, what is the t-ratio?

Treatment group	Control group
$\bar{X}_T = 100$	$\bar{X}_C = 90$
$N = 21$	$N = 11$
$\sigma x^2 = 5000$	$\sigma x^2 = 3000$

A. 1.57
B. 1.65
C. 1.71
D. 2.36

PART B

6. On a test given in a class, the 68 males' mean score was 29.17 with a standard deviation of 3.05, and the 68 females' mean score was 28.52 with a standard deviation of 2.97. Were the male and female means significantly different at the .05 (95% confidence) level?
7. One hundred twenty-five human performance majors were given a pull-up test. Their mean score was 6.5 with a standard deviation of 3.4. After completing a 10-week strength-training program, the students were retested and, on average, completed 7.1 pull-ups with a standard deviation of 4.0. If the correlation coefficient between the pre- and post-training scores was .75, was there a significant difference between the pre- and post-training scores? (Use a confidence level of 95%.)
8. Explain how an independent t-test and a correlation coefficient can lead to the same conclusion about group mean differences.
9. Draw a scattergram figure representing a two-group experiment where the treatment and control groups differ.

COMPUTER ACTIVITIES
t-Test

This activity requires calculating an independent and a dependent t-test and answering several questions based on the results of these tests.

TABLE 7.3 Data set.

SUBJECT	FITNESSLEVEL	VO2MAX
1	F	54
2	F	47
3	F	38
4	F	45
5	F	39
6	F	42
7	F	50
8	F	52

SUBJECT	FITNESSLEVEL	VO2MAX
9	F	37
10	F	36
11	S	42
12	S	37
13	S	29
14	S	34
15	S	36
16	S	34
17	S	22
18	S	18
19	S	27
20	S	32

ACTIVITY 1 Independent *t*-Test

A researcher devised a pencil-and-paper instrument to assess subjects' self-reported amount of time spent in various physical activities. The intent of the instrument was to distinguish between fit and sedentary individuals. To investigate the validity of this instrument (whether or not it correctly identified individuals as being fit or sedentary), the researcher measured the VO2MAX of 10 subjects who had been classified by the written instrument as fit and another 10 subjects classified as sedentary. Enter the data as shown in Table 7.3 and follow the instructions to answer the questions. (Remember to change the type of data when appropriate.)

Using SPSS

1. Based on the descriptive statistics, what are the mean and standard deviation of the VO2MAX for all 20 subjects?
2. Calculate the mean VO2MAX for each group (fit and sedentary) by hand, or by using the SPSS split file command (under the Data menu). In the Split File window select Organize Output by Groups, move fitness level to the Groups Based box. Click OK. Use **Analyze → Descriptive Statistics → Descriptives,** put VO2MAX in the Variables box. What are the two means?
3. Suppose the researcher believes the pencil-and-paper instrument to be valid. State her *research* hypothesis, in terms of the experiment.
4. Also, in terms of the experiment, state the null hypothesis that the researcher would test with an independent *t*-test.

Before continuing, if you used the split file command for question 2 above, you need to undo the split by going to **Data→Split File** and clicking on **Reset.**

Now, go to **Analyze.** Select **Compare Means** and then select **Independent-Sample T Test** . . . . In the Independent-Sample T Test box, move VO2MAX to Test Variable(s), and move Fitness Level to Grouping Variable. Click on **Define Groups** and set Group 1 to be "F" and Group 2 to be "S". Click **Continue**. Click OK.

5. What is the actual difference between the means of the two groups?
6. What should the difference between the means be if the null hypothesis is true?
7. What are the values of the variances for each subgroup of subjects?
8. What t-ratio is calculated for these data?
9. What is the probability of a difference between the means being as large as was obtained, if the null hypothesis is true?
10. What decision should the researcher make with respect to the null hypothesis?
11. What conclusion should she make regarding the validity of the pencil-and-paper instrument?

Using R

Use the following directions to answer items 1 through 11.

Open RStudio.
Install any necessary R packages.
For Chapter 7, we do not require any additional packages.
Set up your RStudio workspace.

In the RStudio menu bar, click on **File → New File → R Script**
On the first line of the R Script (upper left pane of RStudio), type #Chapter 7.

1. Import "fitness" into RStudio (see Chapter 3). Alternatively, get Table 7.3 from the textbook website.

   You should see the fitness (or Table_7_3_Data) data frame listed under the "Environment" tab (upper right pane of RStudio).

2. Enter and run the following RScript commands to obtain descriptive statistics for all 20 subjects:

   - mean(Table_7_3_Data$VO2MAX)
   - sd(Table_7_3_Data$VO2MAX)

3. To calculate the mean and variance of VO2MAX by FITNESSLEVEL, you can use the tapply function.

   - tapply(X= Table_7_3_Data$VO2MAX, INDEX= Table_7_3_Data$FITNESSLEVEL, FUN=mean)
   - tapply(X= Table_7_3_Data$VO2MAX, INDEX= Table_7_3_Data$FITNESSLEVEL, FUN=var)

   When you've finished typing, highlight and Run the two commands you just entered.

The display will provide the mean VO2MAX grouped by FITNESSLEVEL.

4. Now conduct an independent-sample t-test by using the t.test function where we are testing for mean level differences in VO2MAX between Fitness Levels.

   - t.test(Table_7_3_Data$VO2MAX ~ Table_7_3_Data $FITNESSLEVEL)

   When you've finished typing, highlight the entire t.test function and click on **Run**.

   The output will display information on the t-test.

ACTIVITY 2 Dependent *t*-Test

12. Consider the following question: based on what you have learned as a human performance major, do you think that a treatment that consists of having subjects think about fitness for three weeks would increase their fitness scores? Because the researcher already had VO_2max values for 10 sedentary subjects, she asked them to think about fitness for the next three weeks. At the end of this period, she retested the sedentary subjects and obtained the data given in Table 7.4.

Data set. **TABLE 7.4**

SUBJECT	FITNESSLEVEL	VO2MAX	VO2MAXPOST
1	F	54	—
2	F	47	—
3	F	38	—
4	F	45	—
5	F	39	—
6	F	42	—
7	F	50	—
8	F	52	—
9	F	37	—
10	F	36	—
11	S	42	42
12	S	37	38
13	S	29	31
14	S	34	35
15	S	36	37
16	S	34	34
17	S	22	24
18	S	18	20
19	S	27	30
20	S	32	31

Using SPSS

To analyze these data in SPSS, go to **Analyze.** Select **Compare Means** and then **Paired-Samples T Test.** In the Paired-Samples T Test box, click on **VO2MAX** and then on **VO2MAXPOST.** Move both variables into the Paired Variables box. Click OK.

13. State your research hypothesis, based on your answer to question 12.
14. State the null hypothesis in this situation.
15. What are the pre- and post-thinking mean VO_2max values?
16. What is the difference between these means? (Include the sign.)
17. What should this difference be if the null hypothesis is true?
18. What is the value of the dependent t-ratio?
19. If the researcher uses a 95% confidence level, what is her conclusion?
20. List some possible reasons for the difference between the pre- and post-thinking VO_2max means, other than that thinking about fitness caused the fitness scores to improve.

Using R

Use the following directions to answer items 12 through 20.

1. To update your data set with the new VO2MAXPOST measured after three weeks, we could do the following. Be certain that you have imported Table 7.3 which was required for Computer Activity 1 above. Then, create numerical vectors that indicate the sedentary SUBJECT and their VO2MAXPOST after three weeks:

   - SUBJECT <- c(11,12,13,14,15,16,17,18,19,20)
   - VO2MAXPOST <- c(42,38,31,35,37,34,24,20,30,31)

   When you've finished typing, highlight both lines you've just typed in and click on **Run** to create the two numerical vectors SUBJECT and VO2MAXPOST.

2. Next, we can combine the two numerical vectors to create a data frame using the cbind function. The cbind function combines vectors by column.

   - fitnesspost <- cbind(SUBJECT,VO2MAXPOST)

   When you've finished typing, highlight the function you've just typed in and click on **Run** to create the "fitnesspost" data frame.

3. Finally, we can merge the VO2MAXPOST values (after three weeks) by SUBJECT (by=c("STRAIGHT")) for the sedentary individuals with their original VO2MAX values.

   - sedentary <- merge(Table_7_3_Data, fitnesspost, by=c("STRAIGHT"))

   Highlight and click **Run** to create the "sedentary" data frame.

4. Enter and run the following RScript commands to obtain means for the pre and post VO2 values:

   - mean(sedentary$VO2MAX)
   - mean(sedentary$VO2MAXPOST)

5. Now conduct a dependent sample (paired=TRUE) t-test by using the t.test function:

   - t.test(sedentary$VO2MAX, sedentary$VO2MAXPOST, paired=TRUE)

   When you've finished typing, highlight the entire t.test function and click on **Run**

   The output will display information on the paired (dependent sample) t-test.

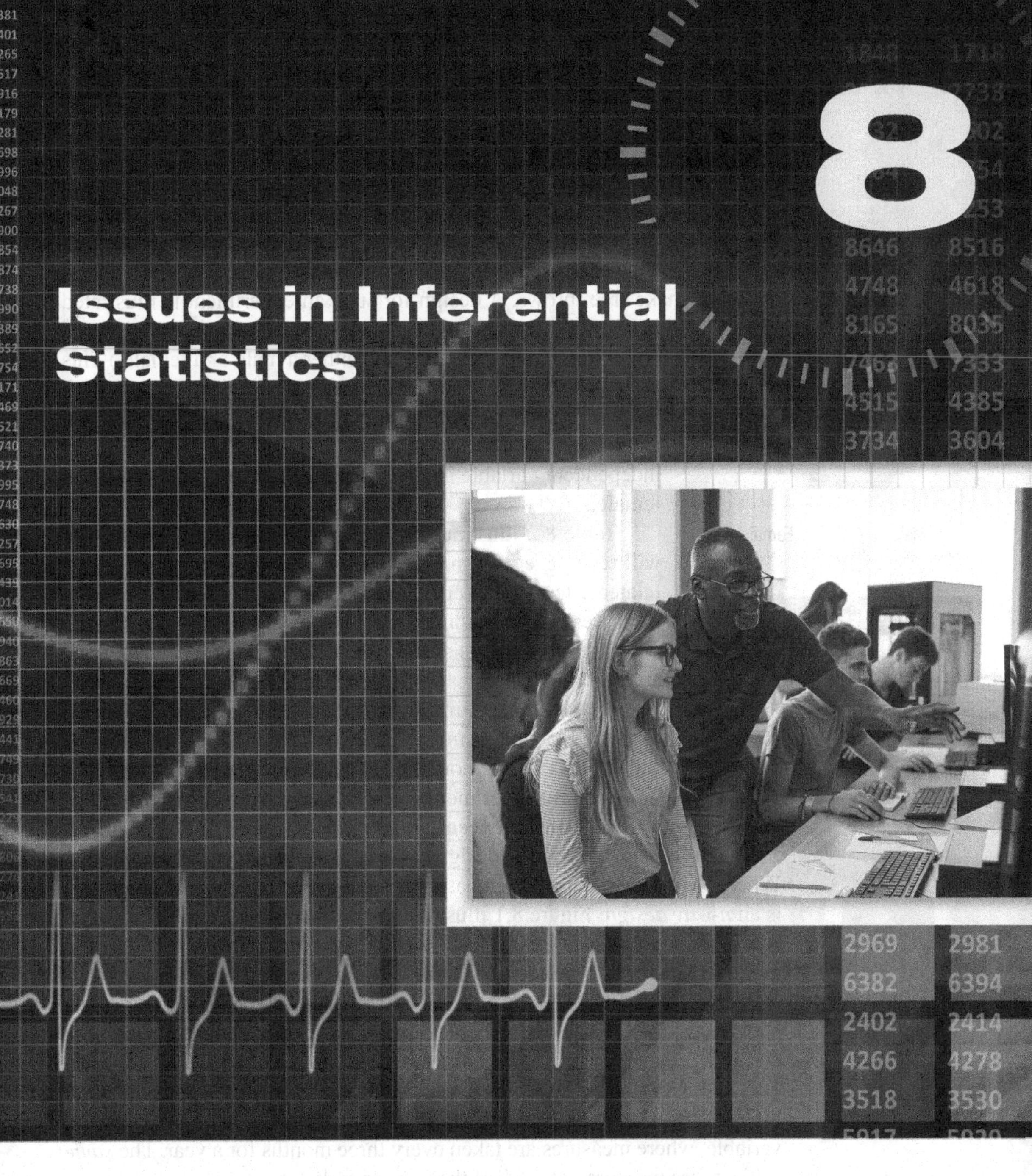

8 Issues in Inferential Statistics

INTRODUCTION

A number of general issues must be considered when we use inferential statistics. Now that you have learned the basic concepts underlying statistical inference, we will examine these general issues. The considerations that we will present in this chapter will continue to be important as we proceed to increasingly complex statistical applications.

BETWEEN-GROUPS AND WITHIN-GROUPS DESIGNS

A key differentiation to be made in experimental designs and statistical inference is that concerning *between-groups* and *within-groups* designs and analyses. These differences are easiest to see in examples. Consider a researcher who wants to investigate the differences in males' and females' ability to make free throws. A design for such an experiment is illustrated in Table 8.1. Notice that each subject can be a member of only one of the groups. (A subject is either a male or a female.) This is an example of a **between-groups design.** In this design, each subject can occur in *one* group of the design only, and individual subjects do not appear across groups. In this example, the independent variable is gender, and it has two levels, male and female.

TABLE 8.1 Between-groups design.

MALES	FEMALES
$Male_1$	$Female_1$
...	...
$Male_N$	$Female_N$

Table 8.2 illustrates a simple **within-groups design**. You will recognize it from Chapter 7, where we presented the correlated *t*-test. The dependent (or correlated, paired, repeated measures, or longitudinal) *t*-test is the simplest illustration of a within-groups analysis. In this example, the independent variable is time, and it has two levels, pre and post.

TABLE 8.2 Within-groups design.

PRETEST	POSTTEST
$Subject_1$	$Subject_1$
...	...
$Subject_N$	$Subject_N$

Both between-groups and within-groups designs can be expanded to include more than two independent variables (also called *factors*) and more than one dependent variable. Table 8.3 illustrates an experimental design where the researcher is interested in two between-groups factors, gender and treatment. Such a design is designated as a 2 x 2 between-groups design. When a between-groups design has *n* between-groups factors, we say that it is an *n-way design*. Figure 8.1 illustrates how a between-groups design can be expanded to a 3 x 2 x 2 design. Here, the factors are Grade, Gender, and Treatment, with three levels for Grade, two for Gender, and two for Treatment. Notice that 12 different cells make up this design. Remember that in a between-groups design, each subject can appear in one and only one of the design's cells.

The within-groups design can be expanded in similar ways. Table 8.4 illustrates a repeated-measures design with four levels of the independent variable, where measures are taken every three months for a year. The *same* subjects are measured at *each* of the time periods.

As you might already have reasoned, an analysis can include both between-groups and within-groups factors. The experimental design illustrated in Table 8.5 is designated as a mixed model ("mixed" in the sense

TABLE 8.3 2 x 2 between-groups design.

		TREATMENT	
		Control	Experimental
GENDER	Males	Male Subject C_1 to Male Subject C_N	Male Subject E_1 to Male Subject E_N
	Females	Female Subject C_1 to Female Subject C_N	Female Subject E_1 to Female Subject E_N

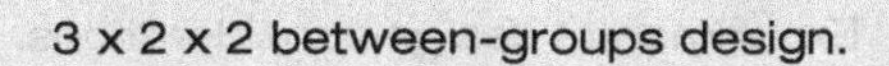
FIGURE 8.1 3 x 2 x 2 between-groups design.

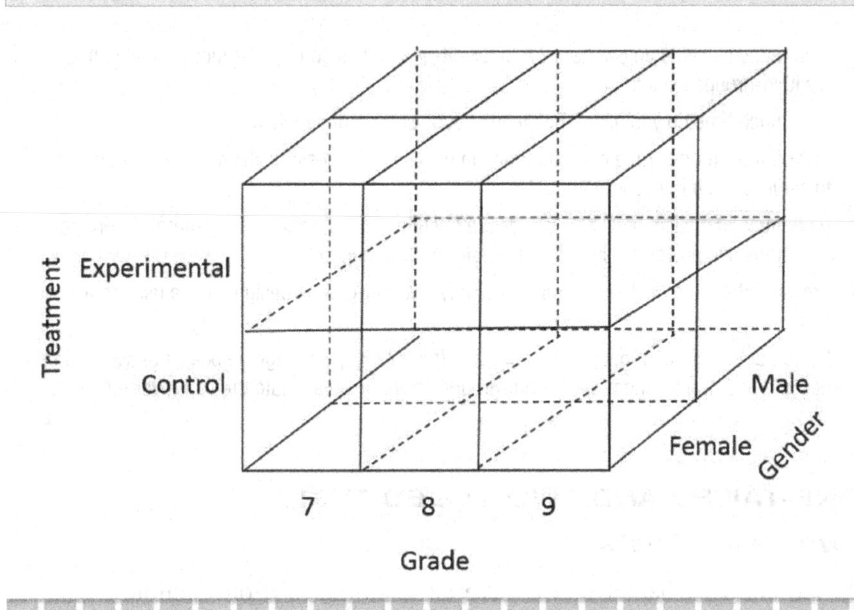

TABLE 8.4 One-way within-groups (repeated measures) design.

BASELINE	3 MONTHS	6 MONTHS	9 MONTHS	12 MONTHS
Subject A	Subject A	Subject A	Subject A	Subject A
Subject B	Subject B	Subject B	Subject B	Subject B
...	...	...	...	...

TABLE 8.5 Mixed-model design.

		BASELINE	6 MONTHS	12 MONTHS
GENDER	Males	Male Subject 1 to Male Subject *n*	Male Subject 1 to Male Subject *n*	Male Subject 1 to Male Subject *n*
	Females	Female Subject 1 to Female Subject *n*	Female Subject 1 to Female Subject *n*	Female Subject 1 to Female Subject *n*

that both between-groups and within-groups factors appear in the design). The design in Table 8.5 is a Gender x Time (2 x 3) design with repeated measures on Time. Note that male subjects appear only within the "male" cells, but they are measured at each level of the Time factor. This pattern is true for females, as well. A mixed-model experimental design can be expanded further, to include multiple between-groups and/or within-groups factors.

The selection of an experimental design depends on many considerations; these are listed in Figure 8.2.

V.8.1 
Between vs. Within Designs

FIGURE 8.2 Experimental design considerations.

1. How many subjects do you have? A between-groups design requires more subjects than does a within-groups design.
2. How much time do you have? Between-groups studies generally require less time.
3. Do you have (or desire) a control group? In within-group designs, the subjects are considered to serve as their own controls.
4. How much statistical power do you desire? Within-groups designs are generally more powerful. (Statistical power will be discussed later in this chapter.)
5. Are you interested in change over time? Only a within-groups design can be used to assess this.
6. Do you anticipate carryover or longitudinal effects? If so, you might choose a between-groups design to control for them or a within-groups design to investigate the carryover effects.

ONE-TAILED AND TWO-TAILED TESTS

Two-Tailed Tests

At the end of Chapter 6, we gave an example of a one-sample *z*-test, in which Professor Kline tested her null hypothesis to determine whether the sample mean score of 80 on the final examination for a class section using a polling application was significantly different from the population mean score of 75, determined from previous classes that did not use the polling application. To test the null hypothesis by using the *z*-test, Professor Kline subtracted the population mean from the sample mean (80 − 75 = 5) and converted this difference to a *z*-score by dividing it by the standard error (1.67) to obtain the value of 2.99. Because she wanted to be 95% confident that her conclusion was correct, she compared 2.99 to 1.96 in the normal distribution. Since 2.99 is greater than 1.96, she knew that this event (a difference of 5 between the means) would occur less than 5% of the time if the two means were equal. Would her decision have been different if she had subtracted the sample mean from the population mean (75 − 80 = −5)? No, because it is the absolute value of the z-score (now −2.99) that determines whether or not the two means are significantly different. This is illustrated in Figure 8.3.

Although the sign of the *z*-score is not a factor in determining whether two means are significantly different, it is important to observe the sign, so that you interpret your results correctly. When Professor Kline began this experiment, she didn't know whether the polling application would improve students' learning or reduce it. This is why she used a **two-tailed test**. In a two-tailed *z*-test, one-half of the critical region occurs at each end of the normal distribution. The researcher doesn't know which mean will be larger than the other one (if they are not equal).

For a two-tailed *t*-test, we use the same procedure as for the *z*-test, except that instead of the critical value being 1.96 for the 95% confidence level, we need to use the appropriate column and *df* row in Appendix A.3 to determine the critical value required to reject H_0. Recall from Chapter 7 that

Locating a z-ratio of ±2.99 on the normal curve in a two-tailed test. FIGURE 8.3

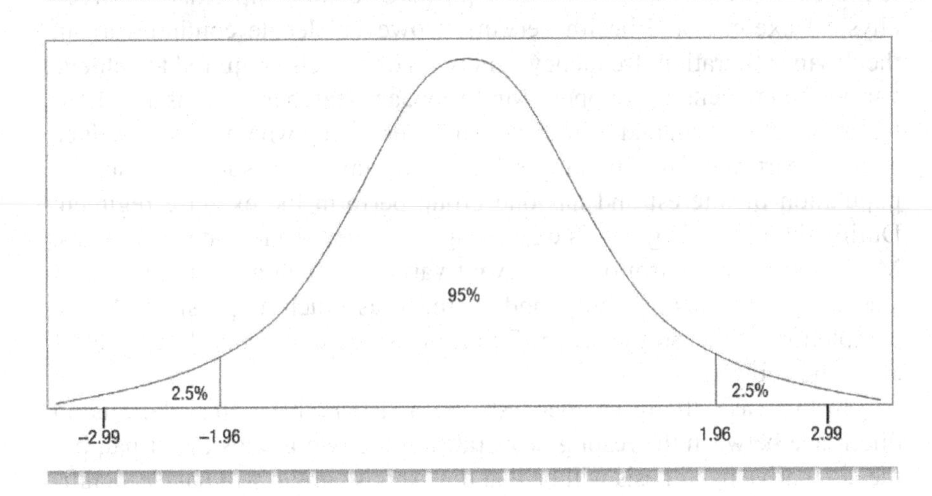

Professor Kline, because she didn't know the population standard deviation, used a t-test for another experiment and found the critical value of 2.042 in Appendix A.3.

Thus, the use of the two-tailed t-test is appropriate when, prior to conducting an experiment, there is no evidence available about which, if either, of the two means might be larger than the other. This condition is often present in exploratory research to compare the effects of two treatment protocols on a dependent variable of interest.

One-Tailed Tests

Often, from knowledge of the relevant research literature, personal experience, or logic, it is possible to predict which treatment should produce the most positive result in the dependent variable. For example, in a common two-sample experimental design, a treatment is applied to one group (called the experimental group), and no treatment is applied to the second group (called the control group). Here, the purpose is to determine whether application of the treatment produces a significant improvement over no treatment.

In experimental designs of this type, the researcher must ensure that all variables except the independent variable are randomly equivalent for both groups (control and experimental), so that only the treatment could possibly be responsible for any detected changes. For example, a strength coach might use a new training regime for one group of athletes but not for another group. The coach needs to ensure that the two groups do not differ in any other way that could affect their strength development. In the **one-tailed *t*-test,** the coach is really only interested in determining whether the new training regime produces significantly *better* results than the old regime (demonstrated by the control group).

Let's use another example to illustrate the use of a one-tailed test. It has been well established that a positive relationship exists between physical exercise and health benefits. However, debate continues about the dosages (duration, frequency, and type) of exercise required to achieve various health benefits. Suppose Mr. Filion, a researcher, designs a particular exercise protocol and is interested in determining whether it can reduce resting heart rate. He randomly selects two samples of subjects from the population of interest and has one group perform the exercise regimen. During the time this group is exercising, the other group watches movies. Mr. Filion must control other relevant variables, such as diet, amount of sleep, other physical activity, and so forth, as much as possible. At the completion of the protocol, Mr. Filion measures the resting heart rate of all of the subjects.

In this case, the researcher tests the null hypothesis (that there is no difference between the resting heart rates of the two groups), as usual, but the alternative hypothesis will be that the exercise group will have a significantly *lower* mean resting heart rate than the control group. In other words, the researcher specifies the *direction* of any difference that might be found. In a two-tailed test, by contrast, the alternative hypothesis is that there will be a significant difference between the means, but the direction of that difference is not specified.

In the two-tailed test, the null and alternative hypotheses are stated, respectively, as follows:

$$H_0 : \mu_E = \mu_C \text{ and } H_1 : \mu_E \neq \mu_C$$

In the one-tailed test, the null and alternative hypotheses are stated, as follows:

$$H_0 : \mu_E = \mu_C \text{ and } H_1 : \mu_E < \mu_C$$

Notice that if the researcher is interested in determining whether the experimental group mean is significantly greater than the control group mean, the alternative hypothesis is H_1: $\mu_E > \mu_C$.

In Mr. Filion's experiment, if the mean resting heart rate of the control (movie watching) group is lower, by any amount, than the mean resting heart rate of the exercising group, he would not perform any statistical tests on the data, because he is only interested in determining whether the exercise protocol is beneficial, and this result would indicate that it is not. However, if the results are in the direction he believes they should be, he would still want to know whether the mean difference is significant or is reasonably attributable to sampling error. He would, therefore, perform a one-tailed *t*-test.

Typically, when higher scores indicate better performance, the control group mean is subtracted from the experimental group mean to produce a positive value. Remember that it is only the absolute value of the calculated *t*-ratio that will be compared to the critical value from the *t*-table. However, in this case, because a lower resting heart rate indicates a more beneficial

Locating the critical value on a *t*-distribution for a one-tailed test. FIGURE 8.4

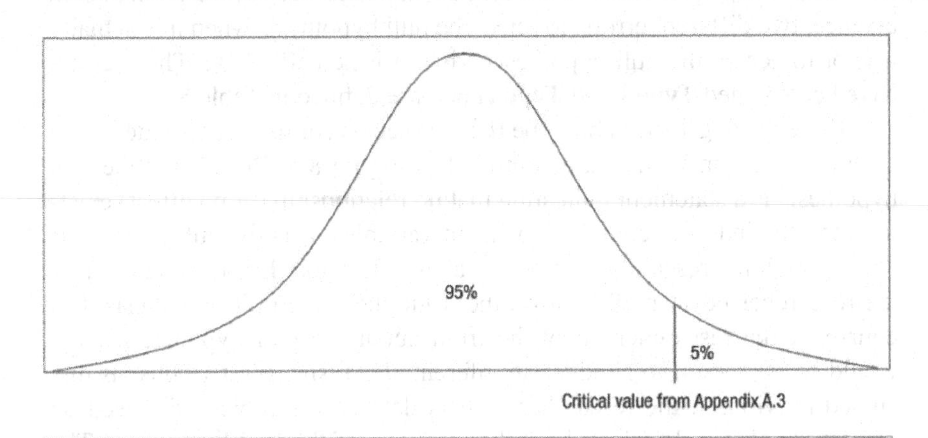

effect than a higher resting heart rate, Mr. Filion would subtract in the opposite direction to obtain a positive mean difference $\left(\overline{X}_C - \overline{X}_E\right)$.

In this case, the researcher is interested only in comparing the calculated *t*-ratio to the positive end of the appropriate *t*-distribution. To accomplish this, he consults Appendix A.3, and selects the appropriate (considering *df* and level of confidence desired) critical value from the one-tailed portion of the table. Figure 8.4 illustrates the concept underlying the one-tailed *t*-test. If the obtained *t*-ratio is to the right of the critical value, the researcher rejects the null hypothesis and concludes that the exercise protocol was beneficial. If the obtained *t*-ratio is to the left of the critical value, he accepts the H_0, because the difference between the means could be explained by sampling error 95% of the time.

Notice that in the one-tailed *t*-test, all 5% of the area that would result in the rejection of the null hypothesis is located at only one end of the distribution. This is the advantage the researcher has gained by deciding *not* to test in the other direction. If this were a one-tailed *z*-test, the critical value line in Figure 8.4 would be at 1.65 for the 95% confidence level. This is less than the 1.96 that is the critical value for a two-tailed *z*-test. Thus, the *z*-ratio required to reject the null hypothesis is smaller in the one-tailed *z*-test than in a two-tailed *z*-test, yet the level of confidence (95%) is the same. This is true with one-tailed *t*-tests, as well. In effect, the one-tailed test increases the probability of being able to reject the null hypothesis and accept the alternative hypothesis, if the data support that decision. In other words, by using a one-tailed test we increase the **power** of the statistical test.

V.8.2 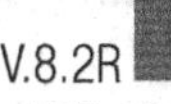
SPSS One-Tailed and Two-Tailed Tests

V.8.2R 
RStudio One-Tailed and Two-Tailed Tests

TYPE I AND TYPE II ERRORS

Inferential decisions are all based on probability. Hence, no conclusion can be stated with *absolute* certainty. In a very real sense, we cannot prove anything with inferential statistics. We can make decisions and draw conclusions based on evidence, but—even with a great deal of evidence—to state that something has been proven with statistical inference would be incorrect.

Since all inferential decisions are based on probability, one always runs the risk of making an error. In inferential statistical testing, we can make, in essence, two different errors: to reject the null hypothesis when it is actually true or to accept the null hypothesis when it is actually false. These errors have been named **Type I** and **Type II** and are defined in Table 8.6.

If identifying Type I and Type II errors seems confusing, consider when each of them can be made, as outlined in Figure 8.5. Recall that the null hypothesis is a statement indicating that no relationship (or no effect) exists between the independent and dependent variables. It is the null hypothesis about which the researcher actually makes a decision. Errors arise because the researcher never really knows the truth about the null hypothesis. (Of course, if the researcher knew the truth about the null hypothesis, there would be no reason to conduct an inferential statistical test.) Thus, as discussed previously, the researcher gathers data on a sample and, based on the data, makes a decision about the accuracy of the null hypothesis. The researcher decides to reject the null hypothesis when there is little probability (often $p < .05$) that the difference (or relationship) would be observed if, in fact, the null hypothesis is true. However, as previously illustrated, the researcher will be wrong 5% of the time, because 5% of the observations will lie in the extremes of the normal distribution when the null is actually true. Thus, the researcher will sometimes reject the null hypothesis when

TABLE 8.6 Type I and Type II inference errors.

TYPE I ERROR	TYPE II ERROR
Concluding that the null hypothesis is false when it is really true	Concluding that the null hypothesis is true when it is really false
■ Probability denoted by α (alpha) ■ A "false positive" ■ Can occur only when one rejects the null hypothesis	■ Probability denoted by β (beta) ■ A "false negative" ■ Can occur only when one *fails* to reject the null hypothesis

FIGURE 8.5 Type I and Type II errors.

YOUR DECISION	TRUE STATE IN POPULATION: H_0 is true, H_1 is false	TRUE STATE IN POPULATION: H_0 is false, H_1 is true
Reject H_0, accept H_1	Type I error (α) No relationship exists, but you concluded there is a relationship.	Correct Decision
Accept H_0, reject H_1	Correct Decision	Type II error (β) A relationship exists, but you concluded there is none.

it should not have been rejected. This is an error of the first type, or Type I error. The probability of a Type I error is symbolized by α (alpha).

Similarly, if the researcher, based on the evidence, concludes that the null hypothesis is the true state of circumstance when in fact it is not, then the researcher has made an error of the second type, or Type II error. The probability of a Type II error is symbolized by β. Key to understanding the potential of a Type I or Type II error is that the researcher *never* knows the true state of the null hypothesis in the population and can only, through inferential techniques, make a "guess" based on probability. This being the case, there is always the risk of the researcher being wrong. In the next section, we will see how researchers can reduce the probability of making Type I and Type II errors.

V.8.3
Type I and Type II Errors

POWER

The level of confidence that we choose is the probability of not committing a Type I error (sometimes called an *alpha error*). By choosing the probability that the null hypothesis is true, we set the probability of making a Type I error. Thus, if we choose to set alpha at .05, then, in essence, the probability of making a Type I error is 5%. We can change the probability of making a Type I error simply by adjusting the alpha level. To be more conservative with regard to making a Type I error, we could set alpha at .01 (or even .001). Conversely, to be more liberal regarding a Type I error, we could set alpha at .10. However, moving alpha (either up or down) has an impact on the probability of making a Type II error (as you will soon see). Note that the level of confidence we have when making a decision about the null hypothesis is equal to 1 − α. For example, if α = .05, the level of confidence is 1 − .05 = .95.

Power is the probability of rejecting the null hypothesis when it is false. Said another way, power is the probability of avoiding a Type II error. The probability of committing a Type II error is called β, so power is equal to 1 − β. It is possible to estimate power for most situations but, to do so, we must know (or estimate) certain facts about the situation. To learn more about power, let's determine the power to reject H_0 correctly in the situation depicted graphically in Figure 8.6. Here, the sample drawn from the population has a mean of 60, and the null hypothesis states that the population mean is 50.

In Figure 8.6, Curve A is the sampling distribution of the population mean when H_0 is true (μ = 50). Curve B is the sampling distribution of the sample, where $\bar{X} = 60$. The researcher has selected α = .05 and is conducting a two-tailed *z*-test. The population standard deviation is known to be 17, and a sample size of 30 was used; thus, the value of the standard error of the mean is $\frac{17}{\sqrt{30}} = 3.1$.

By rearranging the equation for a *z*-test, from

$$z\text{-}ratio = \frac{\bar{X} - \mu}{\sigma_{\bar{X}}} \quad \text{Eq. 8.1}$$

FIGURE 8.6 An illustration of calculating power.

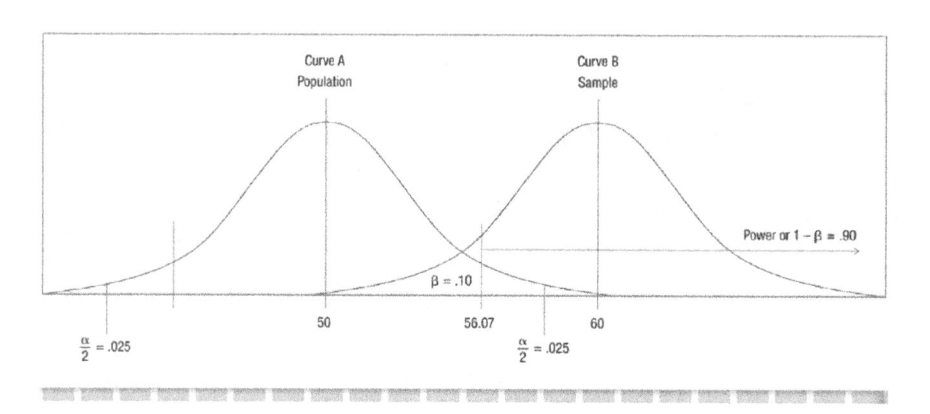

to

$$z\text{-}ratio * (\sigma_{\bar{X}}) = \bar{X} - \mu$$ Please center

and substituting 1.96 for the z-ratio (the critical value from the normal-curve table associated with an alpha of .05), we can determine the location on curve A where the null hypothesis would be rejected.

$$z\text{-}ratio\,(\sigma_{\bar{X}}) = \bar{X} - \mu$$
$$1.96 * (3.1) = \bar{X} - 50$$
$$6.07 = \bar{X} - 50$$
$$\bar{X} = 56.07$$

Thus, if the null hypothesis is true, any sample mean larger than 56.07 would occur by chance less than 2.5% of the time. (Recall that this is a two-tailed test, so 2.5% are in each tail of the distribution.) A sample mean larger than 56.07 would result in rejection of the null hypothesis. In this example, we know this to be a correct decision and not a Type I error because 60 is larger than 56.07.

In this situation, what is the researcher's probability of committing a Type II error? If the mean of the sample is less than 56.07, the researcher would decide to retain the null hypothesis, and this decision would constitute a Type II error. We can find the probability of this occurring by determining the area under Curve B designated as β. To do this, we locate the value of 56.07 on Curve B in standard error units. We assume that the standard error of the mean for Curve B is the same as that of Curve A, that is, 3.1.

Thus, the z-score equals (56.07 − 60)/3.1 = −1.27. Consulting Appendix A.1, we find that approximately 40% of the normal curve is located between this z-score and the mean of 60. Since 50% of the normal curve lies above the mean, the researcher's power in this situation is 40% + 50% = 90%. The area marked as β would then be 10% (i.e., 1 − power = 1 −.90 = .10). This is the probability of committing a Type II error in this situation.

The calculation of power for a *t*-test is identical to the procedure described above, except that the value of 1.96 in Eq. 8.1 is replaced by the value from the correct *df* row and α column in the *t*-table.

In this situation, we calculated that the power to reject the null hypothesis is .90. Note that this result is specific to this situation. Changing any of several factors will change the power value. For example, notice what would happen to β and to power in Figure 8.6 if curve B was moved to the right while everything else remained the same.

Power is an extremely important concept. As researchers, we conduct inferential tests because we believe there is "something going on" (i.e., there is a relationship between variables). Thus, it is important that we gather data and conduct tests in such a way that we will decide that there is something going on only if there *actually is*. That is, we wish to avoid making a Type II error, and, thus, we must have sufficient power when conducting the experiment. Factors that affect the statistical power of an inferential test include our choice of α; the actual difference between the means; and the value of the standard error of the mean, which itself is a function of the variability of the dependent variable; the size of the sample; and the reliability of the measurement of the dependent variable. Below we discuss how each of these issues affects power.

How Does α Affect Power?

The probability of committing a Type I error is totally under the control of the researcher. No one wants to make an error, so why not choose α to be .01 or even .001? Go back to Figure 8.6 and consider what happens to the area labeled β if the critical value line is moved to the right. Doing this would afford greater protection against making a Type I error, but it would increase the probability of making a Type II error and reduce statistical power. For example, assume that the researcher in our discussion of Figure 8.6 chose an α of .01 instead of .05. Confirm that the probability of making Type II error would increase from 10% to approximately 26%, thus reducing power from 90% to 74%.

On the other hand, if the researcher chose an α of .10 instead of .05, this would reduce the probability of a Type II error from 10% to slightly less than 6%, thus increasing statistical power. However, note that the probability of a Type I error would increase from 5% to 10%. The seriousness of making either a Type I or a Type II error is unique to each experiment and is one of the issues that a researcher must consider when designing an experiment and choosing an α level.

Obviously, Type I and Type II errors are related to each other. In any situation, decreasing α (say, from .05 to .01) increases the probability of a Type II error and increasing α (say, from .05 to .10) decreases the probability of a Type II error (and increases power). Keep in mind that Type I and Type II errors can be made only in certain circumstances. Following an inferential test, *all* researchers run a risk (large or small) of making *either* a Type I or Type II error, depending on their decision about the null hypothesis. A Type I error can potentially be made only when the researcher rejects the null hypothesis. A Type II error can potentially be made only when the researcher accepts the null hypothesis. Of course, the researcher also might make a correct inferential decision.

How Does the Actual Difference Between the Means Affect Power?

In the situation represented by Figure 8.6, we know that the actual difference between the means is 10. Of course, in a real-life situation, not only do we not know what the actual difference is, we don't even know whether there is any difference at all. In fact, the null hypothesis, which we will test, states that the means are not different. The researcher must select how big a difference between the means (often called Δ) she is interested in detecting and use that difference in the power estimations.

Consider what happens to the β area in Figure 8.6 if all conditions remain the same except that Curve B moves to the right—say, to center on a mean of 65 instead of 60. Verify that the probability of a Type II error would decrease from 10% to less than 1%. Conversely, if Curve B moves to the left, to center on a mean of 55, the probability of making a Type II error jumps to nearly 64%. Obviously, it is easier to find a significant difference between means that are far apart, and it is more difficult to reject the null hypothesis when the means are only slightly different. The decision about how big a difference really makes a difference is often one of the most difficult for researchers to make.

How Does the Value of the Standard Error of the Mean Affect Power?

The value of the standard error of the mean is a function of two other characteristics, the standard deviation of the population and the sample size. The equation for the standard error of the mean, for a *z*-test, is $\frac{\sigma}{\sqrt{N}}$. For the independent and dependent *t*-tests, the value of the standard error of the mean is more complicated, but it is still a function of the variability of the dependent variable (σ) and N. Examination of the equations for the standard error of the mean (see Chapter 6) reveals that its value is directly proportional to the amount of variance in the dependent variable and indirectly related to the size of N. Since the standard error of the mean is in the denominator in the equations for both the *z*-test and the *t*-test, the probability of rejecting H_0 is higher when the standard error is small than when it is large.

The researcher does not have much control over the amount of variability present in the dependent variable. The variability is affected by

- the breadth of the population being studied (typically, larger variability is associated with diverse populations vs. narrowly defined populations);
- the reliability of the procedure used to measure the dependent variable (as the reliability increases, the individual variability decreases due to the reduction of measurement error); and
- the range of scores produced by the measurement process.

Some variables are simply more variable than others. For example, if you weighed a sample of elementary school students, the standard deviation of the weights would be larger for a sample from the whole school than for a sample from just the third graders. Additionally, the variability of the weights for a sample of subjects will almost always be greater than the variability of the heights of these same subjects.

However, N is often under the control of the researcher. If you examine the equations for the standard error of the mean, you will see why the favorite mantra of statisticians is, "you need to increase your sample size".

How Does the Sample Size Affect Power?

Let's return again to Figure 8.6. Verify that if the researcher could reduce $\sigma_{\overline{X}}$ from 3.1 to 2.5 (perhaps by increasing N from 30 to 36), the probability of a Type II error would decrease from 10% to approximately 3%. This is true if the additional 6 subjects do not affect the standard deviation of the sample.

A question that researchers often pose to statisticians is, "How many subjects do I need?" An easy answer is "lots" or "more than you think". However, a specific number can be estimated for a particular probability of power, if the researcher is able to answer (or estimate the answers to) some questions.

Assume that a researcher designs an experiment such that an independent t-test is the appropriate statistical test. The researcher plans equal Ns for each group. To determine the number of subjects, the researcher must answer or estimate the answers to the following questions:

1. How much protection against a Type I error do I want? (What is my α level?)
2. How much protection against a Type II error do I need? (What is my β probability?)
3. How variable is the dependent variable? (What is the standard deviation of the measurement of the dependent variable?)
4. How big a difference is important for me to detect? (What is Δ?)

The equation to determine the sample size required for a desired amount of power in the specific situation defined by the answers to these questions is:

$$N = \frac{2s^2\left(z_\alpha + z_\beta\right)^2}{\Delta^2} \qquad \text{Eq. 8.2}$$

where

N = the sample size required per group,

s = the standard deviation of the dependent variable,

z_α = the z-value associated with the α level (from Curve A, for example),

z_β = the z-value associated with the β level (from Curve B, for example), and

Δ = the difference between the means the researcher desires to be able to detect

Let's return to Figure 8.6 one more time. We'll assume that Curve A represents a control group, Curve B represents the treatment group, and the four questions above were answered thusly:

1. $\alpha = .05$
2. $\beta = 15$
3. $s = 10$
4. $\Delta = 6$

In this specific situation, then, the researcher wants an alpha level of .05 for protection against making a Type I error and desires the power of the test to be .85 (protection against making a Type II error of .15), the standard deviation of the dependent variable is 10, and the researcher wants to detect any mean difference of 6 units or more. By substituting these values into Equation 8.2, we find that the researcher needs 50 subjects in each group.

$$\begin{aligned} N &= \frac{2s^2\left(z_\alpha + z_\beta\right)^2}{\Delta^2} \\ &= \frac{2(10)^2(1.96 + 1.04)^2}{6^2} \\ &= \frac{(200)(3)^2}{36} \\ &= \frac{(200)(9)}{36} = \frac{1800}{36} = 50 \end{aligned}$$

This equation applies to an independent z-test with equal Ns in each group. It is possible to determine power for unequal Ns and dependent z-tests, as well as for statistics we will introduce later in the textbook. Notice that this equation provides an *estimate* of the sample size needed, not an absolutely correct answer. This is because, typically, the researcher's answers to the questions listed above are estimates and because the equation uses normal-curve values instead of t-distribution values. Computer programs (e.g., G*Power) are available to calculate power for various situations.

Power Simulation

ROBUSTNESS

The concept of robustness, introduced in Chapter 7, is another important issue that we must consider when using inferential statistics. A little history will help you understand the concept. When many of the inferential statistical tests that we employ today were first proposed, there were no computers or even electronic calculators. Thus, early statisticians took it for granted that certain characteristics (i.e., assumptions) about their data were true. Over time, examining the necessity of meeting these assumptions has become possible. With certain important exceptions, many inferential statistical procedures have been determined to be robust, meaning that they are quite accurate even when some of the assumptions are not met. The method used to determine the necessity of meeting specific assumptions is referred to as Monte Carlo, a term from mathematics. Monte Carlo studies are based on probabilities developed from large, repetitive samplings of populations with known characteristics.

Let's consider in more detail the assumptions listed for the *t*-test in Chapter 7.

Normality

It is convenient for researchers to assume that the variable of interest is normally distributed in the population because the mean and variance of samples drawn from a normal distribution are statistically independent. Although many human performance variables are in fact normally distributed, it has been found that the two-tailed *t*-test is quite robust regarding this assumption. Even when skewed and non-mesokurtic distributions are involved, the probabilities of Type I and Type II errors vary little from what would be expected when the normality assumption is met. This is what is meant by "robust".

The condition of robustness for a one-tailed *t*-test requires that the smallest sample size be larger than 20 subjects. Figure 8.7 provides

Sampling distribution from various population shapes. FIGURE 8.7

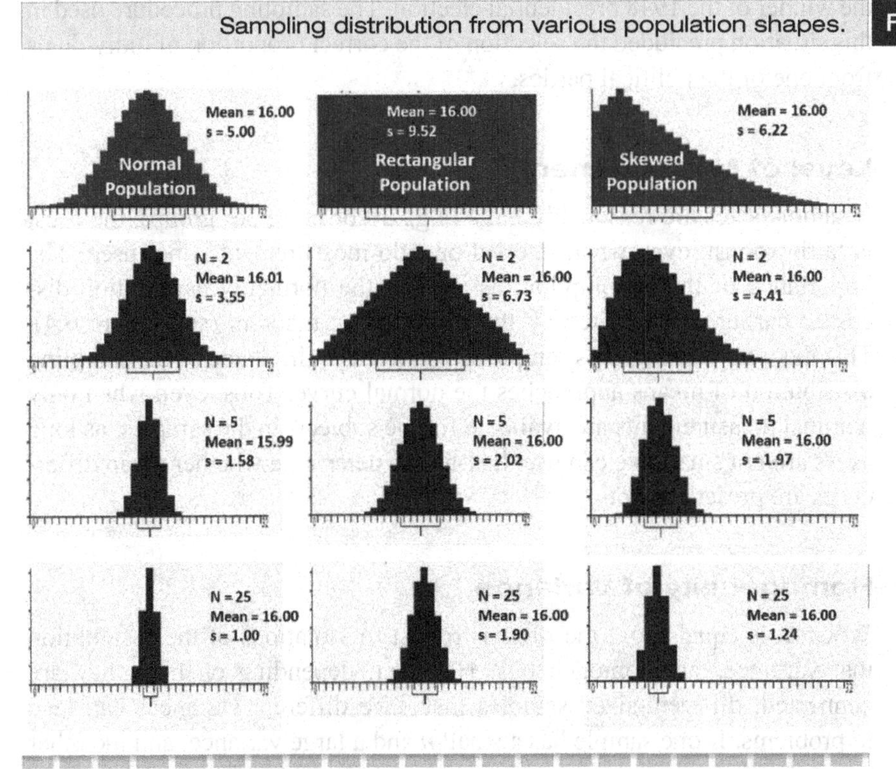

sampling distributions for the mean for various sample sizes and population shapes. Note that, regardless of the shape of the population curve, as the size of the sample drawn to create the sampling distribution of the mean increases, the sampling distribution approaches a normal distribution. This approximation toward normality provides the robustness that allows us to use the normal distribution in inferential tests. Notice also, that as the sample size increases, the standard deviation (standard error of the mean) decreases.

Random Sampling

Both the *z*-test and the *t*-test are based on the assumption that the samples have been randomly drawn from the population. This assumption must be made because the variabilities of both the normal distribution (*z*-test) and the students' *t*-distributions (*t*-test) are due only to chance, a condition that can occur only when every subject in the samples has an equally likely and independent probability of being selected. Unfortunately, random sampling is often difficult to achieve in the real world. With few exceptions, when sampling procedures other than random sampling are used, the results of an inferential statistical procedure cannot legitimately be generalized back to the population.

If biased (i.e., not random) samples are used, and if the biasing factor is correlated with the variable of interest, the probability of a Type I error usually increases. Recall the example of the famous incorrect prediction of the winner of the 1948 presidential election. The sampling procedure used in this situation precluded the selection of the correct proportion of individuals from one of the political parties.

Level of Measurement

If sample sizes are reasonably large (e.g., 30 or more per group), the *t*-test is fairly robust, even when interval or ratio measurement is not used. The importance of this assumption, as well as the normality assumption discussed earlier, is mitigated by the central limit theorem (see Figure 6.4). This theorem states that as sample sizes increase, the shape of the sampling distribution of means approaches the normal curve. Thus, even when only nominal measurements are available for the subjects in the samples, as long as *N*s are not small, we can use the *t*-test to determine whether mean differences are present or not.

Homogeneity of Variance

When n_1 is equal to n_2, the *t*-test is robust to violations of the assumption that variances are homogeneous. However, depending on how they are connected, different-sized samples that have different variances can lead to problems. If one sample has a small *n* and a large variance, and the other sample has a large n but a small variance, the probability of Type I errors will be larger than the α selected for the desired level of confidence. As this discrepancy between the ns and variances increases, so does the problem. For example, if n_2 is five times as large as n_1 and σ_1^2 is five times as large as σ_2^2, the probability of a Type I error will be more than .20 when the researcher selects α to be .05.

Interestingly, when the larger n and larger σ^2 are paired in the same sample (and, necessarily, the smaller sample size and smaller σ^2 *are* paired), the actual probability of a Type I error is slightly *less* than the chosen α. As summarized in Table 8.7 and illustrated in Figure 8.8, this situation slightly increases the chances of committing a Type II error. The general concept, although illustrated here for only two samples and variances, generalizes to multiple samples and variances. Carefully review Figure 8.8 to see how different sample sizes and variances affect the probability of making a Type I error.

STATISTICAL VERSUS PRACTICAL SIGNIFICANCE

We will state again here that statistics can never be used to *prove* anything. Statistical tests can provide very strong evidence about some outcome, but there is always some chance of the occurrence of either a Type I or a Type II error. That chance might be exceedingly small, but it is never zero. When properly used and a significant difference is found, statistical tests allow us to say with a selected level of confidence (short of 100%) that the difference is very unlikely to be attributable to chance (or sampling error).

It is an unfortunate practice to call the difference obtained when the null hypothesis is rejected "statistically **significant**", because the word *significant* has a connotation of importance. It might be better to say that the difference is statistically "reliable", indicating that we are very certain the difference is real and not attributable to chance (or sampling error). The observed difference, although statistically significant, is not necessarily of practical importance.

Under certain conditions, it is possible to find a statistically significant difference that might not be all that important. If the variability of the dependent variable is relatively small and the sample sizes are quite large, a very small difference might be statistically significant but not practically meaningful.

For example, suppose you are an associate dean for student academic affairs of the College of Arts and Sciences in a large public university. One of your tasks is to monitor the overall grade point average for the college from semester to semester. In a communication to department chairs, you note that the GPA has risen from 2.69 to 2.72 from one semester to the next. You receive a letter from a department chair asking whether this increase of .03 is significant. It certainly is statistically significant, because the variability of grades (for scores ranging from 4 for an A to 0 for an F) is very small. If each of the

TABLE 8.7 Homogeneity of variance and sample size.

CONDITION	EFFECT ON α
ns are equal	Selected α = actual α
Large *n* is paired with larger σ^2	■ Actual α is less than the selected α (a conservative test) ■ Type I error probability decreases ■ Type II error probability increases
Large *n* is paired with smaller σ^2	■ Actual α is greater than the selected α (a liberal test) ■ Type I error probability increases ■ Type II error probability decreases

FIGURE 8.8 Relationship between sample size, variance, and α level.

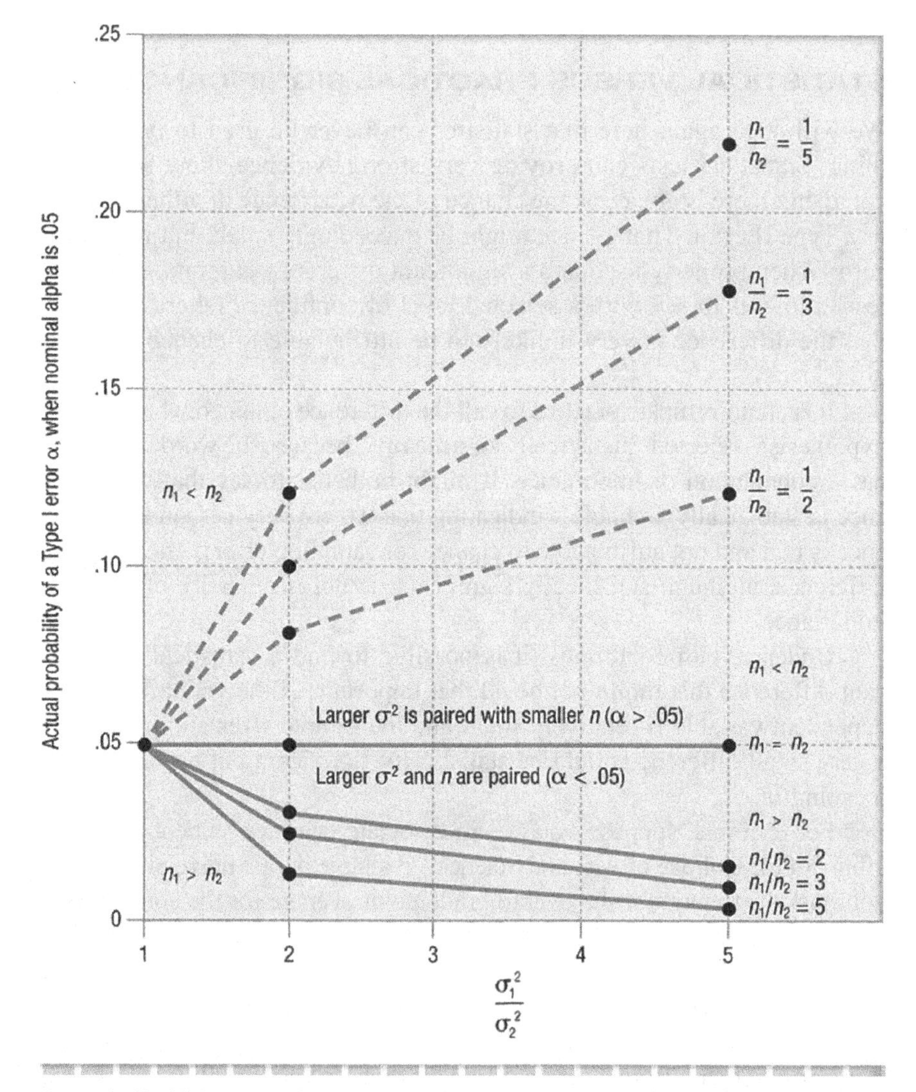

Source: GLASS, GENE V; HOPKINS, KENNETH D., *STATISTICAL METHODS IN EDUCATION AND PSYCHOLOGY*, 3rd Edition, © 1996, p. 294. Reprinted by permission of Pearson Education, Inc., Upper Saddle River, NJ.

17,000 students in the college received a conservative estimate of four grades each semester, the N of the population of the total number of grades received is approximately 68,000. A z-test to determine whether .03 is different from zero would produce a z-ratio far in excess of 1.96, so there is no doubting the statistical significance of the increase. Regardless of these findings, what the department chair needs to know is whether or not the increase is meaningful or of practical importance. This is a fair question, and one that is difficult to answer. Notice that, rounded to the nearest one-tenth, both GPAs are 2.7! You could inform the department chair that the difference is not attributable to chance, but to determine the cause of the increase and whether it indicates a trend or is merely an anomaly would take additional study.

The point is that determining statistical significance is just one part of determining practical significance. To determine practical significance, we must evaluate other statistics. We must compare the amount of difference to norms, consult expert opinion, and consider previous experience to make this judgment. Practical significance is like beauty: it is in the eye of the beholder. How much time, cost, and effort were required to achieve that difference, and what other factors were involved? Finding statistical significance alone cannot answer these questions. However, some other statistics can provide additional information to help us in making this evaluation. Consider the following experiment.

In a study on sleep deprivation, a researcher used an animal model. She required rats to run on a treadmill under two conditions, first as a control condition and then in a sleep-deprived state. The researcher was investigating the effect of sleep deprivation on treadmill time to exhaustion to the nearest minute. The data and summary statistics are given in Table 8.8.

To the researcher's amazement, a dependent t-test (t-ratio = 2.905) revealed the difference between the means (1.10) to be statistically significant at the 95% confidence level, indicating that the sleep-deprived rats ran longer than those in the control condition. Surprised by this finding, the researcher thought to check just how much difference there actually was. Although it was statistically significant, was the 1.10 difference actually meaningful? She calculated three statistics to help make a decision. The three statistics were the percent improvement, effect size, and omega squared.

TABLE 8.8

Data set.

ANIMAL	CONTROL (minutes)	SLEEP DEPRIVED (minutes)
1	53	53
2	48	49
3	40	42
4	45	46
5	47	48
6	45	45
7	33	35
8	29	31
9	38	41
10	43	42
N	10	10
$\overline{X}$	42.1	43.2
s	7.23	6.56
r	0.99	

Percent Improvement

To determine how much the rats had changed from the control to the sleep-deprived condition, the researcher calculated the **percent improvement** with the following equation:

$$PI = \left(\frac{\bar{X}_2 - \bar{X}_1}{\bar{X}_1}\right)100$$ **Eq. 8.3**

where

PI = percent improvement,

$\bar{X}_1$ = control mean, and

$\bar{X}_2$ = sleep-deprived mean.

In her experiment,

$$PI = \left(\frac{43.2 - 42.1}{42.1}\right)100 = 2.6\%$$

This indicates a rather minimal level of improvement.

Effect Size

Next, the researcher calculated the **effect size** (Cohen's *d*), which you recall was briefly introduced in the previous chapter. This statistic reflects the amount of change that took place, in terms of standard deviation units rather than in units of the dependent variable (in this case, treadmill minutes). This standardizes the amount of change that took place so that it can be compared to other experiments in which a different dependent variable might have been used.

If there is a control group, as there was in this experiment, the amount of change is generally expressed in terms of the standard deviation of the control group, because this standard deviation is uncontaminated by any treatment effect. Thus, it is thought to reflect the population variability most accurately. If there is no control group, the denominator for the ES is expressed as a pooled standard deviation derived from both groups. The equation for effect size is:

$$ES = \frac{\bar{X}_2 - \bar{X}_1}{s}$$ **Eq. 8.4**

where

ES = effect size,

$\bar{X}_2$ = largest mean,

$\bar{X}_1$ = smallest mean, and

s = standard deviation of control group or pooled standard deviation from both groups.

In the example,

$$ES = \frac{43.2 - 42.1}{7.23} = 0.15$$

The interpretation of this effect size is that, in this experiment, the sleep deprivation was responsible for changing the dependent variable by only

0.15 of a standard deviation. This would generally be considered to be a very small amount of change. Cohen (1988) has suggested the following guidelines in qualitative terms to define the effect size:

- about .2 = small,
- .5 = modest, and
- .8 = large amount of change.

A tutorial on how to estimate and interpret various effect sizes, is presented in an article by Vacha-Haase and Thompson (2004).

Omega Squared (ω^2)

Finally, the researcher calculated a statistic known as **omega squared** (ω^2). This statistic is an estimate of the percentage of variability in the system that can be attributed to the influence of the independent variable (sleep deprivation, in this example). For the *t*-test, the equation for ω^2 is:

$$\omega^2 = \frac{t^2 - 1}{t^2 + N_1 + N_2 - 1}$$ Eq. 8.5

where

ω^2 = omega squared,

t = the obtained *t*-ratio,

N_1 = the number of subjects in group 1, and

N_2 = the number of subjects in group 2.

In the example,

$$\omega^2 = \frac{2.905^2 - 1}{2.905^2 + 10 + 10 - 1}$$

$$= \frac{8.439 - 1}{8.439 + 10 + 10 - 1} = \frac{7.439}{27.439} = .27$$

We interpret this ω^2 to mean that 27% of the total variability in treadmill minutes is attributable to the sleep deprivation, but 73% of the variability resulted from other factors, such as measurement error, preexisting differences between subjects, and other uncontrolled variables. Omega squared is similar to the coefficient of determination (r^2), introduced in Chapter 5.

None of these statistics (dependent *t*-test, percent improvement, effect size, and omega squared) is more definitive or correct than the others. The dependent *t*-test suggests that the observed change was unlikely to be attributable to chance, but the other three statistics indicate that this change is not particularly meaningful. Although the difference is statistically significant, it might not be a meaningful or practical difference. The researcher's overall decision about the importance of the finding will be a qualitative decision based on a review of the experiment, the data, and relevant literature in the area.

Examination of these statistics can sometimes work in the opposite direction, although this is not common. A difference between means that is found to be close but not quite statistically significant (due to a lack of power) could be judged to be meaningful enough that additional experimentation is warranted.

In the experiment described above, the researcher reflected about the experiment and developed three possible explanations for the outcome:

1. The statistically significant *t*-ratio was an instance of a Type I error, even though the probability of this was less than 5%.
2. There was measurement error in recording the treadmill times.
3. The most likely explanation was that the improvement resulted from a practice effect. The rats' slight average improvement in the sleep-deprived condition could have resulted from their familiarity with treadmill running.

SAMPLE SIZE

We have presented a great deal of information about sample size in this chapter. As you have seen, sample size comes into play in many respects, from power, to assumptions, to statistical significance level, to the critical value. As we have previously suggested, the question that many novice researchers often start with is, "How large should my sample be?" This is an excellent question, but to reach a decision about the specific sample size is difficult. Suffice it to say that the sample size should not be based on convenience but on sound inference practice, as discussed earlier in this chapter. A small sample could result in low power. A sample that is too large could waste time and resources. Choosing the appropriate sample size for your statistical test is the best option.

SUMMARY

In this chapter, we presented an overview of issues related to inference testing. Regardless of the inferential statistical test used, the process, the thinking, and the decision-making are similar. In Chapter 7 we discussed a simple means comparison with the *z*-test and *t*-test. When we deal with higher-level inferential statistics, the steps that we take and the decision we make are similar to those we have already introduced.

Also in this chapter, we presented issues that researchers should consider as they begin to conduct inferential tests. These assumptions, steps, and decisions also apply to increasingly sophisticated hypothesis testing.

Essentially, hypothesis testing is about identifying the relationship between an independent variable *(X)* and a dependent variable *(Y)*. This fundamental concept can be extended to multiple *X*s and multiple *Y*s, and that is exactly what we do in more sophisticated inferential tests. Let's now take the concepts you have learned about hypothesis testing and the relevant issues and, in the following chapters, apply them to some of the more involved statistical tests. The essential concepts will be the same, so if you become confused, take a step back and remember that you are testing to find out whether there is a relationship between *X* and *Y* and attempting to make a decision about what happens to the value of *Y* when you change (or manipulate) the value of *X*.

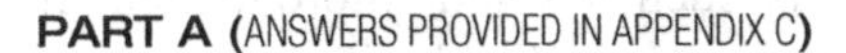

Review Questions and Practice Problems

PART A (ANSWERS PROVIDED IN APPENDIX C)

1. A nationally standardized test measuring positive attitude toward physical activity has a mean of 75. Suppose 25 randomly selected human performance majors take the test. Their mean score is 81, and the standard deviation of their scores is 14. Are they like the population, or do they have a statistically significantly better attitude toward physical activity? (Use $\alpha = .05$.)
2. If your conclusion in question 1 is actually incorrect, what type of an error is this?
3. What percentage of the time would you commit this type of error?
4. Independent *t*-test is to between as dependent *t*-test is to
   A. Degrees of freedom
   B. Correlation
   C. Standard error of the difference
   D. Within
5. Refer back to the example immediately following Equation 8.2, where it was determined that 50 subjects per group would be required to detect a difference of 6 between the means if $\alpha = .05$, the standard deviation of the dependent variable is 10, and the chance of making a type II error is 15%. Use Equation 8.2 to determine how the following changes would affect the number of subjects required:
   A. Change the standard deviation to 12 (more variance in the dependent variable).
   B. Change the alpha level from .05 to .10 (reduce the protection against a Type I error).
   C. Change the protection against a Type II error from 15% to 10%.
   D. Change Δ to 4 (to consider a smaller difference to be significant).

Make each of these changes individually; that is, remember to return previously changed elements back to the original values for parts B, C, and D.

PART B

6. On the final examination in an anatomy class, the mean score for the 68 males was 29.17 with a standard deviation of 3.05, and the mean score for the 68 females was 28.52 with a standard deviation of 2.97. Did the males do significantly better than the females?
7. If your conclusion to question 6 is actually incorrect, what type of an error is this?

Use the following information to answer questions 8 through 15:

One hundred twenty-five human performance majors were given a pull-up test, and their mean score was 6.5 with a standard deviation of 3.4. After completing a 10-week strength-training program, the same subjects were retested and, on average, completed 7.1 pull-ups with a standard deviation of 4.0.

8. The correlation between the pre- and post-training scores was .75. Was the strength-training program effective? (Use $\alpha = .05$ and a one-tailed test.)
9. Why use a one-tailed test in this situation?
10. Assuming the pretest scores to represent a control group, what effect size is associated with the pull-up improvement?
11. What percent improvement was achieved by the strength-training program?
12. What is the probability of committing a Type I error?
13. What is the probability of committing a Type II error?
14. What is the researcher's statistical power?
15. Why is such a small improvement in pull-ups statistically significant?

COMPUTER ACTIVITIES

Sample Size and One-Tailed vs. Two-Tailed Tests

ACTIVITY 1 Sample Size

Ten females and 10 males were randomly selected from a physical education class that had been measured on the FITNESSGRAM PACER Test, with the results given in Table 8.9.

Using SPSS

From the Computer Activities in previous chapters, and with a few hints given in the directions below, you have already learned all the SPSS commands required to answer questions 1 through 11 for Activity 1. Note that Table_8_9_Data SPSS file on the textbook website has some additional columns that are used in these commands.

Go to **Analyze → Compare Means → Independent Samples t-test.** Put PACERLAPS in the Test Variables and GROUP in the Grouping Factor. Click on GROUP and Define Groups. Define your groups as "Females" and "Males" (without the quotes). Click on Continue and OK.

1. What are the means and standard deviations for the two sets of scores?
2. Are the variances significantly different? Hint: You can determine this by examining Levene's test in the SPSS output and the Variance test in RStudio.
3. What is the difference between the means for males and females?
4. Is this difference statistically significant? (Use $\alpha = .05$.)

The investigator decided to add another 10 randomly selected scores for both males and females. The additional scores selected are given in Table 8.10.

Repeat the SPSS commands from above with the new data file with the additional subjects.

5. What are the means and standard deviations for the 20 males and the 20 females?
6. How do these new means compare to the means of the original 10 scores?
7. Are the variances for the two sets of 20 scores significantly different?
8. What is the difference between the means for the 20 males and the 20 females?
9. How does this compare to the original difference?
10. Is this difference statistically significant? (Use $\alpha = .05$.)
11. Why do your answers to questions 4 and 8 differ?

TABLE 8.9

Data set.

FEMALES (0)	MALES (1)
44	20
40	42
18	34
27	35
29	41
21	42
20	31
20	17
25	25
17	38

TABLE 8.10

Data set.

FEMALES (0)	MALES (1)
26	33
25	32
27	31
26	38
30	26
22	32
29	35
23	29
29	37
23	28

Using R

The directions for using R for Activity 1 are given below.

Open RStudio.

Install any necessary RStudio packages.

For Chapter 8, we will not require any additional packages.

Set up your RStudio workspace.

In the RStudio menu bar, click on **File → New File → R Script**.

On the first line of the R Script (upper left pane of RStudio), type #Chapter 8.

Use the RScript commands to answer questions 1 through 4.

1) Using the data from table 8.9, create two numerical vectors—one for females and one for males.

- FEMALES <- c(44,40,18,27,29,21,20,20,25,17)·
- MALES <- c(20,42,34,35,41,42,31,17,25,38)

Highlight both lines and click on **Run** to create the numerical vectors "FEMALES" and "MALES".

2) Conduct an F test (default test for equal variances in R) to determine whether the variance in the PACER test in females is different from that of the males.

- var(FEMALES)
- var(MALES)

- var.test(FEMALES, MALES)

Highlight these three lines and click on **Run** to get and test the equality of the variances.

3) Enter the following commands, highlight and run

- mean(FEMALES)
- mean(MALES)
- sd(FEMALES)
- sd(MALES)
- var(FEMALES)
- var(MALES)
- var.test(FEMALES, MALES)
- t.test(FEMALES, MALES)

4) Make the following changes to your RStudio script to include the combined data from Tables 8.9 and 8.10 and answer questions 5 through 11 from the RStudio output.

- FEMALES <- c(44,40,18,27,29,21,20,20,25,17,26,24,27,26,30,22, 29,23,29,23)
- MALES <- c(20,42,34,35,41,42,31,17,25,38,33,32,31,38,26,32,35, 29,37,28)
- mean(FEMALES)
- mean(MALES)
- sd(FEMALES)
- sd(MALES)
- var(FEMALES)
- var(MALES)
- var.test(FEMALES, MALES)
- t.test(FEMALES, MALES)

ACTIVITY 2 One-Tailed Versus Two-Tailed Tests

The researcher decides to determine whether there is a significant difference between the means for steps taken per week for a group receiving verbal encouragement (Group 2) to increase physical activity as contrasted with a control group (Group 1). Because the researcher expects to find greater physical activity in the encouraged group (and not a decrement), he decides to employ a one-tailed test. He chooses to use an alpha level of .05. *Hint:* The default in SPSS and R is the two-tailed test, so you will need to consult the *t*-distribution table (see Appendix A.3). The data are presented in Table 8.11.

Using SPSS

From the Computer Activities in previous chapters you have already learned all the SPSS independent groups t-test commands required to answer questions 12 through 18 for Activity 2.

12. What are the two means?
13. Was there an improvement?
14. What is the difference between the means?
15. What is the t-ratio?
16. If you use the SPSS/RStudio results (a two-tailed test), is this difference statistically significant at $\alpha = .05$?
17. What t-ratio (from the t-distribution table) is required for significance for a one-tailed test?
18. What is the researcher's decision with a one-tailed t-test?

Using R

1. Import Table_8_11_Data from the textbook website (or create the file yourself).
2. Use the following RStudio commands to conduct a two-tailed t-test:
   - tapply(X=Table_8_11_Data$STEPS, INDEX-=Table_8_11_Data$GROUP, FUN=mean)
   - t.test(Table_8_11_Data$STEPS ~ Table_8_11_Data$GROUP)

   Highlight and run the two commands you have entered.

Answer items 12 through 18 using the RStudio results.

TABLE 8.11

Data set.

SUBJECT	GROUP	STEPS
1	1	28465
2	1	32516
3	1	29468
4	1	34803
5	1	35368
6	1	36592
7	1	41541
8	1	43258
9	1	47142
10	1	54561
11	1	31127
12	1	43562
13	1	54547
14	1	55634
15	1	36167
16	2	54695
17	2	55981
18	2	54326
19	2	49945
20	2	39478
21	2	68124
22	2	61354
23	2	27989
24	2	53154
25	2	28267
26	2	32936
27	2	28549
28	2	56261
29	2	54287
30	2	51365

REFERENCES

Cohen, J. (1988). *Statistical power analysis for the behavioral sciences* (2nd ed.). Hillsdale, NJ: Lawrence Erlbaum.

Glass, G. V, & Hopkins, K. D. (1996). *Statistical methods in education and psychology* (3rd ed.). Upper Saddle River, NJ: Pearson.

Vacha-Haase, T., & Thompson, B. (2004). How to estimate and interpret various effect sizes. *Journal of Counseling Psychology*, *51*, 473–481.

9 One-Way Between-Groups ANOVA

INTRODUCTION

In Chapter 7, we introduced the two-sample *t*-test and identified the dependent variable as the one that we measure to see whether its values change under different conditions of the independent variable. The conditions of the independent variable, which are selected by the investigator, are called *levels*. In all *t*-tests, the independent variable has only two levels. For example, a common design for which the *t*-test is the appropriate statistical test consists of a control group and an experimental group that receives some treatment. The treatment might be, for example, a particular dose of a medication to lower blood pressure. In this situation, blood pressure is the dependent variable, and the two levels of the independent variable are the control group and the medicated group.

What if the investigator wanted to examine three different dosages of the drug, as well as maintain a control group in the experiment? In other words, the investigator would like to change the number of levels of the independent variable from two to four and compare the four resulting means to one another. The independent *t*-test cannot be used now, because it can accommodate only two means. The solution to this problem was determined by Sir Ronald A. Fisher, an English statistician, and published in 1925.

Fisher proposed a method involving the analysis of the variance **(ANOVA)** in a data set to determine whether the differences among several means are greater than would be expected by chance (sampling error) alone. Note that this reasoning and the ultimate decision-making based on it are identical to those of the *t*-test. Variations and extensions of this procedure have resulted in ANOVA becoming one of the most commonly used—if not *the* most commonly used—statistical tests in science. In this chapter, we will examine the simplest application of this procedure: the one-way ANOVA. It involves only one dependent variable and one independent variable, the independent variable having any number of levels. In later chapters, we will expand this procedure to accommodate any number of independent variables and then any number of dependent variables.

MULTIPLE *t*-TESTS

In the example outlined above (a study of three drug dosages, with a control group, to test the efficacy of a drug to lower blood pressure), your first reaction might be to use multiple *t*-tests to compare pairs of means. With three means, there are three possible comparisons; with four means, there would be six possible comparisons. To determine how many possible comparisons of two means exist when K means are available, we use the equation

$$C = \frac{K(K-1)}{2} \quad \text{Eq.9.1}$$

where

C = the number of paired comparisons possible, and

K = the number of means available.

Use the equation to confirm that if the independent variable has six levels, the number of possible comparisons of two means would be 15.

Unfortunately, the probability of committing a Type I error increases above the selected α level when more than one comparison is made. In fact, in the example with four levels of the independent variable, if the researcher used multiple t-tests, the probability of committing at least one Type I error is approximately .26, even though the researcher may have set alpha at .05. The equation for calculating this probability is

$$P = 1 - (1 - \alpha)^C \qquad \text{Eq.9.2}$$

where

P = the probability of making at least one Type I error,

α = the probability of making a Type I error if the null hypothesis is true (i.e., the researcher's selected alpha), and

C = the number of comparisons possible among levels of the independent variable.

Confirm that if the independent variable has six levels, the probability of making at least one Type I error could be as high as approximately .60 when the selected alpha is .05. The exact alpha level depends on the correlations among the various tests being conducted. Table 9.1 gives the potential impact on Type I error for multiple t-tests at α = .05 and α = .01 (Hopkins, Hopkins, & Glass, 1996, p. 272).

TABLE 9.1 Potential Type I errors as a function of number of multiple t-tests.

SELECTED ALPHA	NUMBER OF GROUPS	NUMBER OF POSSIBLE PAIRED COMPARISONS	POTENTIAL ALPHA
.05	2	1	.05
.05	3	3	.14
.05	4	6	.26
.05	5	10	.40
.05	6	15	.60
.05	10	45	.90
.05	15	105	.995
.05	20	190	.999
.01	2	1	.01
.01	3	3	.03
.01	4	6	.06
.01	5	10	.10
.01	6	15	.15
.01	10	45	.36
.01	15	105	.65
.01	20	190	.85

Source: Adapted from Hopkins, K.D., Hopkins, B.R., & Glass, G.V. (1996). *Basic statistics for the behavioral sciences* (3rd ed.), Table 14.1, p. 272. Needham Heights, MA: Allyn & Bacon.

ANOVA NULL HYPOTHESIS

In the *t*-test, the null hypothesis states that there is no difference between the two means ($\mu_1 = \mu_2$ or $\mu_1 - \mu_2 = 0$). If the calculated *t*-ratio leads to the conclusion of retaining the null hypothesis, this signifies that the difference between the means is attributable to chance (sampling error). If the calculated *t*-ratio leads to the decision to reject the null hypothesis, simple inspection of the means reveals that one is significantly higher than the other one, and the probability that the difference is actually attributable to sampling error (i.e., the probability of making a Type I error) is equal to the selected α level.

In the one-way ANOVA, the null hypothesis is

$$H_0 : \mu_1 = \mu_2 = \mu_3 = \ldots = \mu_K \text{ where } K = \text{the number of levels}$$

With an ANOVA analysis, the statistic that corresponds to the *t*-ratio in a *t*-test is termed the *F-ratio*. We will discuss the *F*-ratio at length later in this chapter. For now, just be aware that the *F*-ratio is the final test statistic in an ANOVA analysis. If the *F*-ratio leads to the conclusion of retaining the null hypothesis, this indicates that the pattern of means (even though they might not all be exactly equal) is within the range that could be expected by chance. The conclusion would be that there are no statistically significant differences among the means. If the *F*-ratio leads to the decision to reject the null hypothesis, this signifies that the pattern of the means (i.e., the variability among the means) is too great to be attributed to chance. However, the *F*-ratio does not indicate which means are significantly different from which other means. Essentially, the *F*-ratio provides an "omnibus", or overall test, of whether or not the different means are likely to result from chance (sampling error). If after the omnibus test, the null hypothesis is rejected, the researcher must then decide how to determine which means are different from one another and how they differ. Later in this chapter, in a section titled "Multiple Comparisons", we will discuss methods that have been developed to determine this.

Advantages of ANOVA Over Multiple *t*-Tests

ANOVA offers three major advantages over conducting multiple *t*-tests when the independent variable has more than two levels.

1. As explained above, with ANOVA, the Type I error rate is known (i.e., it is whatever α level the investigator has selected). With multiple *t*-tests, the Type I error rate generally will be greater than the selected α level, and it increases (depending on the correlations among the various levels of the independent variable) with an increasing number of comparisons made.
2. In the *t*-test, the estimate of the population variance that becomes the denominator in the *t*-ratio is based on pooled information from the two groups. In ANOVA, the estimate of the population variance (which becomes the denominator of the *F*-ratio) is pooled information from as many groups (levels) as there are for the independent variable. Since this estimate is based on a larger amount of information, it is more accurate, and the net result is that the *F*-ratio in ANOVA is more powerful than the *t*-ratio. Increased statistical power leads to a greater

probability of rejecting the null hypothesis and a decreased probability of making a Type II error.

3. As we will discuss in Chapter 11, ANOVA can be extended to experiments involving the combined effects of two or more independent variables acting simultaneously. For example, the investigator who is studying the effects of the three drug dosages on blood pressure might, at the same time, be interested in determining whether the pattern of means is the same or different for males and females. The addition of gender is a second independent variable with two levels. Data from such an experiment could be analyzed with a two-way ANOVA. In later chapters, we will expand this concept even further, to include multiple independent and multiple dependent variables. This advantage of ANOVA is what has made it the most ubiquitous statistical procedure in science.

ASSUMPTIONS AND ROBUSTNESS OF ANOVA

Just as with the *t*-test, several assumptions underlie the use of ANOVA. As with the *t*-test, ANOVA is reasonably robust, even when these assumptions are not totally met. The assumptions of ANOVA are as follows:

- *Normality*. It is assumed that the dependent variable is normally distributed in all levels of the independent variable.
- *Homogeneity of variance*. It is assumed that the variance of the dependent variable is the same for all levels of the independent variable.
- *Independence*. The one-way ANOVA discussed in this chapter is based on the assumption that the levels of the independent variable are between groups and that each is an independent sample of subjects from the population. In other words, this chapter discusses an ANOVA that is similar to the independent *t*-test. In the next chapter, we present the repeated-measures ANOVA, which is analogous to the dependent *t*-test (within groups), where the same (or matched) subjects are measured two or more times.
- *Continuous measurement*. It is assumed that the dependent variable is assessed at an interval or ratio level of measurement.

Regarding *robustness:* As long as a reasonably large number of subjects is present in each level of the independent variable, ANOVA is relatively robust against violations of the normality and the homogeneity of variance assumptions. Having equal or nearly equal numbers of subjects in each group also increases the robustness of ANOVA (especially with respect to the homogeneity of variance assumption).

CALCULATIONS

ANOVA's name, which stands for "analysis of variance", is very appropriate. ANOVA partitions the total variation in the dependent variable into various sources. Hence, an "analysis of variance" is exactly what occurs in ANOVA. We will discuss the steps that lead to the final statistic (the *F*-ratio) in more detail for the one-way ANOVA than for the more complex procedures presented later in the text. This will help you continue to build on your knowledge of how statistics works.

Before we begin, let's define some rules about notation and look at a very simple data set that we will use to demonstrate the calculations. This data set, presented in Table 9.2, represents the scores for 9 male subjects randomly selected from a population of interest, who were then randomly assigned to one of three groups and measured on some dependent variable. Sample sizes this small are not typical. We are using them here so that the illustration will not become too cumbersome. Assume that Group 1 engages in moderate aerobic training, Group 2 engages in vigorous aerobic training, and Group 3 engages in both moderate and vigorous aerobic training. The dependent variable is the subjects' attitudes toward the training method, with higher scores indicting a more negative attitude.

TABLE 9.2 Data set.

	GROUP 1	GROUP 2	GROUP 3	TOTAL
	1	3	5	
	2	4	6	
	3	5	7	
ΣX	6	12	18	36
$\bar{X}$	2	4	6	4

Notation

Each subject in the data set is uniquely identified by two subscripts. For example, the first subject in Group 1 is identified by the notation $X_{ij,}$ with the first subscript identifying which individual he is within his group and the second subscript identifying the group of which he is a member. Thus, the first subject in the first group is X_{11}. His score on the dependent variable is 1. The first subject in Group 2 is labeled $X_{12,}$ and his score is 3. The last subject in Group 3 is X_{33} and has a score of 7.

The three group means, $\bar{X}_1$, $\bar{X}_2$, and $\bar{X}_3$, are 2, 4, and 6, respectively. The number of subjects in each group is identified by n_1, n_2, and n_3, and in this case, they are all equal to 3. The total number of subjects in the data set *(N)* is 9, and the grand mean (the mean of all scores in the data set) is 4. The symbol for the grand mean is $\bar{\bar{X}}$.

Calculating Sums of Squares

We will begin the one-way ANOVA analysis by calculating three values called **sums of squares**. To indicate what these values are and where they come from, we will present the definitional equations here, but these equations would be quite cumbersome to use with any but very small data sets. Equivalent equations are available that use raw scores, and we will present some of them here.

Sum of Squares Total (SS_T)

The sum of squares total, or SS_T, is the numerator of the variance, a statistic we first presented in Chapter 3. The variance quantifies the amount of variability that exists in a set of scores.

We obtain the value of the sum of squares total by subtracting the grand mean from every score in the data set, squaring each resulting difference, and then adding them together. Hence, the equation for the sum of squares total in ANOVA is

$$SS_T = \Sigma(X_{ij} - \bar{\bar{X}})^2$$ Eq.9.3

where
Σ = the sum of,
SS_T = sum of squares total,
X_{ij} = each score in the data set, and
$\bar{\bar{X}}$ = the grand mean.

Use the equation to verify that for the data set above, SS_T is equal to 30. The value of 30, although it is not the variance of the total data set (it is only the numerator of the variance), is a composite value representing all sources of variability in these scores.

Our next step is to divide this composite score into the two sources of variability that exist in a one-way between-groups ANOVA. These are

1. variability due to the groups differing from one another (reflected by the fact that their means are not equal). This variability is labeled SS_B, which stands for the sum of squares between; and
2. variability resulting from the fact that the subjects differ from one another within their groups. This variability is labeled SS_W, which stands for sum of squares within.

It is important to recognize that

$$SS_T = SS_B + SS_W$$ Eq.9.4

Sum of Squares Between (SS_B)

If the three means in the data set were all equal to one another, we would not need to proceed any further, because the null hypothesis would be true (there is no difference in attitude toward the exercise program as a result of exercise intensity). However, since the three means of 2, 4, and 6 are not the same, the question becomes, "Could this range (or pattern) of means, suggesting that as exercise intensity increases attitude toward exercising becomes more negative, be reasonably expected to occur by chance (sampling error) or not?" The sum of squares between will reflect how much of the variability in the data set results from the difference in the means.

To obtain the value of SS_B for the data set, we subtract the grand mean from each of the group means, square each difference, add them together, and multiply the total by *n* (the number in any one group). We can use the following equation:

$$SS_B = \Sigma n_j(\bar{X}_j - \bar{\bar{X}})^2$$ Eq.9.5

where
SS_B = sum of squares between,
n_j = the number of subjects in the group,
$\bar{X}_j$ = the mean of the group, and
$\bar{\bar{X}}$ = the grand mean.

Because this data set represents a balanced design (all *n*s are equal), the equation can be modified to read:

$$SS_B = n\Sigma(\bar{X}_j - \bar{\bar{X}})^2$$ Eq.9.6

Verify that for this example $SS_B = 24$.

Sum of Squares Within (SS_W)

The value of SS_W reflects the fact that the subjects within each group differ from one another. In other words, all of the subjects in Group 1 do not have the same score on the dependent variable. This is true for Group 2 and Group 3, as well. This variability occurs simply because subjects differ for many (often unknown) reasons. This variability is labeled *error variance*. This term does not mean that the variance is a mistake, it just means that we can't directly explain its source.

To obtain the value of SS_W for this data set, we subtract the Group 1 mean from each of the scores for all subjects in Group 1, square each difference, and add the squares. We repeat this process for Groups 2 and Group 3. Finally, we add the three obtained sums together. We can calculate the value of SS_W with the following equation:

$$SS_w = \Sigma\Sigma(X_{ij} - \bar{X}_j)^2$$ Eq.9.7

where
SS_W = sum of squares within,
X_{ij} = each score in the group, and
$\bar{X}_j$ = the mean of the group.

Verify that for this example $SS_W = 6$.

As an aside, we could have calculated the SS_W to be 6 by noting that $SS_T = SS_B + SS_W$. It follows that $SS_W = SS_T - SS_B$; thus, $SS_W = 30 - 24 = 6$.

We have now partitioned SS_T (sum of squares total, a quantification of all the variability in the data set) into two parts: the variability resulting from differences in the means and the variability resulting from variance error. (Bear in mind that this is not really "error", but simply unexplained variance.)

Degrees of Freedom for a Between-Groups One-Way ANOVA

In Chapter 3, we divided the sum of squares by its degrees of freedom to obtain the variance. We will follow the same procedure here to obtain what are called **mean squares**. Whenever you hear the term *mean square*, think "variance estimate". The degrees of freedom associated with the three sums of squares for a between-groups one-way ANOVA are as follows:

- For SS_T the $df = N - 1$,
- For SS_B the $df = J - 1$, and
- For SS_W the $df = N - J$,

where
N = the total sample size, and
J = the number of levels of the independent variable.

Thus, for our example, the total degrees of freedom for SS_T are $9 - 1 = 8$, for SS_B are $3 - 1 = 2$, and for SS_W are $9 - 3 = 6$. Notice that $8 = 2 + 6$. We have partitioned the total degrees of freedom into two parts, just as we partitioned the sum of squares for total into two parts. We are now ready to calculate the values of the mean squares and then, finally, the value of the *F*-ratio.

Calculating Mean Squares

The calculation of an *F*-ratio requires only the mean square between (MS_B) and the mean square within (MS_W); hence, we need to calculate only these two mean squares. The general equation for mean squares is

$$MS = \frac{SS}{df} \qquad \text{Eq.9.8}$$

where
SS = the sum of squares, and
df = degrees of freedom.

For our example, $MS_B = 24/2 = 12$ and $MS_W = 6/6 = 1$.

Calculating the F-ratio

We have calculated the needed values and are ready to calculate the *F*-ratio in a one-way between-groups ANOVA. The equation is

$$F\text{-}ratio = \frac{MS_B}{MS_W} \qquad \text{Eq.9.9}$$

For our example, the obtained value for the *F*-ratio is $12/1 = 12$.

As with the calculated *t*-ratio, we need to compare the obtained value of the *F*-ratio with a table of critical values, in this case, an *F*-table. This table is located in Appendix A.4. With the *t*-ratio, we found the appropriate critical value by locating the column for the selected α level and the row for the number of degrees of freedom. The *F*-table is a bit more complicated, because we have two different degrees of freedom values to consider, one associated with the numerator of the *F*-ratio and one associated with the denominator. We indicate this by using the following nomenclature: F-ratio (2,6) = 12, where the first number in the parenthesis is the degrees of freedom for the numerator (between, in our example) and the second number is the degrees of freedom for the denominator (within, in our example). We also consider the selected α level. The different values of alpha appear on different pages of the F-table, so we first locate the appropriate page. Then, we find the degrees of freedom for the numerator in the columns and the degrees of freedom for the denominator in the rows.

Let's go to Appendix A.4 to determine whether the F-ratio of 12 is less or greater than the critical value. Notice that when we compare the calculated F-ratio to the tabled (i.e., critical) value, we are using exactly the same logic that we used with the t-test. That is, if the calculated F-ratio is in the extreme end of the distribution, it is an unlikely occurrence, if the null hypothesis is actually true. Assume the selected α for the example to be .05.

By going to the column for 2 degrees of freedom in the numerator and to the row for 6 degrees of freedom for the denominator, we find the critical value of 5.143 associated with an α level of .05. Hence, we reject the null hypothesis, because our calculated F-ratio of 12 is larger than the critical value of 5.143. Our conclusion is that this pattern of means (2, 4, 6) would occur less than 5% of the time if the null hypothesis is true. In this case, the omnibus ANOVA test gives a significant result, indicating that there is a difference between some of the means. At this point, we do not know where that difference is. Later in this chapter, we will further investigate this difference.

TABLE 9.3

ANOVA source table.

SOURCE OF VARIANCE	*SS*	*DF*	*MS*	*F*
Between	24	2	12	12
Within	6	6	1	
Total	30	8		

It is common practice to summarize an ANOVA by creating an ANOVA source table. Table 9.3 provides an example. Ultimately you would calculate all these values by using a statistical computer package (e.g., SPSS or R).

HOW ANOVA WORKS

A reasonable question to ask at this point is, "How does analyzing the variance in a data set allow us to make a decision about whether the means of the levels of the independent variable are equal or not?" We will answer this question with an illustration that uses the same data set as the example above. By rearranging the data set as shown in Table 9.4, we can see how the total variability in the scores is partitioned into two sources.

When we subtract the grand mean from Subject 1's score of 1, the result is −3. By reading across the row for Subject 1, we see that this value of −3 is composed of

- −2, which is the difference between Subject 1's group's mean (2) and the grand mean (4), and
- −1, which is the difference between Subject 1's score (1) and his group's mean (2).

You can do this for all subjects. For example, for Subject 8, all of the difference between his score and the grand mean comes from the difference

TABLE 9.4 Partitioning of sums of squares.

PERSON	SCORE	$\bar{\bar{X}}$	$Diff_1$	$Diff_1^2$	$Diff_2$	$Diff_2^2$	$Diff_3$	$Diff_3^2$
1	1	−4	−3	9	−2	4	−1	1
2	2	−4	−2	4	−2	4	0	0
3	3	−4	−1	1	−2	4	1	1
4	3	−4	−1	1	0	0	−1	1
5	4	−4	0	0	0	0	0	0
6	5	−4	1	1	0	0	1	1
7	5	−4	1	1	2	4	−1	1
8	6	−4	2	4	2	4	0	0
9	7	−4	3	9	2	4	1	1
				SS_T		SS_B		SS_W
Total			0	30	0	24	0	6

Person = identification number for each subject
Score = subject's score
$\bar{\bar{X}}$ = the grand mean

$Diff_1$ = difference between subject's score and grand mean
$Diff_1^2$ = $Diff_1$ squared
$Diff_2$ = difference between subject's group mean and the grand mean
$Diff_2^2$ = $Diff_2$ squared
$Diff_3$ = difference between subject's score and subject's group mean
$Diff_3^2$ = $Diff_3$ squared
SS_T = sum of all $Diff_1^2$ (i.e., sum of squares for total)
SS_B = sum of all $Diff_2^2$ (i.e., sum of squares between groups)
SS_W = sum of all $Diff_3^2$ (i.e., sum of squares within groups)

between his group's mean (6) and the grand mean (4), because Subject 8's score (6) does not differ from his group's mean (also 6).

When we square and total the results for all nine subjects, we obtain the values for SS_T, SS_B, and SS_W. From here, we can use the associated degrees of freedom to calculate MS_B, MS_W, and the F-ratio. To understand why the F-ratio reveals whether we should retain or reject the null hypothesis, we need to learn about the **expected values** for MS_B and MS_W.

Expected Values of MS_W and MS_B

The **expected value** of a statistic is the mean of its sampling distribution. For example, assume that a very large number of samples are taken from a population and the mean of each of the samples is calculated. If these means are plotted, the result would be a normal distribution with a mean equal to the mean of the population from which the samples were drawn. Thus, the expected value of a sample mean, $\bar{X}$, is μ, the mean of the population.

MS_W

The expected value of the mean square within is the population variance, σ^2, and this is true whether the null hypothesis is true or false. This occurs because MS_W is a function only of the variability of scores within each group. It does not reflect variability around the grand mean.

In the one-way ANOVA, we obtain J independent estimates of the population variance, with J equal to the number of levels of the independent variable. The mean of these J estimates is the mean square within, and its expected value is σ^2.

MS_B

The mean square between is a function of how the sample means fluctuate around the grand mean. If the null hypothesis is true, any differences among the sample means is due to sampling error and, in this situation, the expected value of MS_B is σ^2, the same as the expected value for MS_W. Thus, if the null hypothesis is true in a one-way between-groups ANOVA, the F-ratio (MS_B/MS_W) is approximately equal to 1.0, because both MS_B and MS_W are estimates of the same parameter, namely, σ^2.

However, if the null hypothesis is false, the expected value of MS_B is σ^2 plus a value we will call σ_B^2. This value increases as the differences between the sample means and the grand mean increases. If the value of σ_B^2 is high enough, the value of $\sigma^2 + \sigma_B^2 = MS_B$ will cause the F-ratio to exceed the critical value from the F-table, and the researcher's conclusion will be to reject the null hypothesis.

Put another way, when the null hypothesis is true, the following are true:

$$MS_W \approx \sigma^2$$
$$MS_B \approx \sigma^2$$

$$F\text{-}ratio = \frac{MS_B}{MS_W} = \frac{\sigma^2}{\sigma^2} \approx 1.0$$

However, when the null hypothesis is not true, additional variation occurs in the value of MS_B. This variation comes from the sum of squares calculated earlier. In this case, the following are true:

$$MS_W \approx \sigma^2$$
$$MS_B \approx \sigma^2 + \sigma_B^2$$

where σ_B^2 represents the variance attributable to the differences between the J means.

When the null hypothesis is *not* true and there are differences between the means, we calculate the F-ratio as follows:

$$F - ratio = \frac{MS_B}{MS_W} = \frac{\sigma^2 + s_B^2}{2} > 1.0$$

The F-ratio increases as a result of the increased variance associated with the difference between the means. Thus, the expected value when the null hypothesis is not true is some value > 1.0. We then use the F-table to determine whether the obtained F-ratio is large enough to allow us to decide with some certainty (i.e., the confidence level) that the null hypothesis should be rejected. Notice that if the null hypothesis is true, then σ_B^2 is equal to zero, and the resultant F-ratio will be approximately 1.0.

In summary, in a one-way between-groups ANOVA, MS_W and MS_B both estimate the same parameter, Σ^2, when the null hypothesis is true. When the null hypothesis is false, MS_W is still an estimate of σ^2, but MS_B becomes an estimate of σ^2 *plus* a value that becomes increasing large as the null hypothesis becomes less and less tenable.

Unequal-*n* ANOVA

For ease of illustration and calculation, we have discussed an example with equal *n*s in each level of the independent variable. When n is not equal in each level, we perform the same calculations. However, as we showed in Chapter 8, sample sizes that are vastly different could affect the alpha level. Recall that when the larger sample is matched with the larger variance and the smaller sample is matched with the smaller variance, the actual alpha will be less than the selected alpha. Conversely, if the smaller sample is matched with the larger variance and the larger sample with the smaller variance, then the actual alpha will be greater than the selected alpha.

For an example of an unequal-*n* between-groups ANOVA, suppose an investigator selected 32 subjects randomly from a population and randomly assigned them to four different methods of improving upper-arm strength. Following the treatment, the investigator measured the subjects' upper-arm strength by using pull-ups as the dependent variable. Table 9.5 illustrates the data from an unequal-*n* between-groups ANOVA for this study.

The question to be answered by the one-way between-groups ANOVA is: "Can this pattern of means occur reasonably by chance if there really is no difference in the means?"

Raw Score ANOVA Equations

Rather than use the definitional equations presented above to calculate the necessary sums of squares, we will illustrate equivalent equations that require less tedious mathematical operations. These equivalent equations can be used with data sets that have equal *n*s and data sets with unequal *n*s. We will use the unequal-*n* data in Table 9.5 to illustrate the calculating equations.

Sum of Squares Total

An equivalent calculating equation for $SS_T = \Sigma(X_{ij} - \bar{\bar{X}})^2$ is:

$$SS_T = \Sigma\Sigma X_{ij}^2 - \frac{(\Sigma X_{ij})^2}{N} \quad \text{Eq.9.10}$$

TABLE 9.5 Data set for unequal-*n* ANOVA.

	GROUP 1	GROUP 2	GROUP 3	GROUP 4	TOTALS
	5	4	12	6	
	6	7	9	4	
	7	5	13	5	
	10	8	10	3	
	9	5		4	
	6	9		5	
	8	4		9	
	5	6		5	
	7			4	
				6	
				4	
N	9	8	4	11	32
ΣX	63	48	44	55	210
$\bar{X}$	7.0	6.0	11.0	5.0	$6.56 = \bar{\bar{X}}$

where
SS_T = sum of squares total,
$\Sigma\Sigma X_{ij}^2$ = the sum of the scores in the data set after each score has been squared,
$(\Sigma X_{ij})^2$ = the sum of all the scores, the quantity squared, and
N = the total number of scores in the data set.

Using the equivalent calculating equation for the SS_T we find:

$$SS_T = 1572 - \frac{44{,}100}{32} = 193.9$$

Sum of Squares Between

An equivalent calculating equation for $SS_B = \Sigma nj(\bar{X}_j - \bar{\bar{X}})^2$ is:

$$SS_B = \Sigma\left(\frac{X_j^2}{n_j}\right) - \frac{(\Sigma X_{ij})^2}{N} \quad \text{Eq.9.11}$$

where
SS_B = sum of squares between,
$\Sigma\left(X_j^2 / n_j\right)$ = the sum of the J group scores after each score has been squared and divided by their respective ns,
$(\Sigma X_{ij})^2$ = the sum of all the scores, the quantity squared, and
N = the total number of scores in the data set.

Using the equivalent equation for the SS_B, we find:

$$SS_B = \frac{63^2}{9} + \frac{48^2}{8} + \frac{44^2}{4} + \frac{55^2}{11} - \frac{44{,}100}{32} = 109.9$$

Sum of Squares Within

An equivalent calculating equation for $SS_W = \Sigma\Sigma(X_{ij} - \overline{X}_j)^2$ is:

$$SS_W = \Sigma\Sigma X_{ij}^2 - \Sigma\left(\frac{\Sigma X_j^2}{n_j}\right) \qquad \text{Eq.9.12}$$

where

SS_W = sum of squares within,

$\Sigma\Sigma X_{ij}^2$ = the sum of the scores in the data set after each score has been squared, and

$\Sigma(X_j^2 / n_j)$ = the sum of the J groups' scores after each score has been squared and divided by their respective ns.

Using the equivalent calculating equation, we find:

$$SS_W = 1572 - \left(\frac{63^2}{9} + \frac{48^2}{8} + \frac{44^2}{4} + \frac{55^2}{11}\right) = 1572 - 1488 = 84$$

Once we have found the sums of squares and the degrees of freedom, we can calculate the mean squares and the F-ratio. The degrees of freedom for this example are:

$$df\ total = N - 1 = 32 - 1 = 31,$$
$$df\ between = J - 1 = 4 - 1 = 3, and$$
$$df\ within = N - J = 32 - 4 = 28.$$

Hence,

$$MS_B = 109.9 / 3 = 36.6,$$
$$MS_W = 84 / 28 = 3, and$$
$$F = MS_B / MS_W = 36.6 / 3 = 12.2,$$

which is significant beyond the $\alpha = .01$ level.

On the basis of these results, the investigator of treatments for upper-arm strength would reject the null hypothesis, concluding that this pattern of pull-up mean scores is unlikely to occur by chance if the null hypothesis is true. To determine which means are significantly different from one another requires the use of a multiple comparison technique discussed later in this chapter.

V.9.1 
SPSS ANOVA

V.9.1R 
RStudio ANOVA

ANOVA AND TYPE I AND TYPE II ERRORS

The discussion of Type I and Type II errors in Chapter 8 holds, as well, for ANOVA. When a researcher reaches a conclusion regarding the null hypothesis, either that conclusion is correct, or the researcher has committed an error. If the researcher rejects the null hypothesis, a chance (alpha) exists that she has made a Type I error. If the researcher fails to reject the null hypothesis, she runs the risk of making a Type II error. These concepts will remain with us for all of the inferential tests that we discuss in this textbook.

POWER IN ANOVA

Is previously stated, the power of the ANOVA is generally greater than the power of the *t*-test. The factors presented about power in Chapter 8 remain applicable to ANOVA. Key factors to consider are that power will increase with increasing sample size, with increasing number of levels of the independent variable, with a reduction in error of measurement, and a larger difference among the obtained means. Also recall, power is influenced by the chosen alpha level. Additionally, smaller variation within the respective levels of the independent variable increases power by reducing MS_W (the denominator of the *F*-ratio).

MULTIPLE COMPARISONS

If the *F*-ratio in ANOVA leads to the decision to retain the null hypothesis, no further examination of the data is necessary, because this conclusion indicates that any differences among the means are reasonably explained by sampling error. However, if the null hypothesis is rejected, further investigation is warranted to determine which means are significantly different from one another.

We have already ruled out the use of multiple *t*-tests, due to the associated increase in Type I error probability. Rather, we can employ procedures called **multiple comparisons** to investigate which mean differences are significant. Many multiple comparison (MC) procedures have been developed over the past 60 or more years, and to decide which one is appropriate for a given situation, we must answer a few questions. The answers to these questions are closely tied to the research hypotheses of the experiment. The questions are:

1. Are the desired comparisons simple or complex?
2. Do we wish to use a contrast-based or family-based MC procedure for protection from errors?
3. Do we wish to make planned or *post hoc* comparisons?

After discussing these questions, we will provide examples of multiple comparison techniques.

Simple or Complex Comparisons?

A **simple comparison** is one that involves only two means. For example, we might ask whether there is a significant difference between $\bar{X}_1$ and $\bar{X}_2$

A **complex comparison** involves more than two means. For example, we might wish to determine whether there is a significant difference between the mean of two of the treatment means $\left(\bar{X}_1 + \bar{X}_2\right)/2$ and the control mean $\left(\bar{X}_C\right)$. This would be expressed mathematically as:

$$\frac{(\bar{X}_1 + \bar{X}_2)}{2} \text{ } vs. \text{ } \bar{X}_C$$

Notice that three means are involved in this comparison.

To explore the difference between simple and complex comparisons, consider the following experimental design. The independent variable has four levels, whose means are labeled $\bar{X}_1, \bar{X}_2, \bar{X}_3$, and $\bar{X}_C$. The three numbered means represent three treatments, and $\bar{X}_C$ represents the mean for a control group. If the research hypotheses for this experiment can all be answered by comparing each of the means to each of the other means, choosing a MC procedure for simple comparisons would be appropriate. If each of the treatment means is compared to the control mean, there are three simple comparisons. If each of the means is compared to all other means, there are a total of six simple comparisons. If, however, one of the research hypotheses was that administering any treatment is better than doing nothing (represented by the control group), a complex comparison is required:

$$\frac{(\bar{X}_1 + \bar{X}_2 + \bar{X}_3)}{3} \text{ } vs. \text{ } \bar{X}_C$$

MC procedures that allow only simple comparisons are more powerful than those that allow complex comparisons. However, the decision as to which type to use should be based on the comparisons that must be made to provide evidence for the research hypotheses, not on which comparisons are more powerful.

Contrast-Based or Family-Based Error Rate Protection?

Some MC techniques are designed to protect against a Type I error at the selected α for *each* contrast (comparison) made (**contrast-based error rate**), and others protect against a Type I error for an *entire family* of contrasts (**family-based error rate**). The contrast-based MC techniques are more powerful than the family-based MC techniques. For the example described above, six simple comparisons are possible. If a contrast-based MC technique were used in this situation, the probability of a Type I error would be the selected α for each comparison. However, if a family-based MC technique were used, the probability of one or more Type I errors for the *entire family* of the six comparisons is the selected α.

The decision as to whether to use a contrast-based or family-based MC procedure rests on the researcher's perception of which type of error, Type I or Type II, would be the more serious to make. This is specific to each experiment. The question the investigator has to consider is, would it be worse to declare that a significant difference exists when it actually doesn't (a Type I error), or to accept the null hypothesis when, in fact, there is a significant difference (a Type II error). Contrast-based MC procedures are more prone to Type I errors (and less to Type II errors), and the converse is true of family-based MC procedures.

Planned or *Post Hoc* Comparisons?

In some experiments, to evaluate the research hypotheses, the researcher might not need to make all of the comparisons that are possible; only a smaller number of comparisons might be required. In this situation, the researcher should use a multiple comparison procedure designed to assess only certain comparisons that are decided upon *before* the experiment is started. These are called **planned comparisons**. Some refer to these comparisons as ***a priori* contrasts**. In other situations (e.g., exploratory research), the investigator might like to be able to make all of the possible simple and complex comparisons that are possible. MC procedures for this situation are called ***post hoc* tests** because they allow all possible comparisons to be made following the finding of a significant F-ratio. The difference between *a priori* and *post hoc* tests is that the researcher decides which comparisons to make before *(a priori)* or after *(post hoc)* the experiment.

Planned MC procedures are more powerful than *post hoc* MC procedures in much the same way that a one-tailed test is more powerful than a two-tailed test. When the researcher gives up the possibility of making any and all comparisons, smaller critical values are required to achieve significance for just the planned comparisons. Once again, however, the comparisons that are necessary to evaluate all of the research hypotheses—not the degree of power available—dictate which type of MC procedure is appropriate for a particular study.

Multiple Comparison Procedures

Table 9.6 is an ANOVA source table for a one-way between-groups ANOVA with an independent variable that has three levels. We will examine several MC procedures, using these data for all of the illustrations. Each of the three groups had 10 subjects, and the means were 760.5, 851.5, and 928.5 for Groups 1, 2, and 3, respectively. The selected α level is .05.

TABLE 9.6

ANOVA source table.

SOURCE OF VARIANCE	*SS*	*df*	*MS*	*F*
Between	141,447	2	70,723.5	10.74
Within	177,828	27	6,586.2	
Total	319,275	29		

Scheffé Multiple Comparison Procedure

We will begin by illustrating the most versatile of all MC techniques. It is called the Scheffé test. Because it allows simple and complex comparisons and *post hoc* comparisons and is a family-based technique, it is the most conservative (least powerful) of all the techniques we will present. Any and all comparisons can be made with the Scheffé test. At least one comparison will be found to be significant if the

ANOVA F-ratio has led to rejection of the null hypothesis; however, the significant comparison will not necessarily be relevant to the posited research hypotheses.

If the value of any comparison exceeds the Scheffé value (S), that comparison is declared to be significant. The equation for S is

$$S = \sqrt{(J-1)(F_\alpha)\left(\frac{2MS_e}{n}\right)} \quad \text{Eq.9.13}$$

where
S = the Scheffé critical value,
J = the number of levels of the independent variable,
F_α = the critical value from the F-table used to determine whether the F-ratio from the ANOVA is significant,
$MS_e = MS_W$, from the ANOVA source table, and
n = the number of subjects in each level.

Unequal Ns and the Scheffé test

If the ns are not equal, we determine the Scheffé value by using the following equation:

$$S = \sqrt{(J-1)(F_\alpha)(MS_e)\left(\frac{1}{n_1}+\frac{1}{n_2}\right)} \quad \text{Eq.9.14}$$

where
J, F_α, and MS_e are defined above,
n_1 = the number of subjects associated with $\bar{X}_1$, and
n_2 = the number of subjects associated with $\bar{X}_2$.

If complex comparisons are involved, the ns are the means of the ns involved in obtaining the composite mean(s).

Let's calculate the Scheffé value to determine which means in the example data set are significantly different at $\alpha = .05$. We will use Equation 9.13, because we have equal ns. We use 2 and 26 degrees of freedom from Appendix A.4.

$$S = \sqrt{(J-1)(F_\alpha)\left(\frac{2MS_e}{n}\right)}$$

$$= \sqrt{(3-1)(3.37)\left(\frac{2(6586.2)}{10}\right)}$$

$$= \sqrt{(2)(3.37)\left(\frac{13{,}172.4}{10}\right)}.$$

$$= \sqrt{(2)(3.37)(1{,}317.24)}$$
$$= \sqrt{8{,}878.2} = 94.2$$

Thus, any simple or complex comparison value exceeding 94.2 will be considered to be significant. For the three simple comparisons, therefore, only the difference between $\bar{X}_1$ and $\bar{X}_3\,(760.5 - 928.5 = -168)$ is significant. For an example of a complex comparison that is significant, compare the mean of $\bar{X}_1$ and $\bar{X}_2$ to $\bar{X}_3$. That is, $(760.5 + 851.5)/2 - 928.5 = -122.5$.

Tukey Multiple Comparison Test

If the three answers to the questions discussed earlier are (1) simple, (2) family-based, and (3) *post hoc*, the Tukey MC procedure is appropriate. As with the Scheffé test, we calculate a single critical value and compare this to the difference between any two means. The equation for the Tukey value is

$$T = q\sqrt{\frac{MS_e}{n}} \qquad \text{Eq.9.15}$$

where
T = Tukey critical value,
q = a critical value obtained from the Studentized Range Distribution (see Appendix A.5; the appropriate q value is located on the page with the selected α, at the junction of the column for k [number of levels of the independent variable] and the row for the degrees of freedom associated with the MS_e from the ANOVA source table),
$MS_e = MS_W$, obtained from the ANOVA source table, and
n = number of subjects in the groups.

Unequal Ns and the Tukey test

If the group sizes are unequal, we use the following equation:

$$T = q\sqrt{\left(\frac{MS_e}{2}\right)\left(\frac{1}{n_1} + \frac{1}{n_2}\right)} \qquad \text{Eq.9.16}$$

where
T, q, and MS_e are defined above,
n_1 = number of subjects in the group for the first mean, and
n_2 = number of subjects in the group for the second mean.

Let's use Equation 9.15 to calculate the critical Tukey value for the data set provided in Table 9.6 and determine which mean differences are considered to be significant at $\alpha = .05$ with this procedure. We use $k = 3$ and $df = 24$ from Appendix A.5.

$$T = q\sqrt{\frac{MS_e}{n}}$$

$$T = 3.53\sqrt{\frac{6586.2}{10}} = 3.53\sqrt{658.62} = 3.53(25.66) = 90.6$$

For the three simple comparisons, we find a significant difference between $\bar{X}_1$ and $\bar{X}_3$ $(760.5 - 928.5 = -168)$, as before; however, now the difference between $\bar{X}_1$ and $\bar{X}_2$ $(760.5 - 851.5 = -91)$ is also significant. How can this be, given that the Scheffé critical value procedure found only one significant simple comparison difference? Remember that the Scheffé test is very conservative, because it allows complex as well as simple comparisons. Because the Tukey procedure only allows simple comparisons, it is somewhat more powerful.

Student-Newman-Keuls (SNK) Multiple Comparison Procedure

The Student-Newman-Keuls MC procedure is identical to the Tukey test, except that it uses a contrast-based Type I error rate approach rather than a family-based one. Like Tukey, the SNK procedure can be used only with simple comparisons, and it is also a *post hoc* test. The difference between the two procedures has to do with the number of means *(k)* associated with the independent variable. With the Tukey test, only one critical value is calculated, and it is used for any and all comparisons made. In the SNK procedure, the exact same critical value as calculated for Tukey is compared to the difference between the largest mean and the smallest mean. However, different q values are used to calculate the critical values when means that are closer together are compared. The best way to illustrate this is to use our sample data set.

First, let's arrange the means in order from smallest to largest:

$$\bar{X}_1 = 760.5$$

$$\bar{X}_2 = 851.5$$

$$\bar{X}_3 = 928.5$$

The range (number) of means in this data set *(k)* is 3. So, for the first comparison $(\bar{X}_1 \text{ to } \bar{X}_3)$, the appropriate q value for use in calculating the SNK critical value is found in Appendix A.5 in column 3. The *df* for MS_e are 27, so we use the row for 24 *df*. (If the exact *df* is not listed in the table, we use the q value with the next fewer *df*, to be conservative.) This q value is 3.53. We find the difference between $\bar{X}_1$ and $\bar{X}_3$ $(760.5 - 928.5 = -168)$ and compare this to the SNK critical value:

$$SNK = q\sqrt{\frac{MS_e}{n}}$$

$$SNK = 3.53\sqrt{\frac{6586.2}{10}} = 3.53\sqrt{658.62} = 3.53(25.66) = 90.6$$

which is the same as the Tukey value. The mean difference is significant.

Next, to compare $\bar{X}_1$ to $\bar{X}_2$ and to compare $\bar{X}_1$ to $\bar{X}_3$, we note that the range for these means is only 2, and, thus, the correct q value is located on the same page of Appendix A.5 as before ($\alpha = .05$), but in column 2, row 24. It is 2.92. Thus, the critical value for these two mean differences comparisons is

$$SNK = 2.92\sqrt{\frac{6586.2}{10}} = 2.92\sqrt{658.62} = 2.92(25.66) = 74.9$$

The difference between $\bar{X}_1$ and $\bar{X}_2$ is $760.5 - 851.5 = -91$ and between $\bar{X}_1$ and $\bar{X}_3$ is $851.5 - 928.5 = -77$. Both of these values are compared to 74.9 because in both cases the means are adjacent to each other. Both values are larger than 74.9; therefore, the SNK MC procedure indicates that all three means are significantly different from one another. Although the SNK MC procedure is more powerful than Scheffé and Tukey, it is important to know that if the independent variable has more than three levels, the SNK may lead to an increase in the possibility of Type 1 errors.

It is interesting that as the MC procedures increased in power from Scheffé to Tukey to Student-Newman-Keuls, the number of simple comparisons that were found to be significant increased also, from one to two to three. Perhaps you can see why the decision about which MC procedure to use must be made very carefully and must be based on the purposes of the experiment.

Other Multiple Comparison Procedures

In addition to the MC procedures presented above, a number of others are available in many statistical computer programs. We will describe some of the most common ones here, but we will not present equations for calculating them. Many of these procedures have limitations and are used only under certain conditions.

Planned Orthogonal Comparisons (POC)

Planned orthogonal comparisons is a very powerful MC procedure, but it can be used only when all of the comparisons to be made are independent (orthogonal), meaning that they contain no overlapping information. A mathematical method is available to determine whether the comparisons are independent. We will not present the method, but we will illustrate it here. Consider an independent variable that has means for four levels, and the comparisons that need to be made to examine the research hypotheses are $\bar{X}_1$ with $\bar{X}_2$, $\bar{X}_3$ with $\bar{X}_4$, and $\bar{X}_1$ with $\bar{X}_4$. Are these comparisons all independent? The answer is no. $\bar{X}_1$ versus $\bar{X}_2$ is independent of $\bar{X}_3$ versus $\bar{X}_4$, but the comparison of $\bar{X}_1$ versus $\bar{X}_2$ contains information that overlaps with the comparison of $\bar{X}_1$ versus $\bar{X}_4$. Therefore, these comparisons are not all independent from one another, and POC is not an appropriate choice of procedure.

Dunnett MC Procedure

The Dunnett MC procedure is ideal if the experimental design requires only comparing each treatment group mean with a control group mean. Because it is a planned procedure and accommodates only simple comparisons, it is relatively powerful.

Trend Analysis

Trend analysis is more of an alternative method of analyzing a data set than it is a multiple comparison procedure. In the special situation that the levels of the independent variables are at least ordinal in nature rather than nominal, we can use trend analysis to determine the shape of the curve that expresses the relationship between the independent and dependent variables. For example, if the independent variable is age, with levels of 5, 15, 25, 35, 45, 55, 65, and 75, and the dependent variable is any of several physical measurements, say, grip strength, it is common to find that a parabola best describes the relationship between these two variables (see Figure 9.1).

We will revisit this procedure in the next chapter, because it is common in repeated-measures studies for the levels of the independent variable to be measured on an ordinal scale. Trend analysis can indicate whether the

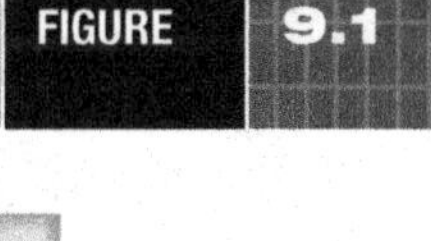

FIGURE 9.1 Example of a quadratic relationship between dependent and independent variables.

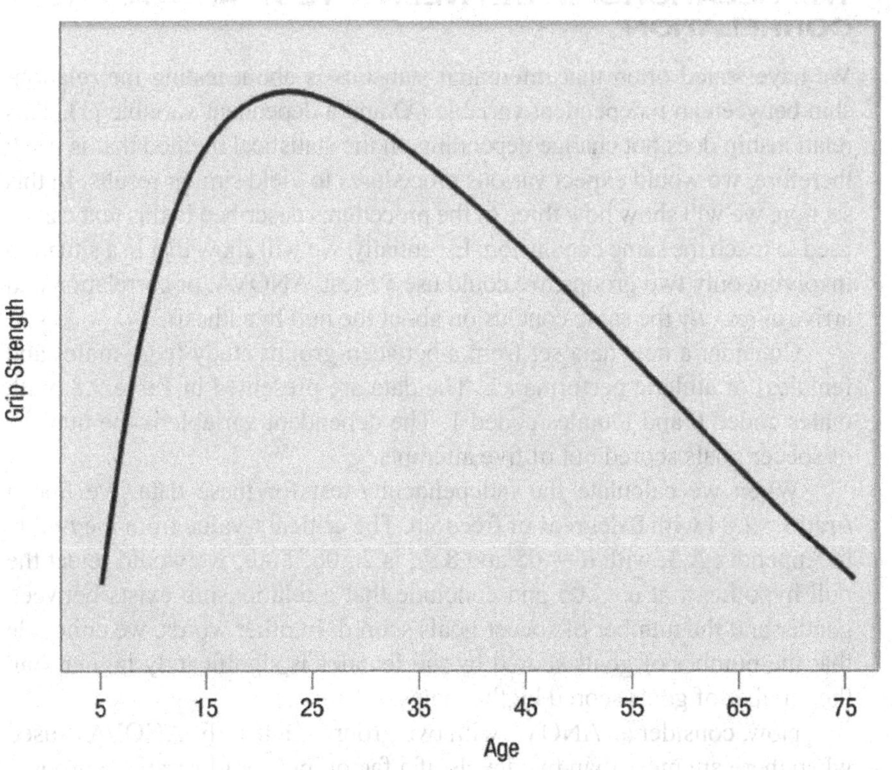

TABLE 9.7 Multiple comparison procedures arranged approximately from most to least powerful and situations where they are appropriate.

	TYPE OF COMPARISON		TYPE I ERROR RATE		BEFORE OR AFTER	
	Simple	**Complex**	**Contrast-based**	**Family-based**	**Planned (a priori)**	***Post hoc***
POC		X	X		X	
SNK	X		X			X
Tukey	X			X		X
Dunnett	X			X	X	
Scheffé	X	X		X		X

V.9.2
SPSS Multiple Comparisons

V.9.2R
RStudio Multiple Comparisons

relationship is linear (a straight line), quadratic (a curve with one inflection point), cubic (a curve with two inflection points), or another shape. Often, this information is of much more interest than knowing which means along the ordinal scale differ from one another.

Descriptions of Multiple Comparison Procedures

Table 9.7 gives a comparison of the various MC procedures presented in this chapter. SPSS and R provide many MC alternatives with ANOVA analyses.

THE RELATIONSHIP BETWEEN *T*-TEST, ANOVA, AND CORRELATION

We have stated often that inferential statistics is about testing the relationship between an independent variable *(X)* and a dependent variable *(Y)*. This relationship does not change depending on the statistical method that is used; therefore, we would expect various procedures to yield similar results. In this section, we will show how three of the procedures described in this text can be used to reach the same conclusion. Essentially, we will show that in a situation involving only two groups, we could use a *t*-test, ANOVA, or correlation and arrive at *exactly* the same conclusion about the null hypothesis.

Consider a new data set from a between-groups study (e.g., males and females) of athletic performance. The data are presented in Table 9.8, with males coded 0 and females coded 1. The dependent variable is the number of soccer goals scored out of five attempts.

When we calculate the independent *t*-test for these data, we find a *t*-ratio = 2.84 with 8 degrees of freedom. The critical t-value from the *t*-table in Appendix A.3, with $\alpha = .05$ and 8 *df*, is 2.306. Thus, we would reject the null hypothesis at $\alpha = .05$ and conclude that a relationship exists between gender and the number of soccer goals scored. In other words, we conclude that the number of goals scored by the females is significantly higher than the number of goals scored by the males.

Now, consider an ANOVA with two groups. Generally, ANOVA is used when there are more than two levels of a factor, but for illustrative purposes we will conduct an ANOVA for the experiment under discussion, with two levels of the independent variable (GENDER). Table 9.9 gives the calculated results.

In Appendix A.4, we find that the critical F-ratio for 1 and 8 degrees of freedom with $\alpha = .05$ is 5.318. Hence, we reject the null hypothesis at $\alpha = .05$. In Table 9.9, notice that the degrees of freedom for the denominator in the F-ratio are 8 (the same as the degrees of freedom for the t-test). If we square the calculated t-ratio ($2.84^2 = 8.067$), the result is equal to the calculated F-ratio. In general, when only two groups are involved (i.e., 1 degree of freedom in the numerator) in a one-way between-groups ANOVA, the exact same results are obtained as when we calculate a t-test ($t^2 = F$).

Finally, let's calculate the Pearson product-moment correlation between gender (X) and the dependent variable (Y). Recall that the null hypothesis in this case states that there is no correlation between X and Y. That is, the null hypothesis states that the correlation coefficient between X and Y is equal to 0.00. When we use the methods presented in Chapter 5, we find that the correlation coefficient between gender and goals scored in this experiment is 0.709. Consulting Appendix A.2, with 8 degrees of freedom, we find that the critical value for rejecting the null hypothesis is .632 with $\alpha = .05$. Hence, we reject the null hypothesis and conclude that a relationship does exist between X and Y. Another way to state this conclusion is to say that the mean number of goals scored was significantly different between the males and females.

Thus, we have used the t-test, ANOVA, and PPM correlation to illustrate that we reach *exactly* the same decision with each statistical procedure. That is, in all cases we find that there is a relationship between X and Y that cannot be fully explained by random variation. Note that we have illustrated this concept *only* for the situation where there is a single independent variable (X) with two levels, but the general concept remains the same for all inferential tests. As the researcher, you are testing whether a relationship exists between the X (independent) and Y (dependent) variables. You test a null hypothesis about a relationship and then make a decision, based on probability, about the likelihood of the null hypothesis being a true statement about the population.

TABLE 9.8

Between-groups example of t-test, ANOVA, and correlation.

X GENDER	Y DEPENDENT VARIABLE
1	5
1	4
1	5
1	3
1	2
0	0
0	2
0	1
0	2
0	3
$\bar{X}_{males} = 1.6$	$s_{males} = 1.1$
$\bar{X}_{females} = 3.8$	$s_{females} = 1.3$

TABLE 9.9

ANOVA results for comparison with t-test.

SOURCE OF VARIANCE	SS	df	MS	F
Between	12.1	1	12.1	8.067
Within	12.0	8	1.5	
Total	24.1	9		

V.9.3 **Canonical Model Comparison**

THE MAGNITUDE OF THE EFFECT

Effect Size

As we have mentioned previously, a statistically significant finding is only one piece of evidence that the relationship between the independent and the dependent variables might be important. When we discussed the t-test, we suggested additional information that could help the investigator decide whether this relationship is practically important (percent improvement, effect size, and omega squared).

When we use ANOVA, we can also examine the effect size of the treatment to help in this decision. The equation for effect size is

$$R^2 = \frac{SS_B}{SS_T}$$ Eq.9.17

where
R^2 = effect size,
SS_B = sum of squares between, and
SS_T = sum of squares total.

We can interpret R^2 in much the same way as the coefficient of determination (r^2) presented in Chapter 5. This effect size is sometimes called eta squared (η^2). It indicates the percentage of the variance in the data set that can be attributed to the treatment. Obviously, the higher the value of R^2, the greater the variation in the measures is associated with the treatment. For the results in Table 9.9, we find

$$R^2 = \frac{SS_B}{SS_T} = \frac{12.1}{24.1} = 50.2\%$$

Therefore, slightly more than 50% of the variability of the soccer scoring ability is attributable to gender, and slightly less than 50% is attributable to other factors.

Omega Squared (ω^2)

Another approach to evaluating the magnitude of the treatment effect is to calculate ω^2. Omega squared also provides an estimate of the proportion of the variance in the dependent variable that is attributed to the effect (i.e., the independent variable). It is calculated from the sums of squares and mean squares.

$$\omega^2 = \frac{SS_B - (df_B)(MS_W)}{MS_W + SS_T}$$ Eq.9.18

where
ω^2 = omega squared,
SS_B = sum of squares between,
df_B = degrees of freedom between,
MS_W = mean square within, and
SS_T = sum of squares total.

When we use the data in Table 9.9, we find

$$\omega^2 = \frac{SS_B - (df_B)(MS_W)}{MS_W + SS_T}$$

$$= \frac{12.1-(1)(1.5)}{1.5+24.1}$$
$$= \frac{12.1-(1.5)}{1.5+24.1} = \frac{10.6}{25.6} = .41$$

Thus, 41% of the variation in the dependent variable (Y, soccer scoring ability) is attributable to differences in levels of the independent variable (X, gender). The value of ω^2 is a slightly more conservative estimate of effect size than R^2 and, thus, will normally be slightly less than R^2.

SUMMARY

The appropriate procedure for comparing differences among more than two interval- or ratio-measured means is analysis of variance (ANOVA). In this chapter, we presented a one-way between-groups ANOVA. Following a significant omnibus test with ANOVA, the researcher can conduct various multiple comparison procedures to investigate mean differences. We also illustrated that the t-test, ANOVA, and PPM correlation calculations result in identical conclusions under certain circumstances. Finally, we discussed how, with ANOVA, we can also examine the effect size to obtain information that will help us determine whether a relationship is practically important.

Review Questions and Practice Problems

PART A (ANSWERS PROVIDED IN APPENDIX C)

Use the following test score data to answer questions 1–10 in PART A and 11–13 in PART B. Use an alpha level of .05.

Control	Treatment A	Treatment B
1	2	3
2	4	6
3	6	9

1. What are the three means?
2. What is the grand mean?
3. What is the null hypothesis?

Using the *definitional* formulas, calculate the F-ratio for these data.

4. What are the SS_T, SS_B, and SS_W?
5. What are the degrees of freedom for each sum of squares?
6. What are the MS_B and MS_W?
7. What is the F-ratio?
8. What is the critical value of F required to reject the null hypothesis?
9. What is your conclusion regarding the null hypothesis?
10. What type of error is possible with this conclusion?

PART B

11. Using the raw score formulas with these data, calculate the sums of squares by hand and then construct a source table for the ANOVA.

Use SPSS or R to confirm the results you obtained in questions 1, 2, 4, 5, 6, 7, and 11.

To use SPSS enter the data above in two columns (TESTSCORE and TREATMENT). Use **Analyze → Compare Means → One-Way ANOVA** and put TESTSCORE in the Dependent List and TREATMENT in the

Factor Box. Click **Options → Descriptive**. Click **Continue**. Click **OK**.

The RStudio commands are:

Enter the data into an Excel file with two columns (TESTSCORE and TREATMENT). Assign the TREATMENT levels as C (Control), TA (Treatment A), and TB (Treatment B). Save the data file as "experiment".

Open RStudio.

Import the Data Set.

To generate the mean for the nine scores enter:

- mean(experiment$TESTSCORE)

To generate the means by TREATMENT enter:

- tapply(X=experiment$TESTSCORE, INDEX=experiment$TREATMENT, FUN=mean)

Now run a one-way ANOVA using the aov (analysis of variance) function where TESTSCORE is the response variable and TREATMENT is the factor of interest.

- treatment_aov <- aov(TESTSCORE ~ TREATMENT, data=experiment)
- summary(treatment_aov)

Highlight all four statements and click on Run.

12. What multiple comparison technique would be appropriate for the results of this ANOVA?
13. Suppose the investigator for this study decided before the data were collected that she would only compare each of the treatment means to the control mean, and assume she wants to be very sure she does not commit a Type I error. How would she answer the three questions given in the chapter that guide the choice of multiple comparison test?

Use the following information to calculate the Scheffé and Tukey values and the answers to questions 14 through 18. Use $\alpha = .05$.

- $k = 6$
- $N = 60$ (10 in each group)
- The six means (in order) are: 10, 11, 12, 13, 14, 20.
- F-ratio = 12
- $MS_B = 84$

14. What is the value required for significance, with Scheffé?
15. What is the value required for significance, with Tukey?
16. With Scheffé, which means are significantly different?
17. With Tukey, which means are significantly different?
18. At $\alpha = .05$, is there a difference between the average of the first three means and the average of the last three means?

COMPUTER ACTIVITIES

One-Way ANOVA and Multiple Comparisons

ACTIVITY 1 One-Way ANOVA

A 60-item knowledge test was administered to a random sample of 15 human performance majors from each class (freshmen, sophomores, juniors, and seniors). The researcher is interested in testing whether a significant difference exists among the mean scores for the classes. The results are given in Table 9.10.

Using SPSS

In SPSS, go to **Analyze→Compare Means→Means**. In the Means box, use the arrow button to put TESTSCORE in the Dependent List box and CLASS in the Independent List box. Click on **Options** and check the box next to "ANOVA table and eta". Click **Continue** and then OK.

Q

1. What is the null hypothesis for this study?
2. What is the research hypothesis for this study?
3. What are the mean and standard deviation for all 60 scores?
4. What are SS_B, SS_W, and SS_T? Hint: $SS_T = SS_B + SS_W$
5. Why are the degrees of freedom 3 and 56?
6. What is the value of the F-ratio?
7. What is the significance value for this F-ratio?
8. Why is the MS_W (Residuals) so relatively small?
9. What is your conclusion regarding the null hypothesis?

Data set. TABLE 9.10

SUBJECT	CLASS	CLASSNUM	TESTSCORE	SUBJECT	CLASS	CLASSNUM	TESTSCORE	SUBJECT	CLASS	CLASSNUM	TESTSCORE
1	F	1	13	21	H	2	14	41	J	3	29
2	F	1	18	22	H	2	19	42	J	3	25
3	F	1	15	23	H	2	18	43	J	3	27
4	F	1	15	24	H	2	19	44	J	3	29
5	F	1	17	25	H	2	19	45	J	3	31
6	F	1	12	26	H	2	20	46	S	4	43
7	F	1	20	27	H	2	20	47	S	4	47
8	F	1	18	28	H	2	18	48	S	4	41
9	F	1	13	29	H	2	15	49	S	4	49
10	F	1	14	30	H	2	17	50	S	4	48
11	F	1	14	31	J	3	28	51	S	4	42
12	F	1	18	32	J	3	26	52	S	4	41
13	F	1	19	33	J	3	31	53	S	4	40
14	F	1	15	34	J	3	23	54	S	4	46
15	F	1	15	35	J	3	26	55	S	4	49
16	H	2	16	36	J	3	29	56	S	4	48
17	H	2	18	37	J	3	28	57	S	4	41
18	H	2	18	38	J	3	27	58	S	4	42
19	H	2	17	39	J	3	32	59	S	4	40
20	H	2	16	40	J	3	30	60	S	4	43

10. If this written test is designed to measure human performance knowledge, does the pattern of the means make sense? (Hint: Think about the typical human performance curriculum.)
11. Why weren't the sophomores labeled S?

Using R

1. Open R and import data from Table 9.10 (Table_9_10_Data) into RStudio.
2. To generate the mean and standard deviation for all 60 scores, type the following:
   - mean(Table_9_10_Data$TESTSCORE)
   - sd(Table_9_10_Data$TESTSCORE)
3. To generate the means by CLASS, type the following:
   - tapply(X=Table_9_10_Data$TESTSCORE, INDEX =Table_9_10_Data$CLASS,FUN=mean)
4. Run a one-way ANOVA using the aov (analysis of variance) function where TESTSCORE is the response variable and CLASS is the factor of interest:
   - class_aov <- aov(TESTSCORE ~ CLASS, data=Table_9_10_Data)
5. To view the relevant information about the ANOVA, we will use the summary function.
   - summary(class_aov)

   Highlight the five lines you've typed and click **Run** to execute and you will see the means, standard deviation, and ANOVA results.

ACTIVITY 2 Multiple Comparisons following One-Way ANOVA

12. Is it appropriate to conduct multiple comparison tests for this study?
13. Which multiple comparison tests might be most appropriate? Answer the three questions given in the chapter to guide the selection of a multiple comparison test. Compare your selection with a classmate's selection. Discuss why you made the selection that you did.
14. Is your ANOVA Source Table the same as what you obtained when you ran the first procedure?
15. Describe what you see in the multiple comparison output.
16. Why do you think the mean values are significantly different in the pattern that you observe?
17. Rerun the ANOVA, this time choosing Scheffé as the *post hoc* procedure. Compare the Scheffé results with the SNK results.

Using SPSS

Rerun an SPSS ANOVA procedure as follows: Go to **Analyze→Compare Means→One-Way ANOVA**. Put CLASSNUM in the Factor box and put TESTSCORE in the Dependent list box. Click on **Post Hoc**. When the Post Hoc Multiple Comparison window opens, click on **S-N-K** and **Scheffe.** Click **Continue** and then OK.

Using R

To conduct Multiple Comparison tests with RStudio we will need to install the "DescTools" package:

In the lower right pane of the RStudio, click on the "Packages" tab and then click on the "Install" button.

In the "Packages" field, type in "DescTools" and click "Install".

[Note that from this point forward, the package "DescTools" will be installed, and you do not have to install it again. However, if you close and reopen RStudio, you will need to run the "Library(DescTools)" command again.]

Add the following three commands to the ANOVA commands to obtain *post hoc* test results:

- library(DescTools)
- PostHocTest(class_aov, method="newmankeuls")
- PostHocTest(class_aov, method="scheffe")

REFERENCES

Fisher, R. A. (1925). *Statistical methods for research workers*. Edinburgh: Oliver and Boyd.

Hopkins, K. D., Hopkins, B. R., & Glass, G. V. (1996). *Basic statistics for the behavioral sciences* (3rd ed.). Needham Heights, MA: Allyn & Bacon.

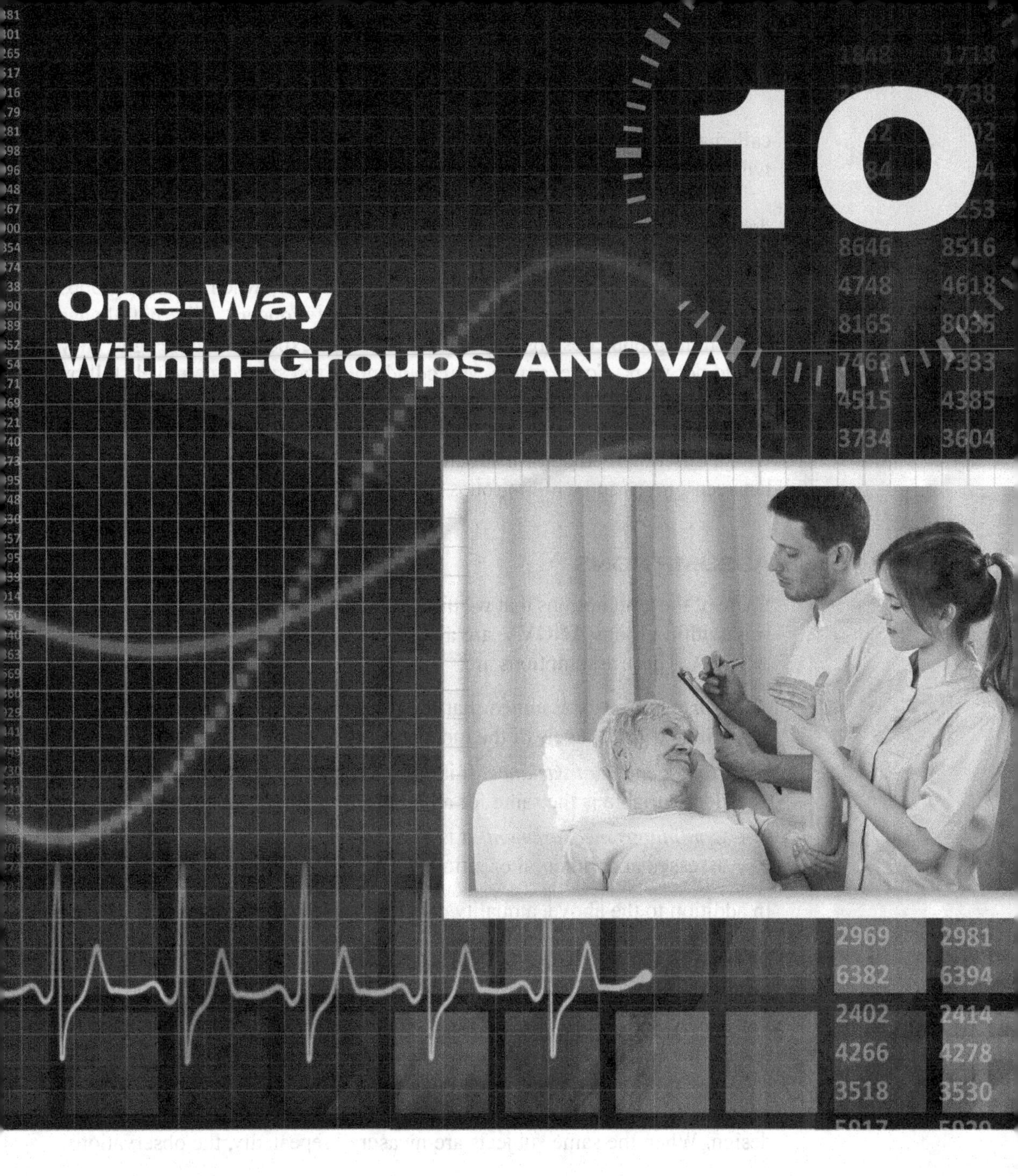

10 One-Way Within-Groups ANOVA

INTRODUCTION

We now turn our attention to a slightly different model. In this chapter, we present information about a one-way within-groups ANOVA. This general topic was introduced in Chapter 8, when we discussed between-groups and within-groups statistical designs. Then, in Chapter 9, we discussed an ANOVA with a between-groups design. In that design, each subject appeared in one and only one level of the independent variable. For within-groups ANOVA, scores for each subject appear in all levels of the independent variable in a one-way design. This design, like the correlated t-test, is often called a repeated-measures design, because the same subjects are measured twice (for a *t*-test) or more than twice (for a repeated-measures ANOVA).

Just as the correlated *t*-test is generally more powerful than the independent *t*-test, the within-groups ANOVA is generally more powerful than the between-groups ANOVA. This is because less unexplained (error) variance results when subjects serve as their own controls than when separate groups of subjects are involved. Given this, you might wonder, "Why not always use within-groups designs?" Because, for some experimental designs, they would not work. For example, you could not teach subjects a novel skill by using one method of instruction and then tell them to forget what they learned because you want to measure them again after they have learned the skill with a different method of instruction. Figure 10.1 illustrates differences for between-groups and within-groups designs with respect to subject placement.

ASSUMPTIONS

Some of the assumptions that we discussed in Chapter 9, with regard to one-way within-groups ANOVA, are also applicable to one-way within-groups ANOVA. These assumptions are:

- *Normality*. It is assumed that the dependent variable is normally distributed in all levels of the independent variable.
- *Homogeneity of variance*. It is assumed that the variance of the dependent variable is the same for all levels of the independent variable.
- *Continuous measurement*. It is assumed that the dependent variable is assessed at an interval or ratio level of measurement.

In addition to the above, a quality called *sphericity* is also assumed.

Sphericity

For a *between-groups* design, we assume independence of observations: that is, the levels of the independent variable are between groups and that each observation is taken from an independent sample of subjects from the population. Obviously, this assumption is not applicable to a within-groups design. When the same subjects are measured repeatedly, the observations,

FIGURE 10.1 Differences for between-groups and within-groups designs.

BETWEEN-GROUPS				WITHIN-GROUPS		
Freshmen	Sophomore	Junior	Senior	Time 1	Time 2	Time 3
Sue	Kelley	Pete	Karen	$Mike_1$	$Mike_2$	$Mike_3$
Kevin	James	Sally	Jamall	Sam_1	Sam_2	Sam_3
...	...	...	...	$\dots_1$	$\dots_2$	$\dots_3$

by their very nature, cannot be independent. In the within-groups ANOVA, the groups are, in fact, not independent but rather are made up of the same subjects. Instead of the independence assumption, within-groups ANOVA requires the **sphericity** assumption, which relates to the pattern of the variances, the correlations, and score differences from all of the levels of the independent variable. Sphericity is not an issue in the correlated *t*-test, because there are only two levels and, thus, only one correlation. When the sphericity assumption is violated, we can correct for the violation by using one of the methods discussed below.

If the assumption of sphericity is not met, the probability of making a Type I error is larger than the selected alpha. It is common for statistical computer packages to test for sphericity and present common adjustments (discussed below), if warranted. If necessary, we can adjust (reduce) the number of degrees of freedom that we use to find the critical value for comparison to the calculated *F*-ratio. Doing this when the sphericity assumption is not met essentially reduces the probability of making a Type I error. However, as always, reducing the degrees of freedom results in a less powerful statistical test.

Two of the most common corrections are the Greenhouse-Geisser and the Huynh-Feldt corrections. The value of epsilon (ε) indicates the degree to which the data deviate from meeting the sphericity assumption.

Epsilon Correction

Both Greenhouse and Geisser (1959) and Huynh and Feldt (1976) estimate the value of epsilon. Essentially, the epsilon value adjusts the degrees of freedom in the numerator and denominator of the critical (tabled) *F*-ratio. For both methods, the degrees of freedom are multiplied by epsilon to reduce the degrees of freedom and the adjusted number of degrees of freedom are used to locate the critical value in the *F*-table (Appendix A.4). If the sphericity assumption is met, the value of epsilon is 1, and thus no correction takes place. As the severity of violation of the sphericity assumption increases, the value of epsilon becomes increasing smaller than 1, but it will never be smaller than $1/(J-1)$, where J is the number of levels of the independent variable.

Lower-Bound Sphericity Correction

In another approach, also suggested by Geisser and Greenhouse (1958), the *df* in the numerator and denominator of the *F*-ratio are both divided by the quantity $J-1$. This approach is more conservative than the epsilon correction because the lowest possible value for epsilon is $1/(J-1)$. Because the *df* for the numerator are $J-1$, this correction reduces the numerator *df* to 1 [i.e., $(J-1)/(J-1) = 1$]. The *df* for the denominator reduce to $(J-1)(N-1)/(J-1) = N-1$. We use these *df* values to locate the critical *F* value in Appendix A.4.

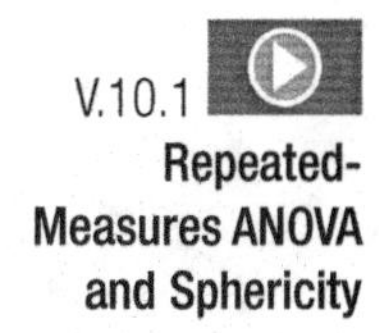

Repeated-Measures ANOVA and Sphericity

This Geisser-Greenhouse correction is referred to as a lower bound or conservative adjustment because it reduces the degrees of freedom as low as would occur if the sphericity assumption was not met. Later in this chapter, we will illustrate adjustments made to correct for lack of sphericity.

CALCULATIONS

Recall that in the one-way between-groups ANOVA, the total sum of squares was partitioned into two parts, the sum of squares between and the sum of squares within (SS_B and SS_W). The sum of squares between (SS_B) could be called the sum of squares "for treatment", because it reflects variance in the system attributable to differences among the means presumably caused by the treatments; similarly, the sum of squares within (SS_W) could be called the sum of squares "for error", because it reflects the unexplained variance in the system.

In the one-way within-groups ANOVA, we will partition the sum of squares into three parts. These are:

1. the sum of squares for columns (SS_C), which reflects the variance in the data attributable to differences in the means;
2. the sum of squares for rows (SS_R), which reflects differences among the subjects; and
3. the sum of squares for error (SS_E), which represents the remaining unexplained error variance.

As you may have guessed, $SS_T = SS_C + SS_R + SS_E$.

For example, measures could be taken at baseline and then repeated on the same subjects at six weeks and at 12 weeks. Such a design might be appropriate when a researcher is interested in how many steps participants in a physical fitness class take weekly, at baseline and then six and 12 weeks into the course. Keep in mind that at baseline, week 6, and week 12, the researcher is measuring the *same* subjects and, thus, this is a repeated-measures (within-groups) design.

The researcher is interested in testing whether the means are significantly different across the three time periods. The null hypothesis is:

$$H_0: \mu_{\text{baseline}} = \mu_{\text{week6}} = \mu_{\text{week12}}$$

The alternative hypothesis (H_1) is that at least one of the means is not equal to the others (i.e., not all means are equal).

We will illustrate the calculations of a one-way within-groups ANOVA, starting with sum of squares and ending with the F-ratio, by using the data presented in Table 10.1. As with the one-way between-groups ANOVA, we use a small data set to simplify the illustration. In reality, it would be unusual to conduct a one-way within-groups ANOVA on such a small amount of data, because it is unlikely that the normality assumption would be met. Following the calculations, we present the output for the same data from SPSS and R for this one-way within-groups design.

Sum of Squares Total

As always, the definitional equation to obtain the total sum of squares requires subtracting the grand mean from every score in the data set, squaring the resulting differences, and summing the squares. An equivalent calculation equation that uses raw scores is:

TABLE 10.1 One-way repeated measures (within-groups) ANOVA with three levels of the independent variable.

	TIME			
SUBJECT	BASELINE	WEEK 6	WEEK 12	ΣX_J
1	1	3	6	10
2	2	4	5	11
3	2	5	4	11
4	3	4	5	12
$\Sigma X =$	8	16	20	$\Sigma\Sigma X = 44$
$\bar{X} =$	2	4	5	$\bar{\bar{X}} = 3.67$
$\Sigma X^2 =$	18	66	102	$\Sigma\Sigma X^2 = 186$

$$SS_T = \Sigma\Sigma X^2 - \frac{(\Sigma X)^2}{nJ} \quad \text{Eq. 10.1}$$

where

SS_T = sum of squares total,
$\Sigma\Sigma X^2$ = the sum of all the scores after each has been squared,
$(\Sigma X)^2$ = the sum of the scores, the quantity squared,
n = the number of subjects, and
J = the number of levels (Columns) of the independent variable.

By using Equation 10.1, we find:

$$SS_T = 186 - \frac{(44)^2}{(4)(3)}$$

$$= 186 - \frac{1936}{12} = 186 - 161.33 = 24.67$$

Sum of Squares for Columns (SS_C)

The sum of squares for columns represents the amount of variance in the system attributable to differences between the means (columns). In this case, it represents the differences that result from the passage of time. The calculating equation is:

$$SS_C = \frac{\Sigma(\Sigma X_J)^2}{n} - \frac{(\Sigma X)^2}{nJ} \quad \text{Eq. 10.2}$$

where

SS_C = sum of squares for columns (Time),
$\frac{\Sigma(\Sigma X_J)^2}{n}$ = the sum of the sum of each column after they have been squared, this sum divided by the number of subjects in each column,

$(\Sigma X)^2$ = the sum of the scores, the quantity squared,
n = the number of subjects, and
J = the number of levels (Columns) of the independent variable.

By using Equation 10.2, we find:

$$SS_C = \frac{8^2 + 16^2 + 20^2}{4} - \frac{(44)^2}{(4)(3)}$$
$$= \frac{64 + 256 + 400}{4} - \frac{1936}{12}$$
$$= \frac{720}{4} - \frac{1936}{12} = 180 - 161.33 = 18.67$$

Sum of Squares for Rows (SS_R)

The sum of squares for rows represents the variance in the system that is attributable to the fact that subjects differ from one another across the time periods. The calculation equation for SS_R is:

$$SS_R = \frac{\Sigma(\Sigma X_R)^2}{J} - \frac{(\Sigma X)^2}{nJ} \quad \text{Eq. 10.3}$$

where

SS_R = sum of squares for rows,
$\frac{\Sigma(\Sigma X_R)^2}{J}$ = the sum of the sum of each row after the row sum has been squared, this sum divided by the number of levels (Columns) of the independent variable,
$(\Sigma X)^2$ = the sum of the scores, the quantity squared,
n = the number of subjects, and
J = the number of levels (Columns) of the independent variable.

By using Equation 10.3, we find:

$$SS_R = \frac{10^2 + 11^2 + 11^2 + 12^2}{3} - \frac{(44)^2}{(4)(3)}$$
$$= \frac{100 + 121 + 121 + 144}{3} - \frac{1936}{12}$$
$$= \frac{486}{3} - \frac{1936}{12} = 162 - 161.33 = .667$$

Sum of Squares for Error (SS_E)

The sum of squares for error represents any remaining variance in the s ystem that cannot be explained. We calculate it by subtraction. Since $SS_T = SS_C + SS_R + SS_E$ then

$$SS_E = SS_T - SS_C - SS_R$$ Eq. 10.4

Thus, $SS_E = 24.67 - 18.67 - 0.67 = 5.33$.

Degrees of Freedom

Degrees of freedom for the various sums of squares are as follows:

df for total = $N - 1$, where N = the total number of observations,
df for columns = $J - 1$, where J = the number of levels of the independent variable (i.e., Time),
df for rows = $n - 1$, where n = the number of subjects, and
df for error = $(J - 1)(n - 1)$.

Verify that for the example, the degrees of freedom for total, columns, rows, and error are 11, 2, 3, and 6, respectively. Notice that 2 + 3 + 6 = 11, which is $N - 1$, the total degrees of freedom.

Mean Squares

As before, the general equation for mean squares is *SS/df*. In our example,

$$MS_C = SS_C / df_C = 18.67 / 2 = 9.33$$
$$MS_R = SS_R / df_R = 0.67 / 3 = 0.22$$
$$MS_E = SS_E / df_E = 5.33 / 6 = 0.89$$

F-ratio and SPSS/RStudio Output

As before, we obtain the *F*-ratio by dividing the mean square representing the explained variability in the system by the mean square for the unexplained (error) variance estimate. Table 10.2 gives the SPSS ANOVA source table for these data. Although, in this design, it is possible to determine whether or not a significant difference exists among the subjects by calculating an *F*-ratio for rows (MS_R/MS_E = .22/.89 = .25 in this case, indicating that there is no significant difference), this ratio is seldom of interest. In many other cases, the *F*-ratio for rows does result in a statistically significant result, but it comes as little surprise that subjects differ. Table 10.2 R contains the RStudio

TABLE 10.2 SPSS ANOVA source table for repeated-measures data in Table 10.1.

SOURCE OF VARIANCE	*SS*	*df*	*MS*	*F*-ratio
Columns (Time)	18.67	2	9.33	10.5
Rows (Subjects)	0.67	3	0.22	0.25
Error	5.34	6	0.89	
Total	24.67	11		

RStudio ANOVA source table for repeated-measures data in Table 10.1. TABLE 10.2R

	$ANOVA							
	Effect	DFn	DFd	SSn	SSd	F	p *p<.05	ges
1	(Intercept)	1	3	161.333	0.666	726.0	0.000112*	0.9641
2	Time	2	6	18.666	5.333	10.5	0.010973*	0.7567

TABLE 10.3 Additional SPSS output for data in Table 10.1.

Tests of Within-Subjects Effects

Measure: Score

Source		Type III Sum of Squares	df	Mean Square	F	Sig.	Noncent. Parameter	Observed Power[a]
Time	Sphericity Assumed	18.667	2	9.333	10.500	.011	21.000	.885
	Greenhouse-Geisser	18.667	1.391	13.417	10.500	.027	14.609	.742
	Huynh-Feldt	18.667	2.000	9.333	10.500	.011	21.000	.885
	Lower bound	18.667	1.000	18.667	10.500	.048	10.500	.591
Error(Time)	Sphericity Assumed	5.333	6	.889				
	Greenhouse-Geisser	5.333	4.174	1.278				
	Huynh-Feldt	5.333	6.000	.889				
	Lower bound	5.333	3.000	1.778				

[a] Computed using alpha = .05

output for the data in Table 10.1. Notice that, although it is formatted differently, the RStudio output contains the same statistical values.

Generally, the *F*-ratio of interest is the one in which we test for significant differences among the means (Columns). In our example, we do find a significant difference: $MS_C/MS_E = 9.33/.89 = 10.5$.

The critical *F*-ratio for 2 and 6 degrees of freedom with an α level of .05 is 5.14. The calculated *F*-ratio (10.5) exceeds this value. Therefore, a significant ($p < .05$) difference exists across the three time periods. The notation $p < .05$ means that the probability of obtaining this pattern of means is less than 5% if the null hypothesis is true. Thus, our conclusion for this example is that the pattern of means (2, 4, and 5) includes at least one significant difference among the means. Thus, the researcher rejects the null hypothesis at $\alpha = .05$.

As we have previously indicated, nearly all of the calculations we are performing in this chapter can be done with statistical software. Let's look at the SPSS and R outputs for the data in Table 10.1. The SPSS output includes a great deal of information, but we will discuss only the results that are most important to us at this point. Table 10.3 includes results for sum of squares, *df*, mean squares, *F*, and significance, and each of these is presented four times. The source labeled "Time" is the effect of the repeated measures, and the source labeled "Error(Time)" is the error or unexplained variance. The value of *F* (i.e., the *F*-ratio) is the result when we divide MS_{Time} by $MS_{\text{Error(Time)}}$ in each row. Notice four different results are listed under "Time", and although the *F*-ratio is equal to 10.500 in each case, the degrees of freedom differ in each case.

SPSS AND R TEST OF SPHERICITY

Earlier in this chapter, we mentioned the sphericity assumption for the within-groups analysis. SPSS tests this assumption for you by using Mauchly's test of sphericity, a test of whether the sphericity assumption has been met. Essentially,

SPSS output for sphericity test of data in Table 10.1. TABLE 10.4

Mauchly's Test of Sphericity[a]

Measure: Score

Within-Subjects Effect	Mauchly's W	Approx. Chi-Square	df	Sig.	Epsilon[b]		
					Greenhouse-Geisser	Huynh-Feldt	Lower bound
Time	.562	1.151	2	.562	.696	1.000	.500

Tests the null hypothesis that the error covariance matrix of the orthonormalized transformed dependent variables is proportional to an identity matrix.

a. Design: Intercept
Within-Subjects Design: Time

b. May be used to adjust the degrees of freedom for the averaged tests of significance. Corrected tests are displayed in the Tests of Within-Subjects Effects table.

the null hypothesis for the test in our example is that the assumption *has* been met. If the assumption is not met, then the result of the Mauchly test will be significant. Table 10.4 gives the SPSS output for Mauchly's test for our example. The actual test statistic used with Mauchly's test is W (which is tested with the chi-square distribution with $(J * [J - 1]/2) - 1$ degrees of freedom). The "Sig" column indicates the probability that the sphericity assumption is being met. The significance is .562, indicating that the sphericity assumption *is* being met. (In other words, no significant difference exists between the pattern of variances and correlations in our data and the pattern assumed by sphericity.)

The three entries on the right in Table 10.4 indicate how we would adjust the degrees of freedom if the sphericity assumption has not been met. Recall that epsilon (ε) is a value that indicates the degree to which the data deviate from meeting the sphericity assumption. Table 10.4 gives the results of three methods of estimating epsilon (see the list on the next page), each indicating a factor by which we would adjust the degrees of freedom to make the test of the null hypothesis for the differences between the repeated measures more accurate in terms of a Type I error.

1. Greenhouse-Geisser provides an estimate of epsilon of .696.
2. Huynh-Feldt estimates epsilon to be 1.00. The Huynh-Feldt adjustment is 1.0 because the sphericity assumption was met. (R reports HFe as 1.1.)
3. The lower-bound estimate of epsilon, .500, results in the greatest possible reduction in the degrees of freedom: $1/(J - 1) = 1/(3 - 1) = 1/2 = .500$. We would use this value of epsilon when the sphericity assumption deviated as greatly as possible from being met.

Let's now look more closely at Tables 10.3 and 10.4/10.4R to see how the results were obtained. In Table 10.3, each line for the Time effect is associated with a line from the Error(Time) source. The obtained F-ratios come from the respective numbers for $MS_{Time}/MS_{Error(Time)}$. Note that the mean squares for Sphericity Assumed and Huynh-Feldt are identical. This is because Table 10.4/10.4R indicates that the sphericity assumption is met, so

TABLE 10.4R RStudio output for sphericity test of data in Table 10.1

$"Mauchly's Test for Sphericity"

	Effect	W	p p<.05
2	TIME	0.5625	.5625

$"Sphericity Corrections"

	Effect	GGe	p[GG] p[GG]<.05	HFe	p[HF] p[HF]<.05
2	TIME	0.6956522	0.02657985*	1.108108	0.01097394*

the Huynh-Feldt calculation makes no adjustment to the degrees of freedom for the tests. Now look at the Greenhouse-Geisser result for epsilon presented in Table 10.4/10.4R. This value, .696, is used to adjust the degrees of freedom for Time and Error(Time). The numbers 1.391 and 4.174 in Table 10.3 are obtained from 2 * .696 and 6 * .696, respectively. The lower-bound epsilon adjustment to degrees of freedom is $1/(J-1)$, where J is the number of levels of the repeated measure. In this example, $1/(3-1) = ½ = .5$. Thus, the degrees of freedom in Table 10.3 are adjusted with the lower-bound estimate of epsilon to 1 and 3. Note that the actual F-ratios in Table 10.3 are identical, regardless of which adjustment is made. However, the degrees of freedom for the numerator (MS_{Time}) and denominator ($MS_{\text{Error(Time)}}$) are adjusted downward based on the value of epsilon. The reduction in the degrees of freedom decreases the test's power, but it also controls against an increase in the Type I error probability if the sphericity assumption is not met. Note that in our example, regardless of which sphericity model is chosen, the results indicate that there is at least one significant difference among the three means ($\alpha = .05$); thus, the null hypothesis can be rejected. Note that Tables 10.4 (SPSS) and 10.4R (RStudio), although formatted differently, provide similar sphericity results. RStudio does not provide the lower bound sphericity correction but you can easily calculate this as described in #3 above.

In our example, all interpretations result in rejecting the null hypothesis for Columns. A problem arises when one of the tests results in rejecting the null and another does not, because of the decreased degrees of freedom. Perhaps the best advice in such cases is to choose the most conservative result. By doing this, you reduce the probability of a Type I error (but obviously increase the probability of a Type II error).

POST HOC PROCEDURES FOR WITHIN-GROUPS DESIGNS

Multiple Comparisons

Multiple comparison tests are appropriate only for **fixed independent variables**. Fixed variables are ones for which the researcher has selected the levels rather than allow the levels to be selected at random. For example, if the independent variable in a one-way repeated-measures ANOVA is Time, and there are four levels (prior to treatment, one week after treatment, two weeks after treatment, and three weeks after treatment), time represents a fixed independent variable. An example of a **random independent variable** is altitude. Researchers might randomly choose four altitudes at which to conduct sport performance tests. They might obtain a significant F-ratio for the altitude factor, but it is not possible to conduct multiple comparisons procedures on this random factor. Another example is the subjects factor. Because the subjects in a study have been chosen at random from the population, this factor is considered a random variable, and multiple comparisons are not available for random factors. However, this is very seldom a problem. For the most common designs for which the one-way repeated-measures ANOVA is appropriate, the researcher typically has no interest in determining whether or not significant differences exist among the subjects. Rather, the goal is to determine whether significant differences exist between the means of the various levels of the independent variable.

If a significant F-ratio is found for the *fixed* independent variable, then we can use a multiple comparisons test to look for significant differences among the levels of the independent variable (e.g., Tukey or Student-Newman-Keuls). To illustrate how this works, let's return to the data presented in Table 10.1.

For these data, we found a significant F-ratio, signifying that some significant differences occur among the means of 2, 4, and 5. Let's assume the investigator decided to use the Student-Newman-Keuls multiple comparison technique to determine which means differ from which other means.

The first step is to compare the largest and smallest means to each other. The equation for the SNK is:

$$SNK = q\sqrt{\frac{MS_e}{n}}$$

where

q = the value from Appendix A.5, with $\alpha = .05$ and $J = 3$,
MS_e = mean square for error, and
n = the number of subjects.

Therefore,

$$SNK = 4.34\sqrt{\frac{0.89}{4}} = 4.34\sqrt{.2225} = 4.34(.472) = 2.04$$

Any difference of means greater than 2.04 is significant. Since the difference between the largest mean (5) and the smallest mean (2) is 3, a significant difference exists between these two means.

The next step is to determine the SNK value for adjacent means. In this case, the q value from Appendix A.5 with $\alpha = .05$ and $J = 2$ is 3.46.

$$SNK = q\sqrt{\frac{MS_e}{n}}$$

$$SNK = 3.46\sqrt{\frac{0.89}{4}} = 3.46\sqrt{.2225} = 3.46(.472) = 1.63$$

In this case, you then compare the differences in the two sets of *adjacent* means to the SNK value above. Since 5 − 4 = 1 is not larger than 1.63, these two means do not differ significantly. However, 4 − 2 = 2 is greater than 1.63; thus, these two adjacent means are significantly different.

Trend Analysis

Although it is sometimes instructive to determine which means are significantly different from other means in a repeated-measures design, if the levels of the independent variable represent an ordinal or continuous level of measurement, we sometimes might prefer to do a trend analysis of the means, as discussed in Chapter 9. This approach allows us to determine whether a line or curve best describes the relationship among the means. For example, if equally spaced means increase at a constant rate, a straight line (linear relationship) best describes the pattern of means. See Figure 10.2. Another possible outcome is a curvilinear relationship, such as the quadratic solution illustrated in Figure 10.3.

As we mentioned in Chapter 9, trend analysis is a relative of the multiple comparison techniques, and we use it in the special situation when the levels of the independent variable are equally spaced on a scale with an ordinal or higher level of measurement. The quadratic relationship among the means depicted in Figure 10.3 is generally more interesting than information about which of the four means are significantly different from one another. Statistical software programs, including SPSS and R, frequently allow calculation of trend analysis.

FIGURE 10.2 Trend analysis illustrating a linear solution.

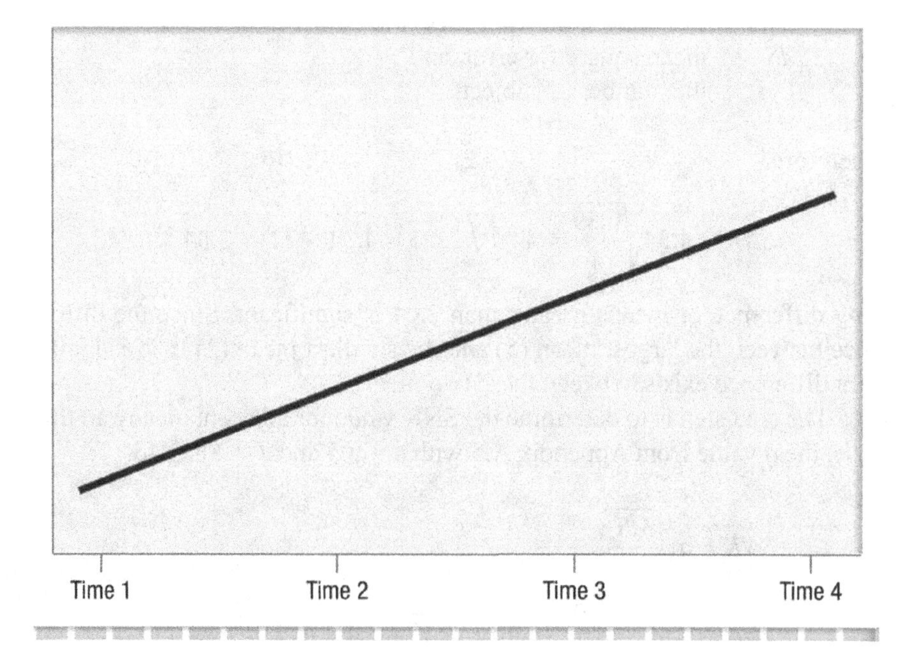

Trend analysis illustrating a curvilinear solution. FIGURE 10.3

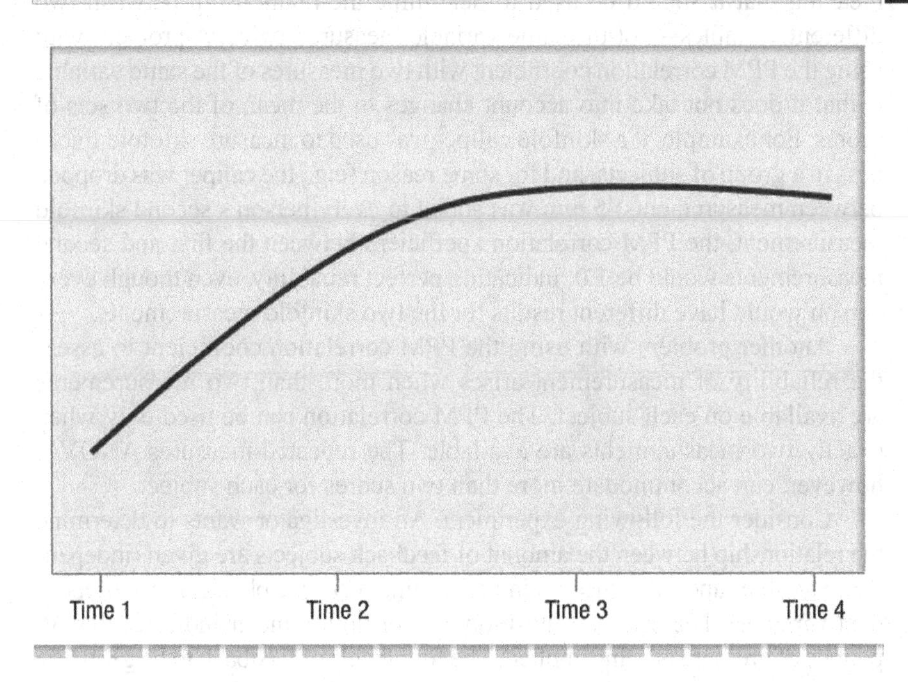

USES OF ONE-WAY REPEATED-MEASURES ANOVA

Inferential Tests

Investigators typically use the one-way repeated-measures ANOVA for one of two different purposes. The first (and perhaps most frequent) use of this design is to determine whether significant differences exist among the means of the levels of the independent variable (or to determine a trend among the means) when subjects are measured repeatedly. The most common independent variables are time, conditions, and exposure. For example, a researcher might be interested in determining how long the effect of a particular medication is detectable, so she might take measurements at regular time intervals. In another example, with conditions as the independent variable, the researcher might want to see how well subjects can perform a particular task under varying conditions of receiving feedback regarding the success of the prior trial. The information we have presented so far in this chapter applies to these types of uses of the repeated-measures ANOVA.

Intraclass Correlation (R)—Reliability Estimation

A second use of the repeated-measures ANOVA is to calculate **intraclass correlation** coefficients. In Chapter 5, we introduced the Pearson product-moment correlation coefficient, a statistic that researchers often use to estimate reliability. At the time, we mentioned that this statistic was not entirely correct, but it could be used to *estimate* the reliability of measurement when two scores were obtained on the same group of subjects and there is no substantial difference in the means between testing periods.

The PPM correlation coefficient is actually an *interclass* correlation, meaning that it should be used to determine the relationship between two different variables—not the same variable measured twice. A problem with using the PPM correlation coefficient with two measures of the same variable is that it does not take into account changes in the mean of the two sets of scores. For example, if a skinfold caliper was used to measure skinfold thickness in a group of subjects, and for some reason (e.g., the caliper was dropped between measurements) 5 mm was added to every person's second skinfold measurement, the PPM correlation coefficient between the first and second measurements would be 1.0, indicating perfect reliability, even though every person would have different results for the two skinfold measurements.

Another problem with using the PPM correlation coefficient to assess the reliability of measurement arises when more than two measurements are available on each subject. The PPM correlation can be used only when exactly two measurements are available. The repeated-measures ANOVA, however, can accommodate more than two scores for each subject.

Consider the following experiment. An investigator wants to determine the relationship between the amount of feedback subjects are given (independent variable) and their success in performing a simple physical task (dependent variable). The task is to push on a lever until a meter indicates that 30 pounds of pressure is being applied. The four levels of feedback to be given are (1) none; (2) if the pressure is within 2 pounds of the target force, the subject is told "close"; if the pressure is more than 2 pounds off (either above or below), the subject is told "not close"; (3) the subject is told by how much the pressure is off but is not told in which direction; or (4) the subject is told by how much and in which direction the pressure is off. Each subject will be measured on six trials under each of the four conditions, and then the researcher will analyze the average for each subject's six trials under each condition.

However, before conducting the experiment, the investigator correctly decides to determine the reliability of the process for recording the pressure exerted. To accomplish this, the investigator randomly selects five subjects from the same population that will be involved in the experiment. After demonstrating the task and allowing a few practice trials with total feedback, she has each of the subjects perform six trials under the no-feedback condition. Table 10.5 gives the results. Normally, a researcher would use a much larger number of subjects to determine reliability than we have included in this example.

TABLE 10.5 Repeated-measures ANOVA for intraclass correlation.

	TRIAL					
SUBJECT	1	2	3	4	5	6
1	29	28	29	31	30	29
2	33	33	32	34	32	31
3	29	28	27	30	31	29
4	30	31	29	28	30	31
5	31	32	29	30	30	28

SPSS intraclass correlation (reliability statistics and ANOVA) output. TABLE 10.6

Reliability Statistics

Cronbach's Alpha		N of Items
.873		6

ANOVA

		Sum of Squares	df	Mean Square	F	Sig
Between People		45.800	4	11.450		
Within People	Between Items	8.667	5	1.733	1.195	.347
	Residual	29.000	20	1.450		
	Total	37.667	25	1.507		
Total		83.467	29	2.878		

Grand Mean = 30.1333

RStudio intraclass correlation (reliability statistics) output. TABLE 10.6R

Reliability analysis

Call: alpha(x = Table_10_5)

raw_alpha	std.alpha	G6(smc)	average_r	S/N	ase	mean	sd	median_r
0.87	0.89	1	0.57	8.1	0.081	30	1.4	0.56

lower	alpha	upper	95% confidence boundaries
0.71	.87	1.03	

We will use SPSS and R to illustrate how we can use a repeated-measures design to determine intraclass reliability for the six trials. With the SPSS **Analyze→Scale→Reliability Analysis** commands, we can obtain the intraclass reliability (also known as **Cronbach's alpha**). Table 10.6 shows part of the output of these commands. The intraclass reliability is .873 for the six trials.

Note that the second box in Table 10.6 looks very similar to some we have seen previously in this chapter. Between Items is the test of differences between trials. The sum of squares, *df*, mean square, and *F* are all calculated as we have discussed previously. However, the purpose here is different: the purpose of the intraclass correlation is to estimate the reliability (i.e., consistency) of measures across multiple measures (i.e., trials). In this case, the intraclass reliability of .873 across six trials is quite high. Although it is not perfect, this level of reliability provides confidence that any differences found between the means at the conclusion of the experiment will not be likely to be the result of unreliable measurement procedures.

Although formatted differently, notice RStudio produces the same value (0.87) for the reliability coefficient.

V.10.2
SPSS Repeated-Measures ANOVA

V.10.2R
RStudio Repeated-Measures ANOVA

V.10.3
SPSS Calculating Reliability with More Than Two Trials

V.10.3R
RStudio Calculating Reliability with More Than Two Trials

The equation to calculate the intraclass reliability coefficient is based on the mean squares between people and trials (obtained from Table 10.6):

$$R = \frac{MS_R - MS_E}{MS_R} \quad \text{Eq. 10.5}$$

where

R = intraclass reliability (Cronbach's alpha) coefficient,
MS_R = mean square for rows (the variation between subjects), and
MS_E = mean square for error (the residual).

$$R = \frac{11.45 - 1.45}{11.45} = \frac{10}{11.45} = .873$$

The obtained Cronbach's alpha coefficient is the reliability across the total number of trials (six, in the current example) and not the reliability of a single trial. Researchers often administer only a single trial and are interested in the reliability of the one trial. SPSS and R can provide an estimate of single-trial reliability. In Table 10.7, the two rows give the reliability for six trials (.873) and an estimate for a single trial (.535) from SPSS. Similar results from R are presented in Table 10.7-R. Note that the estimate of reliability of a single trial is much less than the estimate for all six trials. The researcher in our example plans to administer six trials, so she should be confident in the reliability of the measurement.

TABLE 10.7 SPSS intraclass correlation coefficient.

Intraclass Correlation Coefficient

	Intraclass Correlation	95% Confidence Interval		F Test with True Value 0			
		Lower Bound	Upper Bound	Value	df1	df2	Sig
Single Measures	.535	.172	.917	7.897	4	20	.001
Average Measures	.873	.555	.985	7.897	4	20	.001

TABLE 10.7R RStudio intraclass correlation coefficient.

	type	ICC	F	df1	df2	p	lower bound	upper bound
Single_raters_absolute	ICC1	0.52	7.6	4	25	0.00038	0.17	0.91
Single_random_raters	ICC2	0.53	7.9	4	20	0.00055	0.18	0.91
Single_fixed_raters	ICC3	0.53	7.9	4	20	0.00055	0.17	0.92
Average_raters_absolute	ICC1k	0.87	7.6	4	25	0.00038	0.56	0.98
Average_random_raters	ICC2k	0.87	7.9	4	20	0.00055	0.57	0.98
Average_fixed_raters	ICC3k	0.87	7.9	4	20	0.00055	0.55	0.99

Number of subjects = 5 Number of Judges = 6

Although the RStudio output provides additional information, notice that the values of .53 and .87 appear in both outputs.

ILLUSTRATING THE CANONICAL MODEL

Near the end of Chapter 9, we showed that the independent groups *t*-test and the one-way between-groups ANOVA with two levels of the independent variable *(X)* led to identical decisions. Here, we illustrate that identical decisions are also reached for a paired (correlated) *t*-test and the repeated-measures ANOVA with only two levels of the repeated measures.

Consider a study where the researcher obtains two measures: a pretest and a posttest of ten observations each. Table 10.8 presents the hypothetical data.

TABLE 10.8

Paired *t*-test and repeated-measures ANOVA data.

PRETEST	POSTTEST
1	5
2	3
4	4
3	2
3	5
2	3
4	4
2	3
2	4
3	5

When we use the correlated *t*-test procedures presented in Chapter 7, we find that there are 9 degrees of freedom and the calculated *t*-ratio is 2.714. The critical *t*-value with $\alpha = .05$ and 9 degrees of freedom is 2.262. Thus, the researcher rejects the null hypothesis at $\alpha = .05$.

Table 10.9 gives the results from a repeated-measures ANOVA with the same data. The critical *F*-value for $\alpha = .05$ with 1 and 9 degrees of freedom is 7.21, so the researcher rejects the null hypothesis. As with the independent *t*-test and the between-groups ANOVA, if we square the value of the *t*-ratio, it results in the same *F*-ratio obtained with the repeated-measures ANOVA $(2.714)^2 = 7.36$).

TABLE 10.9

Repeated-measures ANOVA source table.

Source of variation	*df*	Sum of squares	Mean square	*F*
Time (pre/post)	1	7.2	7.2	7.36
Error (within)	9	8.8	.978	

Thus, as with the independent *t*-test and the between-groups ANOVA, the researcher comes to the same conclusion about the null hypothesis, regardless of which procedure she uses. Again, this works only when the repeated measure has exactly two levels. A similar extension does not exist for the Pearson product-moment correlation coefficient where there is only one variable measured twice. This example again illustrates the fact that the null hypothesis is about a *relationship*—in this case, the relationship between time (pre and post), which is the independent variable *(X)*, and the outcome (the dependent variable, *Y*). Thus, as in all inferential testing, the researcher is testing whether a statistically significant relationship exists between *X* and *Y*.

SUMMARY

Beginning with Chapter 6, we have introduced you to the world of inferential statistics. So far, we have covered the methods for testing the null hypothesis between two independent means (independent t-test), between two correlated means (correlated t-test), among more than two independent means (between-groups ANOVA), and, in this chapter, more than two correlated means (within-groups ANOVA). In this chapter, we also considered the repeated-measures ANOVA's assumptions and calculating equations, the sphericity assumption, multiple comparison procedures, and uses of the two-way repeated-measures ANOVA.

All of these designs have one dependent variable and one independent variable. In the next chapter, we will explain how to analyze designs involving more than one independent variable (factor), and in subsequent chapters we will examine the possibility of more than one dependent variable.

From Chapter 6 to this chapter, we have transitioned from presenting calculating equations to examining software output. In the chapters to come, we will continue to rely on software output, because the amount of calculation for the more complex designs becomes onerous. Before proceeding, you might find it helpful to review the meanings of such concepts as independent and dependent variables, fixed factors and random factors, between groups and within groups, and repeated measures.

As we have pointed out many times, in all inferential tests, the logic remains the same. The researcher tests whether a relationship exists between an independent variable (X) and an outcome or dependent variable (Y). The researcher generates null and alternative hypotheses, collects data, and then arrives at a decision regarding the likelihood of the null hypothesis being true.

Review Questions and Practice Problems

PART A (ANSWERS PROVIDED IN APPENDIX C)

1. What is measured more than once in a repeated-measures design?

   A. The reliability
   B. The dependent variable
   C. The total sum of squares
   D. The alpha level

2. Which of the following assumptions is unique to a repeated-measures design?

   A. Homoscedasticity
   B. Normality
   C. Independence
   D. Sphericity

3. Which of the following *least* belongs with the others?

   A. Consistent
   B. Error free
   C. Validity
   D. Reliable

PART B

Use the following information to answer questions 4 through 6.

- Thirteen subjects were measured on four occasions, resulting in a total of 52 measures.
- The means for the occasions were 50, 52, 54, and 61.
- The sum of squares total = 500
- Sum of squares for columns = 165
- Sum of squares for rows = 120

4. Draw a figure illustrating this design.
5. Construct the ANOVA source table for a repeated-measures ANOVA.
6. Use the Tukey multiple comparison procedure to determine which paired means are significantly different from the others (use $\alpha = .05$).
7. Use the following ANOVA source table to calculate Cronbach's alpha (the intraclass reliability coefficient). Forty subjects were measured on five repeated trials.

Source of variation		*SS*	*df*	*MS*	*F*
Between subjects		320	39	8.21	
Within people	Between items	164	4	41.0	21.58
	Residual	297	156	1.90	
	Total	461	160	2.88	
Total		784	199		

COMPUTER ACTIVITIES

Repeated-Measures ANOVA

ACTIVITY 1 Output for a Repeated-Measures ANOVA

A researcher measured 10 subjects performing a hand-eye coordination task under four different conditions. The conditions were: (1) dim light, little feedback; (2) moderate light, little feedback; (3) high light, little feedback; and (4) high light, high feedback. Table 10.10 gives the resulting data. Notice this is a repeated-measures design, because all 10 subjects were tested under all four conditions. The researcher administered the four different conditions in a random order in an attempt to counter any carry-over effects that might result if all subjects completed the conditions in the same order.

Data set. TABLE 10.10

SUBJECT	CONDITION1	CONDITION2	CONDITION3	CONDITION4
1	31	39	14	69
2	40	22	24	100
3	81	18	16	80
4	23	57	33	66
5	11	32	41	45
6	13	77	25	73
7	26	46	77	36
8	29	35	73	81
9	42	62	12	88
10	25	71	77	40

Using SPSS

A. Analyze the data with SPSS by performing the following commands:

- Click on **Analyze→General Linear Model→Repeated Measures**.
- Type in the Within-Subject Factor Name (e.g., CONDITION).
- Type in the Number of Levels (4).
- Click on **Add**.
- Click on **Define**.
- Move the four levels of condition from the left into the Within-Subjects Variables box on the right. Note that this simply indicates that the repeated-measures (within subjects) factor consists of these measures.
- Click on **Options**.
- Click on **Descriptive statistics** in the Display window.
- Click on **Continue**.
- Click **OK**.

B. Notice that the output has seven portions. Some of them are important to our interpretation, and others give very similar information. The portions are:

- *Within-subjects factors:* Confirms your dependent variables across the four conditions.
- *Descriptive statistics:* Lists the means, standard deviations, and sample sizes for each condition.
- *Multivariate tests:* Presents a multivariate analysis of the repeated-measures design. Although this method is sometimes superior for use with repeated-measures designs, it is beyond the scope of this textbook.
- *Mauchly's test of sphericity:* This is an important result.
- *Tests of within-subjects effects:* This is another important result (the outcome of the test of conditions).
- *Tests of within-subjects contrasts:* These are linear, quadratic, and cubic tests of changes across the four conditions. We will not use them here.
- *Tests of between-subjects effects:* No between-subjects effects exist in this example. We will ignore this result.

1. How many degrees of freedom exist for the test of condition?
2. Interpret Mauchly's test of sphericity.
3. Do you have to adjust the within-subjects effects to accommodate the sphericity assumption?
4. Does a significant difference exist across the four levels of condition? (Use $\alpha = .05$.)
5. Why are the "Sig" (probabilities) values different for the sphericity adjusted tests for condition?

6. Show how the degrees of freedom were obtained for each of the sphericity adjustments. (Hint: Epsilon adjustment values are found under the Mauchly test.)
7. What is the null hypothesis for this study?
8. Will you accept or reject the null hypothesis?
9. Assume that the sphericity assumption has not been met. Use the most conservative adjustment for sphericity. What decision would you make about the null hypothesis?
10. What are the means for the four conditions?
11. Draw a figure presenting the means of the four conditions.

Using R

1. Open RStudio.
2. Install any necessary R packages.

   For Chapter 10, we will need to install the "ez" and "reshape2" packages. The "ez" package provides the necessary tools to conduct a repeated-measures ANOVA in R and the "reshape2" package provides the tools to transpose our data set in preparation for a repeated-measures analysis.

   In the lower right pane of the RStudio, click on the "Packages" tab and then click on the "Install" button.

   In the "Packages" field, type in "ez" and click "Install".

   Repeat to install the "reshape2" package.

   Note that from this point forward, the packages "ez" and "reshape2" will be installed and you do not have to install it again. However, once you close out of RStudio, you may need to load them again using the library command described below.

3. Set up your RStudio workspace.

   In the RStudio menu bar, click on **File → New File → R Script**.

   On the first line of the R Script (upper left pane of RStudio), type #Chapter 10.

   On the second line of the R Script, type library(ez).

   On the third line of the R Script, type library(reshape2).

   When you have done that, you should see the following in your new R Script:

   - #Chapter 10
   - library(ez)
   - library(reshape2)

   Highlight and **Run** the three lines that you've typed.

4. Import the data from Table 10.10 (Table_10_10_Data) into R.

   NOTE: In order to conduct a repeated-measures analysis in R, the data are required to be in what is commonly known as the "long" format. This means, there are multiple rows of data for each subject corresponding to the number of measures collected per subject. Table 10.10 was imported in the "wide" format—meaning, there is one row of data for each subject with multiple columns corresponding to the number of measures collected per subject. In order to transpose the data from "wide" to "long" format, we will use the melt function, which is part of the "reshape2" package that we've loaded for this activity.

5. Use the melt function to transpose the data from "wide" format in Table_10_10_Data to "long" ("table_10_10_long"). For these data, our ID variable (id.vars=) is "SUBJECT", and we would like to name the variable (variable.name=)"Condition" that will indicate which condition the measures were collected and finally, name (value.name=) the hand-eye coordination task variable as "TASKSCORE".

   - table_10_10_long <- melt(Table_10_10_Data, id.vars=c("SUBJECT"), variable.name="CONDITION", value.name="TASKSCORE")

   When you've finished typing, highlight the entire function and click **Run**.

   Click on the newly created "table_10_10_long" data frame to see how the data are now structured in the "long" format.

6. Next, we are ready to run our repeated-measures ANOVA using the ezANOVA function from the "ez" package that we've installed and loaded. For this function, we are using the "table_10_10_long" data and are analyzing the dependent variable (dv=) "TASKSCORE" from different individuals (wid=) measured within (within=) a "CONDITION". We also request a detailed output (detailed=TRUE) which provides us additional information about the repeated-measures ANOVA. First, we generate the means:

   - tapply(X=table_10_10_long$TASKSCORE, INDEX=table_10_10_long$CONDITION,FUN=mean)
   - ezANOVA(data=table_10_10_long, dv=.(TASKSCORE), wid=.(SUBJECT), within=.(CONDITION), detailed=TRUE)

   Highlight and click **Run** what you've typed to execute and you will see the information on the repeated-measures ANOVA analysis.

The output is quite extensive and provides the important information we need:

Effect: "Intercept" refers to the between subject whereas "CONDITION" refers to the between condition (or generically, between measures)

DFn/d: degrees of freedom – numerator/denominator

- DFn (CONDITION) = df for columns (conditions)
- DFd (Intercept) = df for rows (subjects)
- DFd (CONDITION) = df for error

SSn/d: Sum of Squares (SS) – numerator/denominator

- SSn (CONDITION) = Sum of squares for columns (SS_C)
- SSd (Intercept) = Sum of squares for rows (SS_R)
- SSd (CONDITION) = Sum of squares for error (SS_E)

ges: Generalized Eta-Squared (ignore)

Sphericity Corrections

- GGe = Greehouse-Geisser
- HFe = Huynh-Feldt

ACTIVITY 2 An Experiment Using a Repeated-Measures Design

A researcher has 10 individuals complete a maximum treadmill run to exhaustion and records heart rates at three times: (a) immediately at the end of the test; (b) one minute after completion; and (c) two minutes after completion. Table 10.11 gives the data that were collected.

TABLE 10.11

Data set.

IMMEDIATE POST	POST1 MINUTE	POST2 MINUTES
194	123	98
182	120	85
168	150	110
202	155	120
175	123	94
185	125	89
164	140	100
191	142	112
180	160	120
192	130	115

Using SPSS

A. Enter the data into SPSS and complete a repeated-measures ANOVA by following the steps suggested above.

12. How many degrees of freedom exist for the test of Time?
13. Interpret Mauchly's test of sphericity.
14. If you decide to conduct the test for Time with the most conservative adjustment to the degrees of freedom which adjustment would you use? What do you conclude about the null hypothesis regarding changes in heart rate across the three time periods?
15. Draw a figure representing the changes in heart rate across the three Time periods. Put Time on the horizontal axis and the heart rate (HR) values on the vertical axis. You obtain descriptive statistics under the Options choice.
16. Interpret the figure you have drawn.
17. Rerun the SPSS analysis, but this time select **Plots** in the Repeated-Measures window. Then move the time factor to the Horizontal Axis box. Click on **Add, Continue**, and **OK**. The RStudio command to obtain this figure is "ezplot". Compare the resulting figure with the one you drew for question 15.
18. What do you conclude from this research?

Using R

Use the following R directions to respond to items 12 through 19

1. Enter the variable names (IMMEDIATE, MINUTE1, MINUTE2) and corresponding values found in Table 10.11 into a spreadsheet and create a csv file named "table_10_11_wide.csv" to be imported into R.
2. Import "table_10_11_wide.csv" into R (see Chapter 3).

   You should see the table_10_11_wide data frame listed under the "Environment" tab (upper right pane of RStudio).

3. Before we transpose our "wide" data set to the "long" format, we will first add a unique identifier to our data set. Data has been collected on 10 subjects, so we will simply provide them with IDs that range from 1 to 10. To accomplish this task, we will instruct R to create a numerical vector of SUBJECT numbers from 1:10 and then combine the SUBJECT vector with our data.

   Hint: Remember to run the following two commands if you are returning to RStudio after having closed RStudio.

   - library(ez)
   - library(reshape2)
   - SUBJECT <- 1:10
   - table_10_11_wide_subject <- cbind(SUBJECT,table_10_11_wide)

   Highlight what you've typed and click **Run**.

   You should now see a new data frame table10_11_wide_subject under the "Environment" tab (upper right pane of RStudio)

4. Use the melt function to transpose the data from "wide" ("table_10_11_wide_subject") to "long" ("table_10_11_subject_long"). For these data, our ID variable (id.vars=) is "SUBJECT" and we would like to name the variable (variable.name=) "TIMEPOINT" that will indicate which time point the measures were collected and finally, name (value.name=) the heart rates measures as "HEARTRATE".

   - table_10_11_subject_long <- melt(table_10_11_wide_subject, id.vars=c("SUBJECT"), variable.name="TIMEPOINT", value.name="HEARTRATE")

   Highlight the entire function and click **Run**.

   Click on the newly created "table_10_11_subject_long" data frame to see how the data are now structured in the "long" format.

5. Next, your data are now ready to run a repeated-measures ANOVA as done for Activity 1 with the following commands:

   - tapply(X=table_10_11_subject_long$HEARTRATE, INDEX=table_10_11_subject_long$TIMEPOINT,FUN=mean)

- ezANOVA(data=table_10_11_subject_long, dv=.(HEARTRATE), wid=.(SUBJECT), within=.(TIMEPOINT), detailed=TRUE)

6. You can plot your data using the ezPlot function included in the "ez" package:

   - ezPlot(data=table_10_11_subject_long, dv=.(HEARTRATE), wid=.(SUBJECT), within=.(TIMEPOINT), x=.(TIMEPOINT))

REFERENCES

Geisser, S., & Greenhouse, S. W. (1958). An extension of Box's results on the use of the *F* distribution in multivariate analysis. *Annals of Mathematical Statistics*, *29*, 885–891.

Greenhouse, S. W., & Geisser, S. (1959). On methods in the analysis of profile data. *Psychometrika*, *24*, 95–112.

Huynh, H., & Feldt, L. S. (1976). Estimation of the Box correction for degrees of freedom from sample data in randomized block and split-plot designs. *Journal of Educational Statistics*, *1*, 69–82.

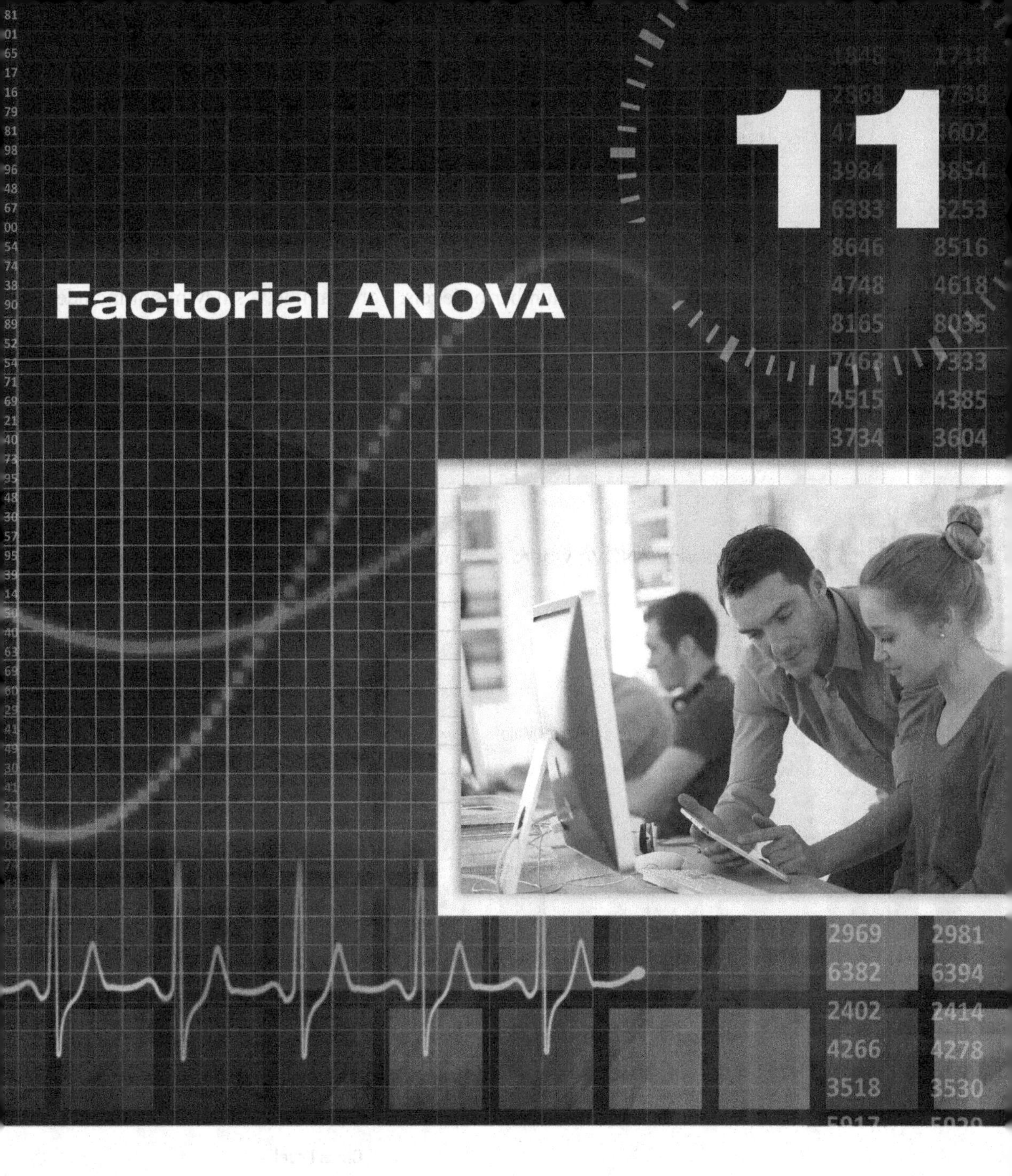

11 Factorial ANOVA

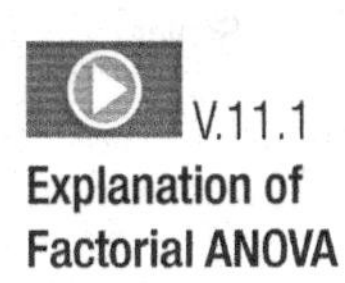
V.11.1
Explanation of Factorial ANOVA

INTRODUCTION

In this chapter, we address the statistical analysis of experimental designs incorporating more than one independent variable. Another name for *variable* is *factor*, hence the term *factorial ANOVA*. For now, we will continue to consider only designs with one dependent (outcome) variable.

FACTORIAL ANOVA DESIGN

The simplest possible factorial ANOVA is a two-way ANOVA where each factor has only two levels. This is called a 2 x 2 design and is illustrated in Figure 11.1.

The 2 x 2 design can be expanded in two ways. First, either of the factors can have any number of levels. For example, a study might include three levels for mode of instruction and four levels for class. This would be a 3 x 4 design. See Figure 11.2.

Another way to expand a 2 x 2 design is to include more than two factors. For example, to form a 2 x 2 x 3 design, we could include age, with two levels (young and old) as another independent variable. See Figure 11.3.

FIGURE 11.1 Two-way ANOVA where each factor has two levels (2 x 2).

		Treatments	
		Experimental	Control
Gender	Male		
	Female		

FIGURE 11.2 Two-way ANOVA with two factors: one factor with three levels and the other with four levels.

		Class Level			
		Freshman	Sophomore	Junior	Senior
Mode of Instruction	Lecture				
	Online				
	Blended				

Three-way ANOVA with two factors having two levels and the third factor having three levels.

FIGURE 11.3

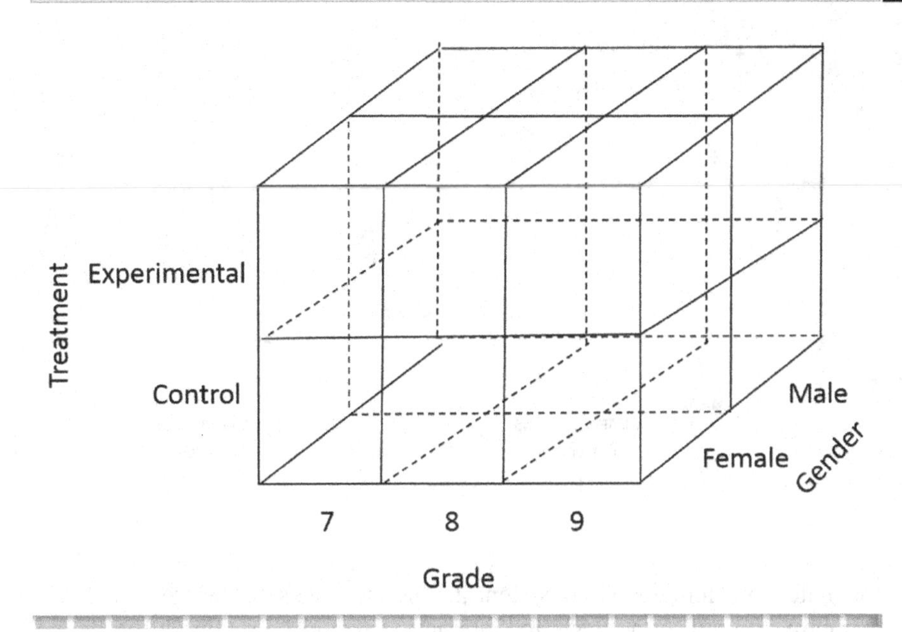

We will address the two-way ANOVA first and then the three-way ANOVA later. To start, let's return to the simplest two-way design, depicted in Figure 11.1: the 2 x 2 design.

TESTABLE HYPOTHESES

In the two-way ANOVA, we formulate three null hypotheses and test them with three separate F-ratios. Two of these tests are called *main effects*, and these are the tests between the means for each of the two factors. With the design in Figure 11.1, we can test to see whether a significant difference exists between the treatment means and to see whether a significant difference exists between the means for males and females. The two null and alternate hypotheses are stated as follows:

For treatments: $H_0: \mu_{T_1} = \mu_{T_2} \quad H_1: \mu_{T_1} \neq \mu_{T_2}$

For gender: $H_0: \mu_M = \mu_F \quad H_1: \mu_M \neq \mu_F$

To state the third testable hypothesis, we need to introduce a new concept, called **interaction**. We state the null hypothesis for the interaction as follows: "The pattern of the means for factor A is consistent across the levels of factor B". If the interaction between the two factors is manifested by an inconsistency in the mean differences for one factor across the levels of the other factor, then we reject the null hypothesis. Figure 11.4 presents the concept of interaction graphically. In outcome 1, treatment 2 appears to be superior to treatment 1 for both males and females. The pattern of the means

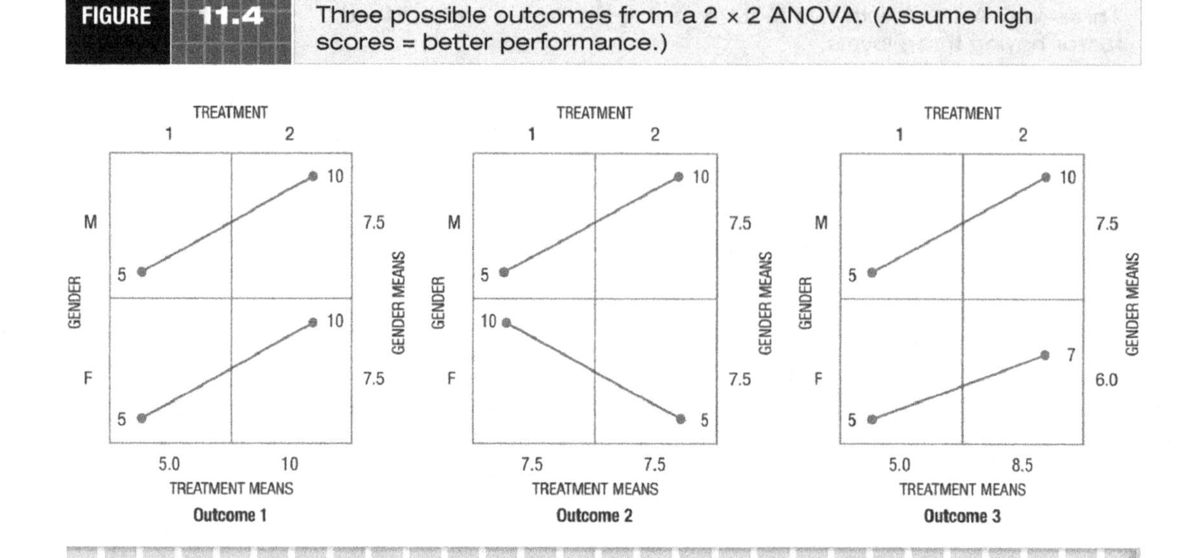

FIGURE 11.4 Three possible outcomes from a 2 × 2 ANOVA. (Assume high scores = better performance.)

for males and females is consistent across the levels of the treatments. In this case, the null hypothesis for interaction would be retained; that is, no interaction is present.

In outcome 2, however, it appears that treatment 2 is superior to treatment 1 for males, but treatment 1 is superior to treatment 2 for females. In this case, when we test the null hypothesis for interaction, the F-ratio would likely indicate that we should reject the null hypothesis. Notice that for each of the main effects in outcome 2, we would retain the null hypothesis (that is, the means are the same for the genders and for the treatments). However, the treatment means are different for each gender, suggesting a possibly significant interaction effect. Examination of the main effects alone leads us to conclude that no difference exists between the treatments or between the genders; however, the significant interaction reflects a significant difference in how the treatments affected the two genders.

In outcome 3, it appears that treatment 2 is superior to treatment 1 for both genders but is not as superior for females as it is for males. To determine whether the pattern of means for males and females is significantly different across the two treatments, we turn to the statistical test for interaction. Just as the main effect F-ratio for the difference between the means of 7.5 (males) and 6.0 (females) indicates whether this difference is statistically significant, the F-ratio for interaction indicates whether the interaction is significant.

Another interpretation of the interaction F-test is that it is determining whether the lines in the figures above are parallel. If they are parallel (or nearly so), we would retain the interaction null hypothesis. If they are not, we would reject the null hypothesis for interaction. As discussed previously, we state the null hypothesis for the interaction as follows: "The patterns of the means for factor A is consistent across the levels of factor B". The decision whether to reject or accept the null hypothesis for interaction will help us decide about the generalizability of the main effects (factors).

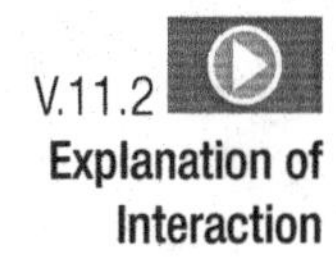

Explanation of Interaction

In essence, with a two-way (or higher) ANOVA, we are partitioning the total variance into the variance attributable to the main effects (factors), the variance attributable to the interaction, and the remaining error (residual) variance. Thus, we are truly "analyzing variance", just as the name Analysis of Variance suggests.

INTERACTION VERSUS GENERALIZABILITY

When interactions are not present in the results of a study, those results can be more readily generalized. For example, in outcome 1 in Figure 11.4, the investigator could generalize by stating that treatment 2 is superior to treatment 1 regardless of the subject's gender. This is a parsimonious finding.

When interactions are present, the results can be more cumbersome to report (e.g., treatment 2 is superior to treatment 1 in some situations, but in other situations this is not the case). However, significant interactions are often very enlightening. For example, outcome 2 in Figure 11.4 reveals that it is necessary to consider a subject's gender when deciding which treatment to apply. Had the researcher not combined these two factors in the experiment, that finding would not have been discovered. In fact, in the field of human performance, examining interactions is the focus of many investigations.

ASSUMPTIONS

The assumptions that were presented in Chapters 9 and 10 for one-way ANOVAs apply to factorial ANOVAs, as well. We will limit the discussion in this chapter to designs with equal numbers of subjects in each cell. This is not a requirement of factorial ANOVA, but it will reduce the complexity of our presentation. Most software programs provide various methods for handling situations involving unequal *n*s. We will not go into detail about these modifications here. However, you should be aware that all of them reduce the power of these designs. All of these more sophisticated designs are based on the same reasoning and logic as the hypothesis testing we have already discussed.

PRELIMINARY INFORMATION

Prior to a detailed examination of the statistical analysis of factorial designs, let's explore how ANOVA works in a more general way. We will focus on a few concepts related to mean squares and their expected values.

The final step in any ANOVA is to calculate an F-ratio and compare the obtained value to an appropriate critical F-value in Appendix A.4. To review, the F-ratio is a ratio between two mean squares, and whenever you hear the term "mean square", you should think "variance estimate". When the null hypothesis is true, a mean square is, in fact, an estimate of the variance (σ^2) of the population from which the sample was selected.

In the one-way between-groups ANOVA presented in Chapter 9, we estimated the population variance in two different ways—in one method we used the differences between the group means (MS_B), and in the other we used the differences between the subjects' scores and their own group mean (MS_W). In this situation, if the null hypothesis is true, the only possible cause for any difference in the group means is chance, and chance is also the only possible cause for any difference between each subject's score and the subject's group mean.

When the null hypothesis is true, the expected value of both of these mean squares is the population variance, σ^2. In this case, the *F*-ratio that results from dividing MS_B by MS_W is expected to be approximately 1.0. In this design, the expected value of MS_W is σ^2 no matter the status of the null hypothesis, because it is purely a function of how subjects differ within their groups. It is not affected by how the groups might differ from each other. However, as the difference between the groups increases (as reflected by larger group mean differences), the null hypothesis becomes increasingly less likely to be true. Correspondingly, the expected value of MS_B changes. It becomes σ^2 *modified by* a value, σ_B^2, that increases in size as the group means get further apart. Therefore, in the one-way between-groups ANOVA, we have three possible outcomes for a decision, based on the value of the *F*-ratio.

First, if group means are identical, the *F*-ratio is expected to be unity:

$$F\text{-}ratio = \frac{MS_B}{MS_W} \approx \frac{\sigma_{est}^2}{\sigma_{est}^2} \approx 1.0$$

The only reason the *F*-ratio may not be exactly equal to 1.0 is that the two mean squares are both *estimates* and, therefore, might not exactly equal each other. Of course, if the group means were identical, there would be no reason to calculate the *F*-ratio to determine whether they are different.

A second possibility is that the group means are not exactly equal but, in fact, differ only by chance. In this case,

$$F - ratio = \frac{MS_B}{MS_W} \approx \frac{\sigma_{est}^2 + \sigma_B^2}{\sigma_{est}^2} \text{ is somewhat} > 1.0$$

The *F*-ratio will be some value larger than 1.0 but not greater than the appropriate critical value from Appendix A.4. The effect of the independent variable (designated by σ_B^2 in the numerator of the *F*-ratio) is explained by the chance differences among the means. In this situation, the investigator would retain the null hypothesis.

The third possibility for the *F*-ratio in the one-way between-groups ANOVA is:

$$F - ratio = \frac{MS_B}{MS_W} \approx \frac{\sigma_{est}^2 + \sigma_B^2}{\sigma_{est}^2} \text{ is much} > 1.0$$

In this case, the *F*-ratio is larger than the appropriate critical value because the contribution of σ_B^2 increases the numerator to the point that the *F*-ratio exceeds the critical value from Appendix A.4. When this occurs, the researcher rejects the null hypothesis and declares that the pattern of means is unlikely to be the result of chance (remembering that the possibility of being incorrect in making this assertion—a Type I error—is equal to the selected α level).

In the one-way between-groups ANOVA, the total sum of squares is partitioned into only two parts, SS_B and SS_W. Thus, only one possible F-ratio can be calculated. In that case, the mean square between is placed in the numerator and the mean square within becomes the denominator, so that the F-ratio can only become larger than 1.0 as the null hypothesis becomes less and less tenable. The denominator of the F-ratio is often called the error term, and, as you might recall, we can call it the mean square for error to indicate this fact.

As the designs we will present become increasingly complex, the number of parts that the total sum of squares will be partitioned into will increase. Thus, we will have more than just two sums of squares, and we will calculate several F-ratios. The decision as to which mean square should be placed in the numerator and which should become the denominator has to do with our ability to determine what makes up their expected values.

We explain the partitioning of the sum of squares into parts in some detail for the designs in this chapter to help you see how the concept works. Although we will rely on computer software to calculate the F-ratios, we will need to be able to indicate certain features of the factors involved before we conduct the calculations. For example, we have to identify whether a factor is a between or a within factor, and we have to indicate whether a factor is a fixed or a random factor.

Although this is certainly not always the case, between factors tend most often to be fixed factors, and within factors are equally likely to be fixed or random factors. By way of review, between factors have unique subjects in each level, whereas, for within factors, each level is populated by the same (or matched) subjects. For fixed factors, the researcher selects the particular levels of the independent variables for which he or she desires information. Examples might be three particular methods of teaching a physical skill or five specific exercises designed to increase flexibility. Random factors are those where the levels of the independent variable(s) are chosen at random from a large number of possibilities. Examples might include altitude or age. Subjects are routinely treated as a random factor. The advantage of random factors is that the results can be applied over a wide range of possibilities, whereas the results for fixed factors apply only to those levels selected for study.

POSSIBLE COMBINATIONS OF FACTORS

In Chapters 9 and 10 we introduced the concept of between and within factors, and we noted that the location of subjects and the statistical analyses differed for between-groups ANOVA and repeated-measures (within-groups) ANOVA. In a two-way ANOVA, three combinations of these two types of independent variables (factors) are possible. These are between-between, between-within, and within-within.

If you examine the examples that we have presented thus far in this chapter, you will find that they are all between-between designs. With a between-between design, it happens that there is a unique set of subjects in every cell of the design. In Figure 11.1, for example, the four cells consist of males receiving treatment 1, different males receiving treatment 2, females receiving treatment 1, and different females receiving treatment 2. See if you can determine the unique characteristics of the subjects in each of the 12 cells of Figure 11.3.

An example of a between-within design might be to repeatedly measure subjects involved in different training regimens (between factor) over several different time periods (within factor), such as one day, one month, and six months after a treatment. The dependent variable might be grip strength. The researcher is interested in examining the change, if any, in grip strength as a result of the different training regimens and at various time points across the six-month period.

An example of a within-within design might be to measure the same subjects at different times as they perform a task under some number of different environmental conditions. For example, a researcher might test a group of subjects in a physical education class to see how many free throws they can make before and after a basketball unit, under the two conditions of shooting in silence or with a recording of crowd noise playing. Notice that in this combination of factors the same subjects appear in every cell of the design.

For all three possible combinations of factors (as long as there is more than one subject in every cell and all factors are fixed), the two-way ANOVA will *always* result in three F-ratios: two main effects (one for each factor) and one for the interaction. However, as we will demonstrate next, the total sum of squares will be partitioned into a varying number of parts, depending on the combination of between and within factors.

To illustrate how the different types of factors (between vs. within) affect the two-way ANOVA results, we will use the same data set under the three possible combinations of between-between, between-within, and within-within. The experiment involved is a 2 x 3 two-way ANOVA, with one factor having two levels of training method (cycle vs. treadmill). We will identify this factor as Method *(M)* and the other factor having three levels of time of measurement (pre, immediately post, and five minutes post). We will identify this factor as Time *(T)*. The dependent variable is systolic blood pressure. Figure 11.5 illustrates the generic data organization.

FIGURE 11.5 Generic data organization.

		TIME		
		Pre	Immediate Post	5 min Post
METHOD	Cycle			
	Treadmill			

Between-Between Fixed-Factors Two-Way ANOVA

Figure 11.6 shows how subjects are placed in the cells and the degrees of freedom, **expected mean squares** (EMS), which error term is used in the calculation of the three *F*-ratios, and the three *F*-ratios. Important aspects of this between-between model include:

1. The total number of participants is 30, with 5 in each of the 6 cells. This results in a total of 30 measures taken.
2. Participants are "nested" within each cell of the 2 x 3 design. That is, each particular participant appears in only one cell of the design.
3. The *F*-ratio is the mean square for the effect to be tested, divided by the error term that includes all of the components *except* the source of variation that is being tested. In the ANOVA source table in Figure 11.6, the EMS for Method is the effect of factor *M* plus error *(E)*. If the null hypothesis is true, then *M* is equal to zero (because there is no Method effect) and the expected *F*-ratio is *E/E* = 1.0, as described above. Similar logic applies to the development of the EMS for the other main effect *(Time)* and the interaction (*M* x *T*).

Statistical Analysis

Tables 11.1 and 11.2 give the SPSS ANOVA between-between output, with descriptive statistics in Table 11.1 and the ANOVA source table in Table 11.2. Tables 11.1R and 11.2R give the R ANOVA between-between output, with descriptive statistics in Table 11.1R and the ANOVA source table in Table 11.2R.

Between-between model and source table. FIGURE 11.6

METHOD		TIME			
		Pre	Immediate Post	5 min Post	*N*= 30
	Cycle	S_1 S_5	S_{11} S_{15}	S_{21} S_{25}	
	Treadmill	S_6 S_{10}	S_{16} S_{20}	S_{26} S_{30}	

Source of Variance	*df*	EMS	Error Term	F-ratio
Method	1	M + s(M x T)	s(M x T)	(M + s(M x T))/s(M x T)
Time	2	T + s(M x T)	s(M x T)	(T + s(M x T))/s(M x T)
Method x Time	2	M x T + s(M x T)	s(M x T)	(M x T + s(M x T))/s(M x T)
s (Method x Time)	24	s(M x T)		
Total	29			

M = Method effect; T = Time effect; M x T = Method by Time interaction; s(M x T) indicates subjects are "nested" within Method and Time cells.

TABLE 11.1 SPSS mean values output for between-between ANOVA model.

Dependent Variable: Systolic Blood Pressure				
			95% Confidence Interval	
Training Method	Mean	Std. Error	Lower Bound	Upper Bound
Cycle	122.867	2.035	118.667	127.067
Treadmill	128.467	2.035	124.267	132.667

Dependent Variable: Systolic Blood Pressure				
			95% Confidence Interval	
Time	Mean	Std. Error	Lower Bound	Upper Bound
Pre BP	119.700	2.492	114.556	124.844
Immediate Post BP	136.800	2.492	131.656	141.944
5-Minute Post BP	120.500	2.492	115.356	125.644

Dependent Variable: Systolic Blood Pressure					
				95% Confidence Interval	
Training Method	Time	Mean	Std. Error	Lower Bound	Upper Bound
I Cycle	Pre BP	119.000	3.525	111.725	126.275
	Immediate Post BP	132.000	3.525	124.725	139.275
	5-Minute Post BP	117.600	3.525	110.325	124.875
I Treadmill	Pre BP	120.400	3.525	113.125	127.675
	Immediate Post BP	141.600	3.525	134.325	148.875
	5-Minute Post BP	123.400	3.525	116.125	130.675

TABLE 11.1R RStudio mean values output for between-between ANOVA model.

Treadmill	Cycle
128.4667	122.8667

Pre BP	Immediate Post BP	5-minute Post BP
119.7	136.8	120.5

	Pre BP	Immediate Post	BP 5-minute Post BP
Treadmill	120.4	141.6	123.4
Cycle	119.0	132.0	117.6

TABLE 11.2 SPSS ANOVA source table for between-between fixed-factors two-way design.

Dependent Variable: Systolic Blood Pressure					
Source	Type III Sum of Squares	df	Mean Square	F	Sig.
Corrected Model	2181.867[a]	5	436.373	7.025	.000
Intercept	473763.333	1	473763.333	7626.992	.000
Training Method	235.200	1	235.200	3.786	.063
Time	1862.467	2	931.233	14.992	.000
Training Method * Time	84.200	2	42.100	.678	.517
Error	1490.800	24	62.117		
Total	477436.000	30			
Corrected Total	3672.667	29			

TABLE 11.2R R ANOVA source table for between-between fixed-factors two-way design.

Anova Table

Response: BP

	Sum Sq	Df	F value	Pr(>F)
(Intercept)	473763	1	7626.9922	<2.2e-16***
Training Method	235	1	3.7864	0.06347
Time	1862	2	14.9917	5.962e-05***
Training Method:Time	84	2	0.6778	0.51721
Residuals	1491	24		

Signif. codes: 0*** 0.001** 0.01* 0.05 0.1 1

The output from the SPSS program is very complete and contains some information that is not directly relevant to our goal of testing the hypothesis. For example, in Table 11.2 the intercept information simply indicates whether or not the overall mean involved is significantly different from zero (which, of course, it is). The corrected model line shows how the sums of squares and 5 degrees of freedom associated with the two factors and interaction have been partitioned. Notice that the sums of squares for Training Method, Time, and Training Method x Time add up to 2181.867, and the Corrected Total includes the sum of squares for Error and the Corrected Model Total (i.e., 1490.8 + 2181.867 = 3672.667). The difference between Total and Intercept is also 3672.667. The RStudio output in Table 11.2R presents information similar to that provided with SPSS but not in the same format. You should confirm that both Tables 11.2 and 11.2R contain identical relevant information (i.e., the F-ratios and statistical decisions made in both analyses are identical).

Our concern here is how the total sum of squares in a between-between fixed-factors ANOVA design is partitioned into four parts. In this illustration, the four parts are: (1) Training Method (factor *M*), (2) Time (factor *T*), (3) Training Method by Time (or interaction; *M* x *T*), and (4) error. The statistical analyses of interest are three *F*-ratios. The first *F*-ratio, the main effect for *A*, leads to a conclusion about the difference between the means of the two Training Methods (factor *A*): 122.87 and 128.47 in our example. The second *F*-ratio is available to examine the other main effect (factor *B*), which involves the differences among the means of the three levels of the Time factor: 119.7, 136.8, and 120.5. The third *F*-ratio allows us to look at whether or not a significant interaction exists between factors *M* and *T*. The interaction *F*-test determines whether the differences between the two Training Methods are the same across the Time factor levels: pre, immediate post, and five minutes post. Of course, none of the means are identical. ANOVA helps us determine whether these mean values are randomly equivalent or are disparate enough for us to conclude that the differences are not attributable to chance (we would, thus, reject the null hypothesis for that effect).

Total Sum of Squares (SS_T)

As always, we can obtain the value of SS_T—called the Corrected Total in SPSS—by subtracting the overall grand mean of the systolic blood pressure scores (125.667) from each of the 30 systolic blood pressure scores, squaring the resulting differences, and summing these squares. The software will do this for us. The degrees of freedom associated with SS_T is $N - 1$; hence, for our example, the degrees of freedom are $30 - 1 = 29$.

Sum of Squares for Training Method (SS_M)

The sum of squares for Training Method is a function of the number of subjects in each cell (5), the number of levels of the Time factor (3), and a calculation involving the differences between the means for the levels of the *M* factor and the grand mean (125.667). The value of SS_M represents the portion of the variability in the data that is attributable to the differences in the means for the levels of Training Method. In our example, these means are 122.867 and 128.467, and $SS_M = 235.200$. The degrees of freedom for the sum of squares for Training Method are J–1, where J is equal to the number of levels for this factor. In our example, this value is $2 - 1 = 1$.

Sum of Squares for Time (SS_{Time})

The sum of squares for Time is a function of the number of subjects in each cell (5), the number of levels of the Training Method factor (2), and a calculation involving the differences between the means of the levels of the

Time factor and the grand mean (125.667). The value of SS_{Time} represents the portion of the variability in the system that is attributable to differences in the mean systolic blood pressures at the three different levels of Time. In our example, these means are 119.7, 136.8, and 120.5, and $SS_{Time} = 1862.467$. The degrees of freedom for the sum of squares for Time are $K - 1$, where K is equal to the number of levels of the Time factor. In our example, this value is $3 - 1 = 2$.

Sum of Squares for Interaction (S_{MxT})

The sum of squares for interaction is a function of the number of subjects in each cell (5) and a calculation involving the differences between the grand mean and the cell means after they have been adjusted to remove the effects of the two main effect factors (Method and Time). In our example, $SS_{MxT} = 84.200$. The sum of squares for interaction represents the portion of the variability in the system that is attributable to any remaining cell differences that result from a combination of the two factors. The degrees of freedom for the sum of squares for interaction are $(J - 1)(K - 1)$. For our illustration, this value is $(2 - 1)(3 - 1) = 2$.

Sum of Squares for Error (SS_E)

The sum of squares for Error is a function of two values: the difference between each individual score and the mean of the specific cell in which the individual score resides, and the number of cells in the design. The sum of squares for Error—also called the error variance—represents the remaining variability in the data set that is not attributable to the two main effect factors (M and T) or to their interaction (M x T). It is the unexplained variability. The degrees of freedom for the sum of squares for Error are $JK(n - 1)$, where n is the number of subjects in each cell. In our example, this value is $(2)(3)(5 - 1) = 24$.

Degrees of Freedom

Notice that, just as we have partitioned the total sum of squares into four parts, we have also partitioned the total degrees of freedom into four parts. Table 11.3 summarizes the partitions and the results for our example.

Degrees of freedom in a between-between two-way ANOVA. TABLE 11.3

SOURCE OF VARIANCE	*SS*	*df*	OUR EXAMPLE
Training Method	SS_M	J – 1	2 – 1 = 1
Time	SS_T	K – 1	3 – 1 = 2
Interaction	SS_{MxT}	(J – 1)(K – 1)	1 x 2 = 2
Error	SS_E	JK(n – 1)	6 x 4 = 24
Total	SS_T	N – 1	30 – 1 = 29

Mean Squares and Expected Values

To obtain the mean squares required for the *F*-ratios, we divide the appropriate sum of squares by its degrees of freedom. Again, when you hear the term "mean square", you should think "variance estimate". In the between-between two-way design, MS_M, MS_T, MS_{MxT}, and MS_E are all estimates of the population variance, σ^2, *when the null hypotheses are true*. In other words, all of the mean squares estimate the same parameter, σ^2, when the H_0s are true.

In our example, the *F*-ratio that we use to determine whether the means for the two levels of the Training Method factor are or are not equal is *MSM/MSE*. For the three means of the time factor, it is MS_T/MS_E. For interaction, it is MS_{MxT}/MS_E. If any of the null hypotheses are true, for the Training Method factor, the Time factor, and the interaction factor, we would expect their respective *F*-ratios to be approximately equal to 1.0.

However, if any of the three null hypotheses are false, the numerator of the respective *F*-ratio not only estimates Σ^2 (i.e., *E*) but also includes an additional value, which becomes greater as the null hypothesis becomes less and less tenable. When the numerator becomes so large that the obtained *F*-ratio exceeds the critical *F*-value found in Appendix A.4, we reject the null hypothesis and declare that the pattern of means is unlikely to be attributable to chance. The probability of our being wrong in this assertion is equal to the α level that we selected.

We repeat Table 11.2 and Table 11.2-R above, for convenience.

Results and Multiple Comparisons

The critical value from Appendix A.4 for the Training Method factor *(M)* at $\alpha = .05$ and $df = 1, 24$ is 4.26. Because the obtained *F*-ratio is 3.786, which is less than the critical value, we conclude that the mean of 122.867 for the subjects training on the cycle is not significantly different from the mean

TABLE 11.2 SPSS ANOVA source table for between-between fixed-factors two-way design (repeated).

Dependent Variable: Systolic Blood Pressure					
Source	Type III Sum of Squares	df	Mean Square	F	Sig.
Corrected Model	2181.867[a]	5	436.373	7.025	.000
Intercept	473763.333	1	473763.333	7626.992	.000
Training Method	235.200	1	235.200	3.786	.063
Time	1862.467	2	931.233	14.992	.000
Training Method * Time	84.200	2	42.100	.678	.517
Error	1490.800	24	62.117		
Total	477436.000	30			
Corrected Total	3672.667	29			

R ANOVA source table for between-between fixed-factors two-way design (repeated). **TABLE 11.2R**

Anova Table

Response: BP

	Sum Sq	Df	F value	Pr(>F)
(Intercept)	473763	1	7626.9922	<2.2e-16***
Training Method	235	1	3.7864	0.06347
Time	1862	2	14.9917	5.962e-05***
Training Method:Time	84	2	0.6778	0.51721
Residuals	1491	24		

Signif. codes: 0*** 0.001** 0.01* 0.05 0.1 1

of 128.467 for the subjects training on the treadmill. Both the SPSS and RStudio outputs indicate that the means of the two Training Methods do not differ significantly at $\alpha = .05$.

V.11.3 **SPSS Factorial ANOVA and Multiple Comparison**

The obtained *F*-ratio for the Time factor *(T)* of 14.992 is larger than the critical value of 3.40 (from Appendix A.4), so we reject this null hypothesis at $\alpha = .05$ and $df = 2, 24$. In this situation, we must use a multiple comparison technique to determine which of the three Time level means (119.7, 136.8, and 120.5) is or are significantly different from the others. Alternatively, the researcher might choose to conduct a trend analysis across the Time factor means.

V.11.3R **RStudio Factorial ANOVA and Multiple Comparison**

For practice, use the SNK to verify that the mean of 136.8 is significantly different from 120.5 and 119.7 but the means of 120.5 and 119.7 are not significantly different from each other. *Hint:* When calculating the SNK (see Chapter 9), set *n* equal to the number of subjects involved in obtaining the means being compared. In this case, the number is 10.

The final *F*-ratio of 0.678 indicates that we should retain the null hypothesis for interaction. A graph of the cell means reveals that the pattern of means is nearly identical across time levels for both of the two training methods. See Figure 11.7.

The finding of no significant interaction allows us to generalize the findings. We can state that the systolic blood pressure pattern is the same for subjects training on a cycle as that for subjects training on the treadmill when measured before, immediately post, and five minutes post exercise. However, significant differences exist in the recorded systolic blood pressures across the time periods.

Between-Within Fixed-Factors Two-Way ANOVA

Next, we will analyze the same data set but treat it as a between-within 2 x 3 two-way ANOVA, with the Training Method being the between variable and

FIGURE 11.7 Graph of interaction between training methods and time.

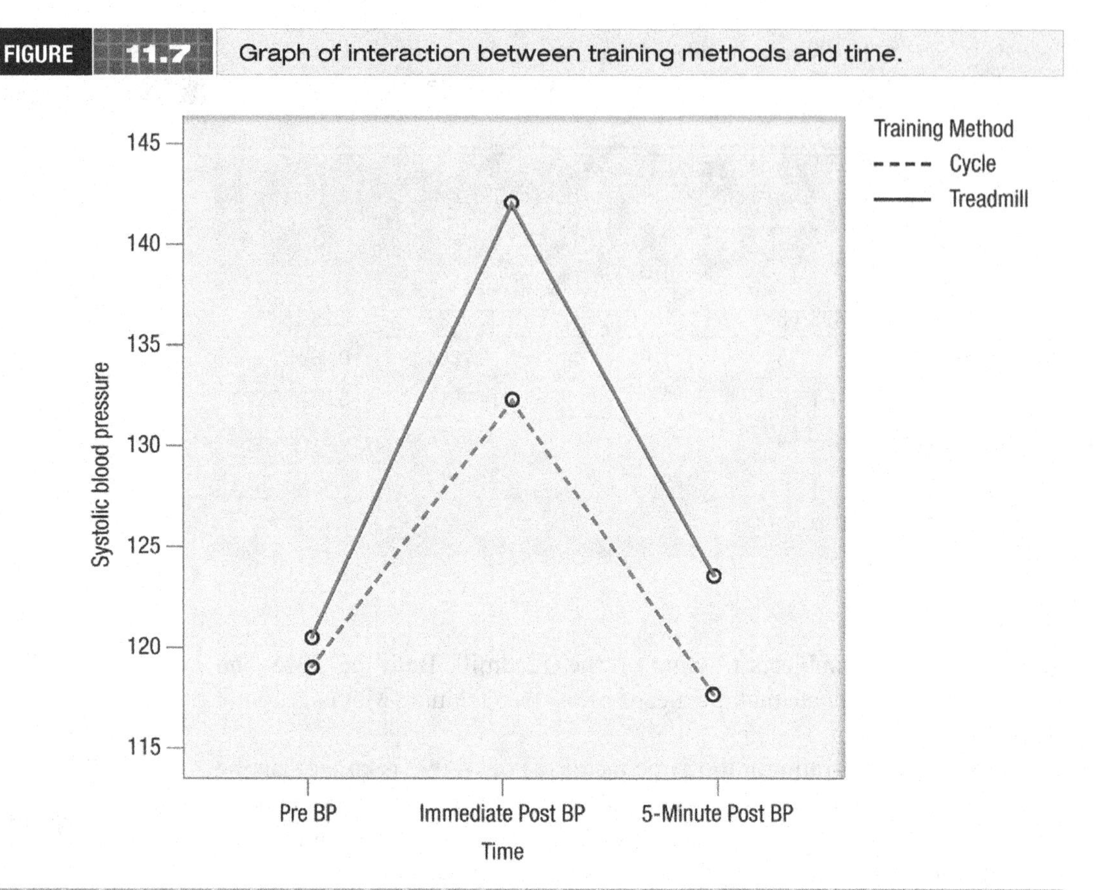

the Time measurements being the within factor. For the experiment that is serving as our example, this would actually be a better design than the previously illustrated between-between design, for two reasons. First, it makes more economical use of the subjects; second, it is a more powerful design for the Time factor, which has changed from a between factor to a within factor. See Figure 11.8 for details of the between-within design.

Notice the following in the between-within model:

1. The total number of participants is 10, with 5 in each training method. The total number of measures taken remains 30.
2. Participants within each Training Method "cross" (i.e., appear) in *each* level of Time.
3. The source table provides expected mean squares, along with the error term in each case. Contrast the values of the expected mean squares and error terms with those of the between-between design presented in Figure 11.6. Note that the degrees of freedom for the various F-ratios are substantially different than those in the between-between model.
4. In this model, we use two different error terms (not one, as in Figure 11.6) to obtain the F-ratios. One is for Training Method, and the other is for Time and the Training Method by Time interaction.

Between-within design and source table. FIGURE 11.8

METHOD		TIME		
		Pre	Immediate Post	5 min Post
	Cycle	S_1 S_5	S_1 S_5	S_1 S_5
	Treadmill	S_6 S_{10}	S_6 S_{10}	S_6 S_{10}

N= 10

Source of Variance	*df*	EMS	Error Term	*F*-ratio
Method	1	M + s(M)	s(M)	(M + s(M))/s(M)
s(Method)	8	s(M)		
Time	2	T + s(M) x T	s(M) x T	(T + s(M) x T)/(s(M) x T))
Method x Time	2	M x T + s(M) x T	s(M) x T	(M x T + s(M) x T)/(s(M) x T)
s(Method) x Time	16	s(M) x T		
Total	29			

M = Method effect; T = Time effect; s(M) indicates subjects are "nested" within Method only but then cross Time.

SPSS Output for Between-Within Design

Table 11.4 gives the SPSS between-within ANOVA mean values output. (Compare Table 11.4 with Table 11.1 and Table 11.1-R.) These values are identical to those obtained with the between-between analysis, because the data are the same, but differences occur in the ANOVA source table results, presented in Table 11.5 and Table 11.5R, because the experimental model is different.

SPSS presents the results of this model in two different source tables. The first (Table 11.5) is for the between-groups factor (Training Method in our example). The obtained *F*-ratio of 1.655 for training method indicates that no significant difference exists between the means of 128.467 and 122.867. The RStudio results presented in Table 11.5R contains both the between-groups and within-groups results.

You might recall that when we have within factors, we must check the assumption of sphericity (see Chapter 10). The sphericity adjustments are presented in Table 11.6 and Table 11.6R. Recall that the obtained *F*-ratio (42.09, in this case) does not change, but the critical *F*-value to which we compare it changes. The results show that when we apply any of the possible sphericity correction techniques, we find at least one significant difference among the three Time means of 136.8, 120.5, and 119.7. Notice that the *F*-ratio of 42.1 for Time is much larger than the critical *F*-ratio of 14.9 that we found when we treated the time variable as a between variable (see Table 11.2). This demonstrates the increased power of the within factor in the between-within design.

TABLE 11.4 SPSS mean values output for between-within ANOVA model.

Dependent Variable: Systolic Blood Pressure				
			95% Confidence Interval	
Training Method	Mean	Std. Error	Lower Bound	Upper Bound
Cycle	122.867	2.035	118.667	127.067
Treadmill	128.467	2.035	124.267	132.667

Dependent Variable: Systolic Blood Pressure				
			95% Confidence Interval	
Time	Mean	Std. Error	Lower Bound	Upper Bound
Pre BP	119.700	2.492	114.556	124.844
Immediate Post BP	136.800	2.492	131.656	141.944
5-Minute Post BP	120.500	2.492	115.356	125.644

Dependent Variable: Systolic Blood Pressure					
				95% Confidence Interval	
Training Method	Time	Mean	Std. Error	Lower Bound	Upper Bound
I Cycle	Pre BP	119.000	3.525	111.725	126.275
	Immediate Post BP	132.000	3.525	124.725	139.275
	5-Minute Post BP	117.600	3.525	110.325	124.875
I Treadmill	Pre BP	120.400	3.525	113.125	127.675
	Immediate Post BP	141.600	3.525	134.325	148.875
	5-Minute Post BP	123.400	3.525	116.125	130.675

TABLE 11.5 SPSS ANOVA source table for between-groups factor.

Source	Type III Sum of Squares	df	Mean Square	F	Sig.
Intercept	473763.333	1	473763.333	3334.014	.000
Training Method	235.200	1	235.200	1.655	.234
Error	1136.800	8	142.100		

As previously, we find the interaction not to be significant, even with the increase in power to detect it. Notice, also, that because the interaction F-ratio uses the error term associated with the within factor, the software checks for sphericity here, as well.

R ANOVA source table. TABLE 11.5R

$ANOVA								
	Effect	DFn	DFd	SSn	SSd	F	p p<.05	ges
1	(Intercept)	1	8	473763.3	1136.8	3334.0	8.987e-12*	
2	Training Method	1	8	235.2	1136.8	1.66	2.342e-01	0.136
3	Time	2	16	1862.5	354.0	42.09	4.234e-07*	
4	Training Method:Time	2	16	84.2	354.0	1.90	1.814e-01	0.053

SPSS ANOVA source table for within-groups factor. TABLE 11.6

Tests of Within-Subjects Effects						
Measure: MEASURE_1						
Source		Type III Sum of Squares	df	Mean Square	F	Sig.
Time	Sphericity Assumed	1862.467	2	931.233	42.090	.000
	Greenhouse-Geisser	1862.467	1.822	1022.180	42.090	.000
	Huynh-Feldt	1862.467	2.000	931.233	42.090	.000
	Lower bound	1862.467	1.000	1862.467	42.090	.000
Time * Training Method	Sphericity Assumed	84.200	2	42.100	1.903	.181
	Greenhouse-Geisser	84.200	1.822	46.212	1.903	.186
	Huynh-Feldt	84.200	2.000	42.100	1.903	.181
	Lower bound	84.200	1.000	84.200	1.903	.205
Error(Time)	Sphericity Assumed	354.000	16	22.125		
	Greenhouse-Geisser	354.000	14.576	24.286		
	Huynh-Feldt	354.000	16.000	22.125		
	Lower bound	354.000	8.000	44.250		

RStudio tests for sphericity. TABLE 11.6R

$"Mauchly's Test for Sphericity"					
	Effect	W	p p<.05		
3	Time	0.902	0.698		
4	Training Method:Time	0.902	0.698		
$"Sphericity Corrections"					
	Effect	GGe	p[GG] p[GG]<.05	HFe	p[HF] p[HF]<.05
3	Time	0.911	1.245e-06*	1.165	4.234e-07*
4	Training Method:Time	0.911	1.861e-01	1.165	1.814e-01

Within-Within Fixed-Factors Two-Way ANOVA

Finally, we will analyze the same data set one more time, this time assuming that both of the independent variables (Training Method and Time) are within factors. The dependent variable remains systolic blood pressure. Although this design provides the most powerful way to analyze the data, it is a less desirable design, because each subject has to experience both Training Methods. This would present problems in interpreting the results, because the order in which the subjects experience the two protocols might affect the results.

Figure 11.9 gives the details of the design, including placement of subjects in the cells, degrees of freedom, expected mean squares, and the *F*-ratios for a 2 x 3 within-within two-way ANOVA.

In the within-within model, notice:

1. The total number of subjects is now only 5.
2. *Each* participant is measured in *each* Training Method and at *each* Time. The total number of measures taken remains 30.
3. When we contrast the expected mean squares and error terms with the results in the previous two models, we find, again, that the denominators of the *F*-ratios contain all of the elements of the numerators, *except* for the effect that is being tested. Note that the degrees of freedom for the various *F*-ratios are substantially different from those in the between-between and the between-within models.

FIGURE 11.9 Within-within model and source table.

METHOD	TIME: Pre	TIME: Immediate Post	TIME: 5 min Post	
Cycle	S_1 … S_5	S_1 … S_5	S_1 … S_5	*N*= 5
Treadmill	S_1 … S_5	S_1 … S_5	S_1 … S_5	

Source of Variance	*df*	EMS	Error Term	*F*-ratio
Method	1	M + M x S	M x S	(M + M x S)/(M x S)
Time	2	T + T x S	T x S	(T + T x S)/(T x S)
Subject	4	S		
Method x Time	2	M x T + M x T x S	M x T x S	(M x T + M x T x S)/(M x T x S)
Method x Subject	4	M x S		
Time x Subject	8	T x S		
Method x Time x Subject	8	M x T x S		
Total	29			

M = Method effect; T= Time effect; S = subjects. Note all factors cross all other factors. Subjects in this design are considered a factor, thus represented by capital S.

4. With the within-within model design, we are using three different error terms to obtain the F-ratios—a different error term for each F-ratio.

We will not repeat the tables for descriptive statistics here; these statistics remain identical to the other two designs, since we are working with the same data.

SPSS Output for Within-Within Design

Tables 11.7 and 11.7R give the SPSS and RStudio outputs for the within-within design. Note that we no longer have any output for a between-groups factor.

SPSS output for within-within design. TABLE 11.7

Tests of Within-Subjects Effects

Measure: Blood Pressure

Source		Type III Sum of Squares	df	Mean Square	F	Sig.
Method	Sphericity Assumed	235.200	1	235.200	2.654	.179
	Greenhouse-Geisser	235.200	1.000	235.200	2.654	.179
	Huynh-Feldt	235.200	1.000	235.200	2.654	.179
	Lower bound	235.200	1.000	235.200	2.654	.179
Error(Method)	Sphericity Assumed	354.467	4	88.617		
	Greenhouse-Geisser	354.467	4.000	88.617		
	Huynh-Feldt	354.467	4.000	88.617		
	Lower bound	354.467	4.000	88.617		
Time	Sphericity Assumed	1862.467	2	931.233	40.518	.000
	Greenhouse-Geisser	1862.467	1.311	1421.061	40.518	.001
	Huynh-Feldt	1862.467	1.693	1100.108	40.518	.000
	Lower bound	1862.467	1.000	1862.467	40.518	.003
Error(Time)	Sphericity Assumed	183.867	8	22.983		
	Greenhouse-Geisser	183.867	5.242	35.073		
	Huynh-Feldt	183.867	6.772	27.151		
	Lower bound	183.867	4.000	45.967		
Method* Time	Sphericity Assumed	84.200	2	42.100	1.980	.200
	Greenhouse-Geisser	84.200	1.257	67.004	1.980	.224
	Huynh-Feldt	84.200	1.561	53.930	1.980	.214
	Lower bound	84.200	1.000	84.200	1.980	.232
Error(Method*Time)	Sphericity Assumed	170.133	8	21.267		
	Greenhouse-Geisser	170.133	5.027	33.847		
	Huynh-Feldt	170.133	6.245	27.243		
	Lower bound	170.133	4.000	42.533		

TABLE 11.7R RStudio output for within-within design.

$ANOVA

	Effect	DFn	DFd	SSn	SSd	F	p p<.05	ges
1	(Intercept)	1	4	473763.3	782.3	2422.31	1.019e-06*	0.997
2	Training Method	1	4	235.2	354.5	2.65	1.786e-01	0.136
3	Time	2	8	1862.5	183.9	40.52	6.517e-05*	0.555
4	Training Method:Time	2	8	84.2	170.1	1.98	2.002e-01	0.053

$"Mauchly's Test for Sphericity"

	Effect	W	p p<.05
3	Time	0.474	0.326
4	Training Method:Time	0.408	0.261

$"Sphericity Corrections"

	Effect	GGe	p[GG] p[GG]<.05	HFe	p[HF] p[HF]<.05
3	Time	0.655	0.0009*	0.846	0.0002*
4	Training Method:Time	0.628	0.2243	0.781	0.2144

Because both factors are within factors, SPSS and RStudio provide possible adjustments for sphericity. Even with this most powerful design, we obtain basically the same results as before. The means for the two Training Methods are still not significantly different, although the significance value in this design is the smallest of all three designs, because of the increased power. The *F*-ratio for the Time factor is significant, and the *F*-ratio for interaction is still not significant.

Two-Way Between and Within ANOVA Model Summary

We have illustrated the expansion of the one-way ANOVA into a two-way ANOVA. The logic remains the same: we partition the total variance into sources associated with different effects (or factors) in the model. The concepts of the null and alternative hypotheses remain the same. Based on the obtained data and the computed *F*-ratios, we make decisions about rejecting or retaining each null hypothesis. An important advantage of the two-way ANOVA is that it permits us to test both main effects and interaction effects. Theoretically, we can include any number of independent variables in the model, and each can have any number of levels.

Fixed Versus Random Factors

V.11.4
Between, Within, Fixed and Random Factors

The three designs presented above varied in their combinations of between and within factors, but in all cases the independent variables (factors; in our example Training Method and Time) were fixed rather than random. That is, the investigator selected the levels of both of the independent variables. We limited this discussion to fixed factors to keep it relatively simple. It would be onerous to illustrate all of the possible combinations of two-way ANOVA designs with between versus within factors and fixed versus random factors. As long as you learn how to input the data and identify the characteristics of each factor, the software will do the work of determining the expected mean squares and the proper F-ratios.

THREE-WAY FIXED-EFFECTS BETWEEN-GROUPS ANOVA

V.11.5
SPSS Three-Way ANOVA

V.11.5R
RStudio Three-Way ANOVA

Finally, to see how we can expand ANOVA to include more than two independent variables, consider the following experiment. The design was presented earlier in generic form in Table 11.3. The investigator is testing whether VO_2max is a function of gender (male and female), age (young and old), and type of training completed (group 1, group 2, or group 3), for a 2 x 2 x 3 model. In this design, all three factors are fixed and all factors are between groups. Obviously, a wide variety of different models are possible with these three factors, as the researcher modifies the fixed or random and between or within nature of each factor.

SPSS Output for Three-Way ANOVA

Table 11.8 gives the cell sizes and mean values for this fixed, fully between-groups 3-way design.

In Table 11.8 panel (a) illustrates the design with 12 cells (2 x 2 x 3) and two observations per cell, resulting in a total $N = 24$. Panel (b) gives the means and standard deviations. The ANOVA output (Tables 11.9 and 11.9R) appears similar to the output of designs with fewer factors. The important points to note are the significance values for all three of the main effects and the interactions—three two-way interactions and one three-way interaction. None of the interactions is statistically significant, and only one of the main effects (age) is significant. It is not necessary to conduct a multiple comparison procedure for the age factor, as only two groups are involved. The ANOVA results in Tables 11.9 and 11.9R indicate the mean VO_2max for the young subjects is 36.17 and the mean VO_2max for the older group is 28.08, values that are significantly different at $p < .001$. Notice the similarity of these results to the previous examples in this chapter.

TABLE 11.8 Cell sizes and means for three-way ANOVA output.

Between-Subjects Factors			
		Value Label	N
Gender	1	Female	12
	2	Male	12
Group	1	Group 1	8
	2	Group 2	8
	3	Group 3	8
Age	1	Young	12
	2	Old	12

(a) Observations per cell

Descriptive Statistics					
Dependent Variable: V02					
Gender	Group	Age	Mean	Std. Deviation	N
Female	Group 1	Young	35.00	.000	2
		Old	25.00	1.414	2
		Total	30.00	5.831	4
	Group 2	Young	33.50	.707	2
		Old	29.50	7.778	2
		Total	31.50	5.066	4
	Group 3	Young	39.00	4.243	2
		Old	26.00	.000	2
		Total	32.50	7.895	4
	Total	Young	35.83	3.189	6
		Old	26.83	4.119	6
		Total	31.33	5.867	12
Male	Group 1	Young	34.50	2.121	2
		Old	28.50	2.121	2
		Total	31.50	3.873	4
	Group 2	Young	36.50	.707	2
		Old	29.50	2.121	2
		Total	33.00	4.243	4
	Group 3	Young	38.50	2.121	2
		Old	30.00	2.828	2
		Total	34.25	5.315	4
	Total	Young	36.50	2.258	6
		Old	29.33	1.966	6
		Total	32.92	4.252	12
Total	Group 1	Young	34.75	1.258	4
		Old	26.75	2.500	4
		Total	30.75	4.652	8
	Group 2	Young	35.00	1.826	4
		Old	29.50	4.655	4
		Total	32.25	4.400	8
	Group 3	Young	38.75	2.754	4
		Old	28.00	2.828	4
		Total	33.38	6.301	8
	Total	Young	36.17	2.657	12
		Old	28.08	3.343	12
		Total	32.13	5.076	24

(b) Cell means

SPSS output for 3-way ANOVA. TABLE 11.9

Tests of Between-Subjects Effects					
Dependent Variable: V02					
Source	Type III Sum of Squares	df	Mean Square	F	Sig.
Corrected Model	485.125[a]	11	44.102	4.923	.005
Intercept	24768.375	1	24768.375	2764.842	.000
Gender	15.042	1	15.042	1.679	.219
Group	27.750	2	13.875	1.549	.252
Age	392.042	1	392.042	43.763	.000
Gender * Group	.083	2	.042	.005	.995
Gender * Age	5.042	1	5.042	.563	.468
Group* Age	27.583	2	13.792	1.540	.254
Gender* Group* Age	17.583	2	8.792	.981	.403
Error	107.500	12	8.958		
Total	25361.000	24			
Corrected Total	592.625	23			

[a]. R Squared = .819 (Adjusted R Squared = .652).

RStudio output for 3-way ANOVA. TABLE 11.9R

Anova Table				
Response: V02				
	Sum Sq	Df	F value	Pr(>F)
(Intercept)	24768.4	1	2764.842	1.472e-15***
Gender	15.0	1	1.679	0.2194
Group	27.7	2	1.549	0.2521
Age	392.0	1	43.763	2.482e-05***
Gender:Group	0.1	2	0.005	0.9954
Gender:Age	5.0	1	0.563	0.4676
Group:Age	27.6	2	1.540	0.2540
Gender:Group:Age	17.6	2	0.981	0.4030
Residuals	107.5	12		

SUMMARY

Prior to this chapter, we examined statistical tests that involved a single independent variable and a single dependent variable. In the one-way model, the independent variable could have any number of levels. In this chapter, we discussed more complex designs—designs that include more than one independent variable. With the expansion to factorial ANOVA, we introduced the concept of interaction. When we deal with combinations of between and within and fixed and random factors, we need to work with multiple error terms to test main effects and interaction hypotheses. In this chapter, we also discussed the role of expected mean squares in the numerator and denominator of the *F*-ratio. A good understanding of expected mean squares will allow you to expand your use of ANOVA to many different models.

As we have pointed out many times, the logic remains the same in all of these inferential tests. The researcher tests whether a relationship exists between an independent variable *(X)* and an outcome or dependent variable *(Y)*. The researcher generates null and alternative hypotheses, collects data, and then makes a decision regarding the likelihood that the null hypothesis is true. In the next chapter, we will consider ANOVA models that include more than one *dependent* variable.

Review Questions and Practice Problems

PART A (ANSWERS PROVIDED IN APPENDIX C)

1. How many independent variables are in a 2 x 4 x 6 ANOVA?

   A. 3

   B. 12

   C. 48

   D. Depends on how the subjects are assigned to the levels

2. Which of the following terms is most closely associated with accepting the null hypothesis for the *F*-ratio for interaction?

   A. Exceptions

   B. Generalizability

   C. Main effect

   D. Type I error

3. In a between-between ANOVA (3 x 5), how many cells would have different subjects in them?

   A. 3

   B. 5

   C. 8

   D. 15

Use the following source table to answer questions 4 through 7. It is a between-between groups design.

Source	*SS*	*df*	*MS*	*F*-ratio
Factor A	A	2	10.5	E
Factor B	I	4	J	K
Interaction	H	B	G	2.5
Error	225	C	D	
Total	330	89		

4. How would you describe this design?
5. What are the values of A, B, C, D, E, G, H, I, J, and K?
6. Which *F*-ratios are statistically significant? (Use $\alpha = .05$.)
7. If the levels of Factor *A* are children, teenagers, and adults, and the levels of Factor *B* are five different methods of teaching the tennis serve, what does the significant interaction indicate?

PART B

8. In a 4 x 4 between-within ANOVA, what is the total number of subjects that would be required if the researcher wanted 10 subjects in each level of the between factor?

   A. 10

   B. 16

   C. 40

   D. 160

9. Which of the following assumptions is relevant only to a within-groups design?

   A. Homoscedasticity

   B. Interval measurement

   C. Normality

   D. Sphericity

10. If the mean square for Factor *A*, which has four levels, is 120, the mean square for error is 7, and the sum of squares total is 500, how much of the variability in the data can be attributed to Factor *A*?

11. What is the total number of subjects required for a 2 x 4 within-within ANOVA design with five subjects per cell?

    A. 5

    B. 8

    C. 30

    D. 40

12. In a 2 x 2 within-within ANOVA design, which assumption is of no concern?

    A. Interval measurement

    B. Homoscedasticity

    C. Normality

    D. Sphericity

13. If the value for epsilon is .5, where in the *F*-table would you look to find the critical *F*-ratio, if the original degrees of freedom were 2 and 20?

    A. 1, 20

    B. 2, 20

    C. 1, 10

    D. 2, 10

14. How many different error terms do we use to calculate the three *F*-ratios in a within-within ANOVA design?

    A. 1

    B. 2

    C. 3

    D. 4

15. How many main effects and how many interactions are tested, respectively, in a 2 x 3 x 4 ANOVA, if all of the independent variables are between factors?

    A. 1, 4

    B. 1, 6

    C. 3, 4

    D. 3, 6

16. How many degrees of freedom exist for interaction in a 3 x 4 factorial design?

    A. 2

    B. 6

    C. 7

    D. 12

17. What information is *most* helpful when we are determining which statistical test to use?

    A. Number of subjects and population mean

    B. Number and type of variables and levels of measurement

    C. Alpha level and power

    D. Population mean and type of sample

COMPUTER ACTIVITIES

Two-way ANOVA

Forty-eight college students participated in a study to examine the effects of two different diets and three dosages of a drug developed to curb hunger. Twenty-four of the subjects were randomly assigned to diet 1, and the other 24 were assigned to diet 2. Within each diet condition, 8 subjects were randomly assigned to each of the three drug dosages. The sensation of hunger was measured with a questionnaire at the end of five weeks. The questionnaire was designed so that high scores indicated that the sensation of feeling hungry was low. Table 11.9 gives the resulting data.

Using SPSS

To analyze these data, we will use a two-way ANOVA, since we have two independent variables. Be sure to enter the data correctly, based on the level of measurement of each variable. Your SPSS Data View file will contain three variables: Dosage with three levels, Diet with two levels, and Questionnaire scores.

- Enter the data, and then go to **Analyze→General Linear Model→Univariate**. This will bring up the Univariate window.
- Move the Diet and Dosage variables into the Fixed-Factors box, and move the Questionnaire score variable into the Dependent Variable box.
- Click on **Options** and check the Descriptive statistics box under Display.
- Click **Continue**.
- Click on **Plots**, and move Diet to the Horizontal Axis box and Dosage to the Separate Lines box. Click **Add**. This graph will be required to answer a later question.
- Click **Continue** and **OK**.

TABLE 11.10 Data set.

	DOSAGE					
Diet	**.1 gram**		**.2 gram**		**.3 gram**	
Diet 1	26	14	41	82	36	87
	41	16	26	86	39	99
	28	29	19	45	59	122
	92	31	59	37	27	104
Diet 2	51	35	39	114	42	133
	96	36	104	92	52	124
	97	28	130	87	156	68
	22	76	122	64	144	142

Using R

1. Open RStudio.
2. Install any necessary R packages.

   For Chapter 11, we will need to install the "car" package which will generate Type II and Type III sum of squares.

   In the lower right pane of the RStudio, click on the "Packages" tab and then click on the "Install" button.

   In the "Packages" field, type in "car" and click "Install".

   [Note that from this point forward, the package "car" will be installed, and you do not have to install it again.]

3. Set up your RStudio workspace.

   In the RStudio menu bar, click on **File -> New File -> R Script**.

   On the first line of the R Script (upper left pane of RStudio), type:

   - #Chapter 11
   - library(car)

   Highlight and **Run** these two lines.

4. Import the data from Table_11_10_Data.
5. To find the mean of the hunger scores (SCORES) by diet (DIET) and dose (DOSAGE) and diet by dosage type the following:

   - Table_11_10_Data$DIET <- as.factor(Table_11_10_Data$DIET)
   - Table_11_10_Data$DOSAGE <- as.factor(Table_11_10_Data$DOSAGE)
   - tapply(X=Table_11_10_Data$SCORES, INDEX=Table_11_10_Data$DIET, FUN=mean)
   - tapply(X=Table_11_10_Data$SCORES, INDEX=Table_11_10_Data$DOSAGE, FUN=mean)
   - tapply(X=Table_11_10_Data$SCORES, INDEX=list(Table_11_10_Data$DIET,Table_11_10_Data$DOSAGE), FUN=mean)

6. To conduct a two-way ANOVA using these data, we will use the linear model (lm) function. We first must set the default contrast to obtain the correct sum of squares.

   - options(contrasts = c("contr.helmert", "contr.poly"))
   - twowayanova <- lm(SCORES ~ DIET + DOSAGE + DIET*DOSAGE, data=Table_11_10_Data)

   Anova(twowayanova, type= 3)

7. To generate two-way interaction plots of the data, the interaction.plot function can be used. In this first example, we will first look at DIET (x-axis) against SCORES (y-axis) plotted by DOSAGE

- interaction.plot(Table_11_10_Data$DIET,Table_11_10_Data$DOSAGE,Table_11_10_Data$SCORES)

Highlight the nine commands above and click **Run** to generate the output required to answer the questions 1 through 20.

Use the resulting analysis to answer the following questions:

1. What are the two independent variables in this experiment?
2. What is the dependent variable?
3. What are the levels of measurement for each of the three variables?
4. What type of design is this?
5. State the null hypothesis and an alternative hypothesis for each of the main effects and the interaction in terms of the experiment.
6. What are the six cell means? Be sure to identify which mean is in which cell.
7. What are the two diet means?
8. What are the three dosage means?
9. What is the sum of squares total? (Hints: In the SPSS two-way ANOVA, output the total sum of squares is not in the row labeled Total. In the RStudio output you will need to add the sum of squares not including the Intercept.)
10. What is total of the sums of squares for the three explained effects?
11. What is the F-ratio for diet? Is it statistically significant?
12. What is the F-ratio for dosage? Is it statistically significant?
13. What is the F-ratio for interaction? Is it statistically significant?
14. State your answer to question 11 in terms of the experiment.
15. State your answer to question 12 in terms of the experiment.
16. State your answer to question 13 in terms of the experiment.
17. For which independent variable in this experiment is it necessary to perform a multiple comparisons calculation?
18. Describe the Profile Plots graph briefly. What does it indicate about any interaction between diet and dosage?
19. Which diet best curbs appetite?
20. Which drug dosage best curbs appetite?

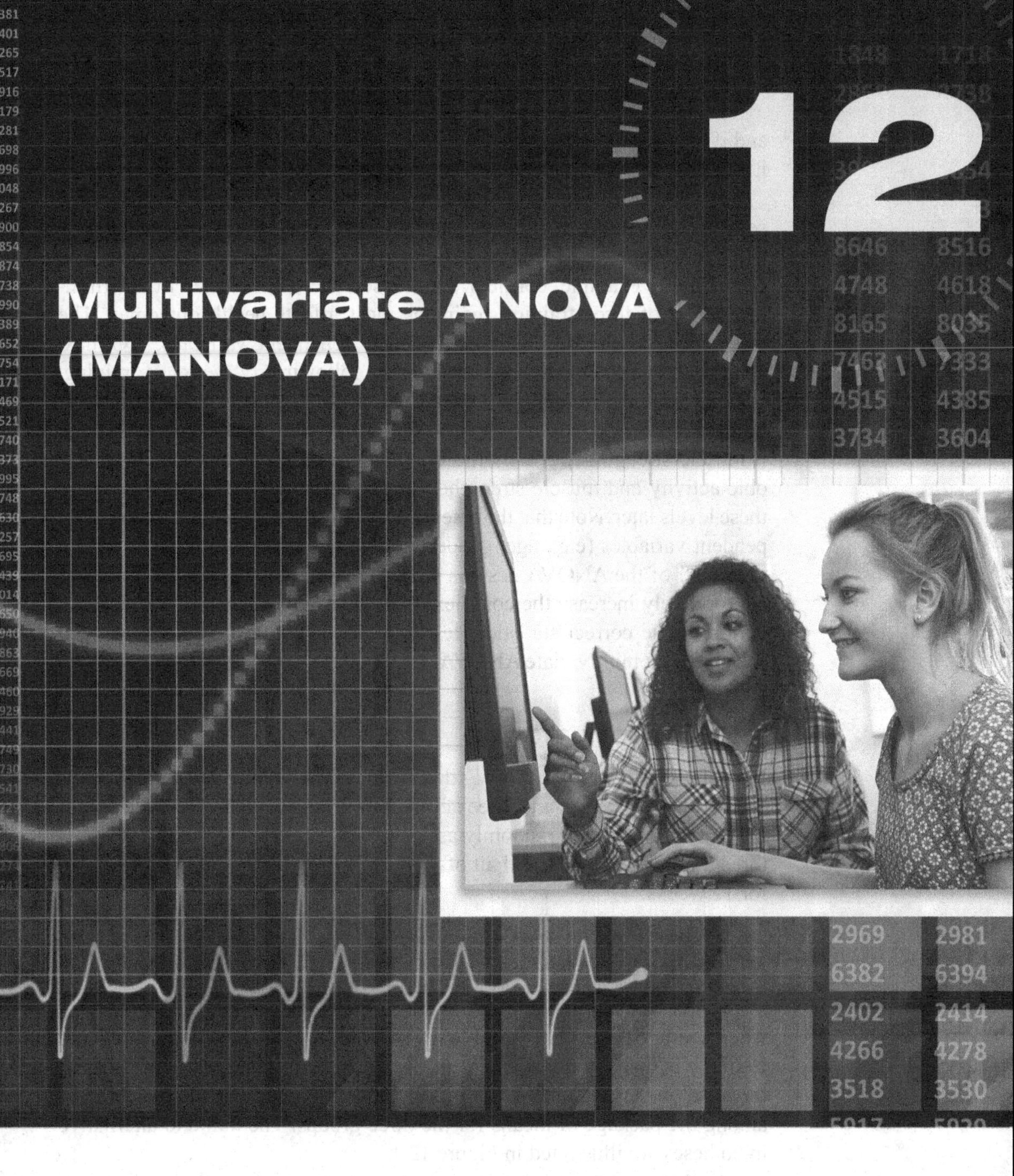

12 Multivariate ANOVA (MANOVA)

INTRODUCTION

The next logical step as we consider increasingly complex experimental designs is to include more than one dependent variable. Up to this point, all of the statistical tests we have examined have involved only one dependent variable.

In previous chapters, we increased the number of means from two (t-test) to any number for a single independent variable (one-way ANOVA), increased the number of independent variables from one to any number (factorial ANOVA), and considered all manner and types of independent variables (varying in how they were measured, if they were between or within, and if they were fixed or random). We considered how all of these possibilities affect only one dependent variable.

As you can imagine, researchers are often interested in detecting how some combinations of independent variables relate to multiple characteristics (outcomes or dependent variables) of the subjects in an experiment. For a simple example, consider that a researcher is interested in determining which of three training regimens most affects health outcomes. The researcher has decided to measure three dependent variables: VO_2max, skinfold thickness, and cholesterol levels, following 12 weeks of training. For this example, we will limit the design to a single independent variable—training method—with three levels. Level 1 is aerobic training only; level 2 is muscle-strengthening activities only; and level 3 is a combination of aerobic activity and muscle-strengthening activities. We will further describe these levels later. Note that the researcher could also include additional independent variables (e.g., age, gender, duration of the activity). In fact, with nearly all of the ANOVA designs we have discussed, the researcher could appropriately increase the complexity by having more than one dependent variable. The correct statistical test in situations with multiple dependent variables is a multivariate ANOVA (**MANOVA**).

HOW MANOVA WORKS

The relatively simple MANOVA described above has one independent variable, with three levels, and three dependent variables. In this example, imagine that the researcher randomly assigns 10 different subjects to each of the three training regimens, and all of them complete the training requirements for 12 weeks.

Null Hypotheses

With three dependent variables, the null hypothesis is a bit different than when a single dependent variable is involved. Essentially, the researcher is testing whether the null hypothesis is true regarding the "package" of the three dependent variables. The null hypothesis is that no differences exist among the package of means for the three groups. The null and alternative hypotheses are illustrated in Figure 12.1.

In these equations, the three means (μ) represent the three dependent variables. The null hypothesis states that the three means are randomly equivalent across the three groups, and any differences are attributable to sampling error. The alternative hypothesis states that the means are sufficiently different, and it can be concluded that the null hypothesis is not correct and should be rejected. Note that this is exactly the logic we have been using for hypothesis testing throughout this textbook. The concepts of an alpha (α) level and power are also appropriate here.

With a single dependent variable, it is easy to see that the researcher is testing whether the means differ. It is possible to visualize the means along a number line, with the means as points along the line. With multiple dependent variables, we must begin to think of a "space" rather than a number

MANOVA null and alternative hypotheses. FIGURE 12.1

Independent Variable

	Level 1		Level 2		Level 3
H_0:	$\begin{pmatrix}\mu_1\\ \mu_2\\ \mu_3\end{pmatrix}$	$=$	$\begin{pmatrix}\mu_1\\ \mu_2\\ \mu_3\end{pmatrix}$	$=$	$\begin{pmatrix}\mu_1\\ \mu_2\\ \mu_3\end{pmatrix}$
H_1:	$\begin{pmatrix}\mu_1\\ \mu_2\\ \mu_3\end{pmatrix}$	$\neq$	$\begin{pmatrix}\mu_1\\ \mu_2\\ \mu_3\end{pmatrix}$	$\neq$	$\begin{pmatrix}\mu_1\\ \mu_2\\ \mu_3\end{pmatrix}$

line. In the example of three Training Methods, think about how the package representing the means for each of the levels of the independent variable might be "observed" in a three-dimensional space. If the means are all the same across the three levels, then within this three-dimensional space, the means would be "hanging in the space" very close to each other. However, if the null hypothesis is not correct, then the three points representing the three means would be located in vastly different places within the three-dimensional space.

Fortunately, it is easier to visualize where the means might be in space when the experimental design has only three dependent variables than when more than three dependent variables are involved. When more than three (e.g., p) means are involved, then the space has p dimensions. We cannot draw p-dimensional space, but it does exist theoretically.

Another way to think of this is in terms of regression. For each subject, SPSS and R mathematically create a "new" variable from a linear combination of the three dependent variables (VO_2max, skinfold thickness, and cholesterol levels) in such a way as to maximize the differences among the three levels of the independent variable. It is this "new" variable that we then analyze, as in an ANOVA.

The results of a MANOVA start with an omnibus (overall) test of significance. The omnibus test indicates whether the researcher rejects the null hypothesis for the package of dependent variables. If the conclusion is that no significant difference exists among the means, the researcher retains the null hypothesis, and analysis ends. However, if the analysis reveals that the pattern of the means across the levels of the independent variable is unlikely to have occurred by chance alone, the researcher rejects the null hypothesis at the selected alpha level, and further analyses are warranted.

More Complex MANOVA Designs

To illustrate how MANOVA could be applied to a slightly more complex design, let's assume that the researcher wants to include another independent

FIGURE 12.2 Two-way MANOVA example.

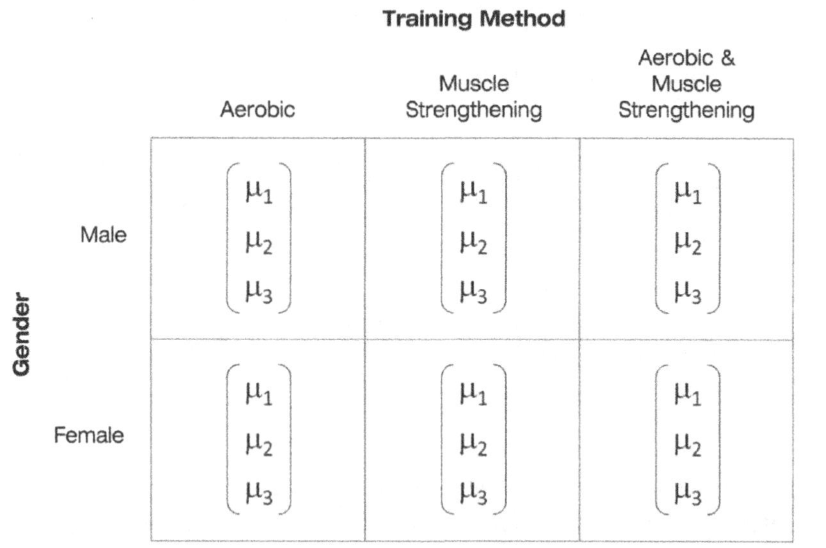

Gender	Aerobic	Muscle Strengthening	Aerobic & Muscle Strengthening
Male	$\begin{pmatrix}\mu_1\\ \mu_2\\ \mu_3\end{pmatrix}$	$\begin{pmatrix}\mu_1\\ \mu_2\\ \mu_3\end{pmatrix}$	$\begin{pmatrix}\mu_1\\ \mu_2\\ \mu_3\end{pmatrix}$
Female	$\begin{pmatrix}\mu_1\\ \mu_2\\ \mu_3\end{pmatrix}$	$\begin{pmatrix}\mu_1\\ \mu_2\\ \mu_3\end{pmatrix}$	$\begin{pmatrix}\mu_1\\ \mu_2\\ \mu_3\end{pmatrix}$

(Training Method)

variable, gender. The researcher wants to determine not only whether the training regimens differ but also whether any differences exist between the results for males and females. Figure 12.2 illustrates this design.

In this factorial model, the MANOVA would create three new linear combinations of the three dependent variables for each subject. One of these combinations would weigh the scores for the variables in such a way as to maximize the differences among the three Training Methods; another combination would maximize the differences between the males and females; and the third combination would maximize the interaction between the two independent variables (Gender and Training Method). Just as in a two-way between-groups fixed-factors ANOVA, the analysis includes tests for the two main effects (Training Method means and Gender means) and a test for the interaction effect.

V.12.1 **Explanation of MANOVA**

We can apply MANOVA to designs employing any number and type (fixed or random, between or within) of independent variables with any number of levels. In these complex designs, it may be difficult or impossible to satisfy many conditions (such as the number of subjects required), and the interpretation of results can become exceedingly difficult.

Why Not Multiple ANOVAs?

You might be asking, "Why not run separate ANOVAs on each dependent variable of interest rather than employing MANOVA?" There are several reasons to use MANOVA. First, just as when a two-way ANOVA can reveal information that we do not find when we conduct two separate one-way ANOVAs, analyzing two or more dependent variables simultaneously with MANOVA might reveal something unexpected that separate ANOVAs would not reveal. Sometimes MANOVA is more powerful than separate ANOVAs.

Also, just as conducting several *t*-tests inflates the possibility of committing a Type I error, conducting multiple ANOVAs can lead to the same problem. Consider an example where we have four dependent variables and just two levels of a single independent variable. If we conducted four *t*-tests (at $\alpha = .05$ for each one), the overall probability of committing a Type I error would be somewhere between .05 and .19. The exact alpha level depends on the correlations among the various dependent variables. Conducting a MANOVA instead would protect us against a Type I error at whatever alpha level we selected. The alpha level for the entire package of dependent variables is the level that the researcher selects (e.g., .05 or .01).

Disadvantages of MANOVA

Besides being more complicated and thus more difficult to interpret than ANOVA, MANOVA involves many conditions that, if not met, can result in loss of power for this statistical test. It is even difficult, sometimes, to assess whether or not some of the conditions *are* being met.

In some situations, MANOVA is less powerful than ANOVA. For example, if two of three dependent variables are in fact not related to an independent variable, but the third one is somewhat related, this fact could be masked by the combination of the three dependent variables into the "new" variable.

Test Statistic

For ANOVA, the final test statistic is the *F*-ratio. For MANOVA, several test statistics result. The most common one is known as Wilks' lambda (λ). Others include Pillai's trace, Roy's largest root, and Hotelling's trace. Since Wilks' lambda is the most commonly used and reported test statistic for MANOVA, we will refer to it for the most part in this chapter. The possible values for Wilks' lambda range from zero to 1.0. Because of the way this statistic is calculated, *lower* lambda values signify significance in the relationship, suggesting rejection of the null hypothesis.

Procedures

As we mentioned at the beginning of this discussion, with MANOVA the researcher tests the overall null hypothesis. If this hypothesis is retained, no further analyses are performed. If the null hypothesis is rejected, the researcher will probably conduct further analyses to identify exactly which dependent variables are most affected by the independent variable. Morrow and Frankiewicz (1979) present strategies for the *post hoc* procedures in studies of human performance. Often, the researcher next conducts separate ANOVAs on each of the dependent variables. However, to mitigate the inflated chance of a Type I error that is possible when we perform multiple tests of significance, researchers often use a procedure called the **Bonferroni**

V.12.2
SPSS MANOVA Example

V.12.2R
RStudio MANOVA Example

adjustment. In our example, with three dependent variables and a desire to be protected against making a Type I error at an alpha of .05, the Bonferroni adjustment has us divide .05 by 3 and set a new alpha level at .016 (.05/3 = .016). Other *post hoc* procedures that a researcher might pursue after finding a significant lambda value include discriminant analyses and stepdown analyses. We will discuss discriminant analysis in the next chapter. Stepdown analysis is beyond the scope of this textbook.

MANOVA ASSUMPTIONS

The assumptions for multivariate ANOVA are analogous to those for univariate ANOVA (one dependent variable), except that they apply to the multivariate situation. Because of this difference, the mathematics involved in explaining the significance tests require the use of matrix algebra and are beyond the scope of this introductory textbook. For the sake of completeness and so that we can discuss some of the implications of using MANOVA, we present the assumptions in a simplified manner.

- *Random sampling and independence*. The subjects are assigned randomly to each group and are independent from one another. As long as the researcher employs random sampling to place subjects into the various groups required by the experimental design, this assumption is generally not an issue. An alternative to meeting the assumption of independence is to employ a repeated-measures design, but it is unlikely in this situation that the remaining assumptions would be met, and the interpretation of the results becomes difficult. See Schutz and Gessaroli (1987) for a description of this "doubly multivariate" model in exercise and sport.
- *Normality*. The dependent variables conform to a multivariate normal distribution. This assumption means that all of the dependent variables are normally distributed in each package of variables being tested in each level of the independent variables. This assumption is both difficult to explain and difficult to assess. However, recall that ANOVA is relatively robust when it comes to the normality assumption, as long as the number of subjects is reasonably large. MANOVA is also robust, especially if the number of dependent variables involved is not large relative to the number of subjects.
- *Homogeneity*. With univariate ANOVA, we assume that the dependent variable displays homogeneity of variance for all levels of the independent variable. In MANOVA, instead, we assume homogeneity of the covariance matrix. The covariance matrix consists of the variance for each of the dependent variables and the bivariate correlations among all of the dependent variables. Essentially, this matrix is assumed to be similar for each level of the independent variable(s). Most statistical software programs provide some help in evaluating the degree to which this assumption is met. It is possible to make some corrections, called transformations of the data, if this assumption is violated too severely. Another approach is to use Pillai's trace as the test statistic

in this situation, because it is more robust for this assumption than is Wilks' lambda.

- *Correlations among dependent variables.* Linear correlations among all pairs of dependent variables are assumed. We typically use scattergrams to assess the condition of this assumption, although instead of straight lines one hopes to see ovals, due to the nature of the data. Violations of this assumption that are severe may, again, require data transformations.

OTHER ISSUES

Testing of MANOVA Effects

If the MANOVA design includes only one independent variable with two levels, the F-tests for all of the test statistics mentioned previously are identical. If two or more independent variables are involved, the researcher must examine interactions among the independent variables *before* interpreting main effects. If no significant interactions are present and the omnibus F-test indicates rejection of the null hypothesis, running separate univariate ANOVAs for each dependent variable (remembering to apply the Bonferroni adjustment) is common.

Number of Subjects

Statisticians disagree about the number of subjects required for MANOVA. Having too few subjects not only reduces power but also increases the probability that the assumptions will not be met. The number of subjects per group must be greater than the number of dependent variables, and many statisticians recommend that, at a minimum, the number of subjects per group should be at least three times the number of dependent variables.

Selection of Dependent Variables

The researcher must select the dependent variables with care. It is not appropriate to measure everything that comes to mind as a dependent variable in the hopes of finding something. Reasoning should underlie the inclusion of any dependent variable. Use of dependent variables that are highly correlated to each other is not recommended, because they will be measuring much of the same variance among the subjects and thus are a waste of degrees of freedom.

Outliers

Finally, as with any other statistical test involving regression, outliers represent a potential problem. Take care when entering data for MANOVA, and examine the entered data for errors, because the existence of an outlier or two can seriously affect the results. Any outliers that are the result of errors of measurement should be eliminated. If they are the result of an atypical subject (someone whose score seems to have been measured correctly but is very different from all the other scores), the researcher faces a possibly difficult decision as to what to do about such a subject. At a minimum, if such outliers are eliminated, the researcher should report this decision, along with the reasoning.

MANOVA EXAMPLE

The following example of MANOVA is based on the three dependent variables previously mentioned in this chapter. We will use a one-way MANOVA for the sake of simplicity. Figure 12.3 presents the design.

In this example, the researcher has considered the 2018 Physical Activity Guidelines from the U.S. Department of Health and Human Services, which encourage adults to engage in aerobic and muscle-strengthening activities to achieve health benefits. The independent variable has three levels (the Training Methods). The researcher is interested in knowing whether aerobic activity alone differs from muscle-strengthening activities alone and from the combination of aerobic and muscle-strengthening activities. The dependent variables of interest are VO_2max, skinfold thickness, and total cholesterol. Ten subjects are randomly assigned to each cell of the design (see Table 12.1). Thus, MANOVA is appropriate, because the design includes multiple (three) dependent variables.

Table 12.2 provides the descriptive statistics for each dependent variable by levels of the independent variable. Tables 12.3 through 12.5 give the SPSS results. Note in Table 12.3 that the multivariate (omnibus) test is significant (Wilks' lambda = .405; $F(6,50) = 4.760$; Sig. < .001) for the package of three dependent variables. Tables 12.3R through 12.5R present the MANOVA results from RStudio.

Table 12.4 presents the univariate ANOVA results for each dependent variable. The important results in this table are those for Training Method. Note that VO_2max is the only significant univariate dependent variable. Look back at Table 12.2 to see the mean values for VO_2max for the three Training Methods. Table 12.5 presents the *post hoc* Student-Newman-Keuls tests for VO_2max.

FIGURE 12.3 One-way ANOVA with three dependent variables.

Training Method

Aerobic	Muscle Strengthening	Aerobic & Muscle Strengthening
$\begin{pmatrix}\mu_1\\ \mu_2\\ \mu_3\end{pmatrix}$	$\begin{pmatrix}\mu_1\\ \mu_2\\ \mu_3\end{pmatrix}$	$\begin{pmatrix}\mu_1\\ \mu_2\\ \mu_3\end{pmatrix}$

TABLE 12.1 Subject placement in a MANOVA design.

Between-Subjects Factors

		Value Label	N
Training Method	1	Aerobic Only	10
	2	Muscle Strengthening Only	10
	3	Aerobic and Muscle Strengthening	10

TABLE 12.2 Descriptive statistics for three dependent variables in MANOVA.

Descriptive Statistics

	Training Method	Mean	Std. Deviation	N
VO2MAX	Aerobic Only	57.80	6.663	10
	Muscle Strengthening Only	38.50	5.482	10
	Aerobic and Muscle Strengthening	47.40	8.566	10
	Total	47.90	10.499	30
SKINFOLDS	Aerobic Only	62.80	7.465	10
	Muscle Strengthening Only	58.80	9.897	10
	Aerobic and Muscle Strengthening	61.60	5.379	10
	Total	61.07	7.719	30
CHOLESTEROL	Aerobic Only	196.70	94.950	10
	Muscle Strengthening Only	164.70	12.248	10
	Aerobic and Muscle Strengthening	169.30	17.237	10
	Total	176.90	56.064	30

TABLE 12.3 SPSS MANOVA results.

Multivariate Tests[a]

Effect		Value	F	Hypothesis df	Error df	Sig.
Intercept	Pillai's Trace	.991	956.453[b]	3.000	25.000	.000
	Wilks' Lambda	.009	956.453[b]	3.000	25.000	.000
	Hotelling's Trace	114.774	956.453[b]	3.000	25.000	.000
	Roy's Largest Root	114.774	956.453[b]	3.000	25.000	.000
Training Method	Pillai's Trace	.603	3.737	6.000	52.000	.004
	Wilks' Lambda	.405	4.760[b]	6.000	50.000	.001
	Hotelling's Trace	1.450	5.799	6.000	48.000	.000
	Roy's Largest Root	1.436	12.449[c]	3.000	26.000	.000

a. Design: Intercept + Training Method

b. Exact statistic

c. The statistic is an upper bound on F that yields a lower bound on the significance level.

TABLE 12.3R RStudio MANOVA results.

	Df	Pillai	Approx. F	Num Df	Den Df	Pr(>F)
(Intercept)	1	0.99136	956.45	3	25	<2e-16***
TRAINING METHOD	2	0.60261	3.74	6	52	0.00364**
Residuals	27					

Signif. codes: 0*** 0.00** 0.01* 0.05 0.1 1

	Df	Wilks	Approx. F	Num Df	Den Df	Pr(>F)
(Intercept)	1	0.00864	956.45	3	25	<2.2e-16***
TRAINING METHOD	2	0.40508	4.76	6	50	0.0006589***
Residuals	27					

Signif. codes: 0*** 0.00** 0.01* 0.05 0.1 1

	Df	Hotelling-Lawley	Approx. F	Num Df	Den Df	Pr(>F)
(Intercept)	1	114.77	956.45	3	25	<2.2e-16***
TRAINING METHOD	2	1.45	5.80	6	48	0.0001332***
Residuals	27					

Signif. codes: 0*** 0.00** 0.01* 0.05 0.1 1

	Df	Roy's	Approx. F	Num Df	Den Df	Pr(>F)
(Intercept)	1	114.774	956.45	3	25	<2.2e-16***
TRAINING METHOD	2	1.436	12.45	3	26	3.086e-05***

Signif. codes: 0*** 0.00** 0.01* 0.05 0.1 1

TABLE 12.4 SPSS Univariate ANOVA results.

Tests of Between-Subjects Effects

Source	Dependent Variable	Type III Sum of Squares	df	Mean Square	F	Sig.
Corrected Model	VO2MAX	1866.200[a]	2	933.100	18.936	.000
	SKINFOLDS	84.267[b]	2	42.133	.692	.509
	CHOLESTEROL	5986.400[c]	2	2993.200	.949	.400
I Intercept	VO2MAX	68832.300	1	68832.300	1396.822	.000
	SKINFOLDS	111874.133	1	111874.133	1837.796	.000
	CHOLESTEROL	938808.300	1	938808.300	297.634	.000
Training Method	VO2MAX	1866.200	2	933.100	18.936	.000
	SKINFOLDS	84.267	2	42.133	.692	.509
	CHOLESTEROL	5986.400	2	2993.200	.949	.400
Error	VO2MAX	1330.500	27	49.278		

Continued. TABLE 12.4

Tests of Between-Subjects Effects

Source	Dependent Variable	Type III Sum of Squares	df	Mean Square	F	Sig.
	SKINFOLDS	1643.600	27	60.874		
	CHOLESTEROL	85164.300	27	3154.233		
Total	VO2MAX	72029.000	30			
	SKINFOLDS	113602.000	30			
	CHOLESTEROL	1029959.000	30			
Corrected Total	VO2MAX	3196,700	29			
	SKINFOLDS	1727.867	29			
	CHOLESTEROL	91150.700	29			

a. R Squared = .584 (Adjusted R Squared = .553)

b. R Squared = .049 (Adjusted R Squared = −.022)

c. R Squared = .066 (Adjusted R Squared = −.004)

RStudio Univariate ANOVA results. TABLE 12.4R

VO2MAX:

	Df	Sum Sq	Mean Sq	F value	Pr(>F)
(Intercept)	1	68832	68832	1396.822	<2.2e-16***
TRAINING METHOD	2	1866	933	18.936	7.256e-06***
Residuals	27	1330	49		

Signif. codes: 0*** 0.001** 0.01* 0.05 0.1 1

SKINFOLDS:

	Df	Sum Sq	Mean Sq	F value	Pr(>F)
(Intercept)	1	111874	111874	1837.7961	<2e-16***
TRAINING METHOD	2	84	42	0.6921	0.5092
Residuals	27	1644	61		

Signif. codes: 0*** 0.001** 0.01* 0.05 0.1 1

CHOLESTEROL:

	Df	Sum Sq	Mean Sq	F value	Pr(>F)
(Intercept)	1	938808	938808	297.6344	4.162e-16***
TRAINING METHOD	2	5986	2993	0.9489	0.3997
Residuals	27	85164	3154		

Signif. codes: 0*** 0.001** 0.01* 0.05 0.1 1

TABLE 12.5 SPSS *Post hoc* test of VO_2max.

VO2max

Student-Newman-Keuls[a,b,c]				
Training Method	**N**		**Subset**	
		1	**2**	**3**
Muscle Strengthening Only	10	38.50		
Aerobic and Muscle Strengthening	10		47.40	
Aerobic Only	10			57.80
Sig.		1.000	1.000	1.000

Means for groups in homogeneous subsets are displayed.

Based on observed means.

The error term is Mean Square(Error) = 49.278.

a. Uses Harmonic Mean Sample Size = 10.000.

b. The group sizes are unequal. The harmonic mean of the group sizes is used. Type I error levels are not guaranteed.

c. Alpha = .05.

TABLE 12.5R RStudio *Post hoc* test of VO_2max.

$TRAINING METHOD				
	Diff	**lwr.ci**	**upr.ci**	**pval**
Muscle Strength Only-Aerobic Only	−19.3	−27.08	−11.52	4.2e-06***
Aerobic and Muscle Strength-Aerobic Only	−10.4	−16.84	−3.96	0.003**
Aerobic and Muscle Strength-Muscle Strength Only	8.9	2.46	15.34	0.009**

Signif. codes: 0*** 0.001** 0.01 0.05 0.1 1

In Table 12.5, note that each of the three Training Methods differs significantly for VO_2max from the other two methods. This is indicated by the fact that each Training Method mean resides in a separate subset. The univariate ANOVA results are not significant for skinfolds and cholesterol. The *post hoc* procedure indicates that all three groups are within chance for these two dependent variables. The differences in VO_2max are substantial enough to result in a significant MANOVA result. The interested reader should see Church and colleagues (2010), a study of training methods very similar to this example, although the researchers used different statistical procedures.

SUMMARY

The expansion from a single dependent variable to multiple dependent variables necessitates a MANOVA analysis. In this chapter's discussion of MANOVA, we explained why using multiple ANOVAs is not appropriate, some common disadvantages of MANOVA, the test statistics involved, and the assumptions of MANOVA. We then presented an example and the associated SPSS and RStudio outputs. MANOVA can include multiple levels of independent variables, which may be fixed or random and between or within. As a result, the researcher can use MANOVA for a multitude of complex designs and analyses. The underlying logic remains the same as that in earlier chapters: the researcher is investigating the relationship between the *X* and *Y* variables.

Review Questions and Practice Problems

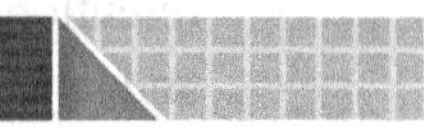

PART A (ANSWERS PROVIDED IN APPENDIX C)

1. What statistical test often follows the finding of a non-significant MANOVA result?

   A. Chi-square

   B. Factor analysis

   C. Univariate ANOVAs

   D. Nothing

2. What is the difference between ANOVA and MANOVA?

   A. The number of dependent variables

   B. The number of independent variables

   C. The number of both independent and dependent variables

3. A researcher is interested in testing the effect of three different training methods on strength and aerobic capacity. The researcher is also interested in the effects on males and females. Is this a MANOVA?

   A. No

   B. Yes, a 3 x 2 x 2 design with multiple dependent variables

   C. Yes, a 3 x 2 design with multiple dependent variables

   D. Yes, a 2 x 2 design with multiple dependent variables

4. What alpha level would a researcher use to test for significance, if she desired protection at the .05 level but applied the Bonferroni adjustment with four dependent variables in a MANOVA?

   A. .0400

   B. .0160

   C. .0125

   D. .0100

PART B

5. What types of tests are often conducted following a significant MANOVA?

   A. Univariate ANOVA

   B. Correlation

   C. Chi-square

   D. All of the above

6. A researcher is interested in testing the effect of three different training methods on the resulting maximal oxygen uptake. Is this a MANOVA?

   A. No

   B. Yes

7. Why would one conduct a MANOVA rather than multiple *t*-tests?

   A. To reduce the need for conducting *post hoc* procedures

   B. To reduce the probability of a Type I error

   C. To test for interactions

   D. To test multiple independent variables

8. Which of the following is true regarding the t-ratio, the F-ratio, and Wilks' lambda?

   A. Larger values of the t-ratio, the F-ratio, and Wilks' lambda lead to rejection of the null hypothesis.

   B. Larger values of the t-ratio lead to retention of the null hypothesis, and smaller values of the F-ratio and Wilks' lambda lead to rejection of the null hypothesis.

   C. Smaller values of the t-ratio, the F-ratio, and Wilks' lambda lead to rejection of the null hypothesis.

   D. Larger values of the t-ratio and the F-ratio lead to rejection of the null hypothesis, and smaller values of Wilks' lambda lead to rejection of the null hypothesis.

COMPUTER ACTIVITIES

MANOVA

Assume that you have completed your degree in exercise science and opened your own fitness center. You want to specialize in strength training for young athletes. You want to base your programs on sound scientific principles, so you design a strength-training program that you believe will improve overall body strength. Before you advertise your "Body Works Strength Training" program, you decide to conduct a research study so that you can advertise that your program has scientific evidence suggesting its effectiveness.

You identify 50 young athletes who desire strength training. You randomly assign them to one of two groups: (1) the strength-training group, in which the subjects complete your "Body Works" program three days per week for 12 weeks, or (2) a control group, in which the subjects continue with their current workout (with the understanding that they will be permitted to enroll in your program at a future date).

You consider what measure(s) you might take to provide evidence of the program's effectiveness. You want to choose measures that are easy to take, so that individuals can conduct self-tests periodically to see how they are progressing. You also want to use measures of core, upper-body, and lower-body strength. You decide to assess total-body strength with three measures: (1) timed 2-minute sit-ups; (2) maximum pull-ups; and (3) a vertical jump.

From your measurement course, you recall that it is unwise simply to add the scores from these three tests, because they involve different units of measurement (i.e., the number of repetitions for sit-ups and pull-ups, and inches for the vertical jump). You also recall from your introduction to statistics in human performance course that you should conduct a MANOVA. Table 12.6 gives your results, with the control group coded 0 and the treatment group coded 1. Enter these data into SPSS or R and answer the questions that follow. Note the data are entered in four columns on the SPSS Data View screen. Alternatively, you can save the data in an Excel file or simply download the table from the textbook website.

Data set. TABLE 12.6

GROUP	SITUPS	PULLUPS	VERTICAL JUMP	GROUP	SIT-UPS	PULLUPS	VERTJUMP
0	42	2	19	1	50	3	23
0	53	3	13	1	74	8	33
0	50	5	28	1	47	7	27
0	46	5	19	1	47	4	28
0	38	6	21	1	60	5	21
0	32	6	17	1	56	5	29
0	45	7	21	1	65	6	34
0	35	7	12	1	51	6	28
0	35	8	27	1	60	6	29
0	42	8	23	1	42	6	32
0	36	0	27	1	56	7	26
0	42	0	20	1	51	7	29
0	51	0	21	1	61	0	24
0	37	4	19	1	45	5	27
0	64	10	15	1	56	8	27
0	50	3	23	1	43	0	20
0	48	2	15	1	46	9	37
0	45	6	25	1	57	1	28
0	43	9	21	1	44	4	36
0	55	5	20	1	72	1	27
0	38	7	28	1	62	1	20
0	49	8	25	1	42	2	24
0	64	10	21	1	43	2	22
0	58	7	26	1	56	2	23
0	45	5	24	1	46	4	25

Using SPSS

Use SPSS to create a correlation matrix for the three dependent variables. Use **Analyze→Correlate→Bivariate**.

Use SPSS to conduct a MANOVA analysis.

- Click on **Analyze→General Linear Model→Multivariate**.
- Move the three dependent variables into the proper window on the right.
- Enter Group as a fixed factor (independent variable).
- Click on **Options** and check the boxes for (1) Descriptive statistics and (2) Estimates of effect size.
- Click **Continue** and **OK**.

Using R

1. Open RStudio.
2. Install any necessary R packages.

   For Chapter 12, we will need to install the "car" package.

   In the lower right pane of the RStudio, click on the "Packages" tab and then click on the "Install" button.

   In the "Packages" field, type in "car" and click "Install".

   [Note that from this point forward, the package "car" will be installed, and you do not have to install it again.]

3. Set up your RStudio workspace.

   In the RStudio menu bar, click on **File** -> **New File** -> **R Script**.

   On the first line of the R Script (upper left pane of RStudio), type #Chapter 12.

   On the second line of the R Script, type "library(car)".

   When you have done that, you should see the following in your new R Script:

   - #Chapter 12
   - library(car)

   Highlight these two lines that you've typed and click **Run**.

   [Note that the second line is where we tell R that we will require the functions that are included in the "car" package.]

You are now ready to conduct the MANOVA analysis.

4. Import the data from Table 12.6.
5. Before starting our analysis, we can label GROUP (1="Strength", 0="Control") by first instructing R to treat GROUP as a factor and then assigning the appropriate level.

   - Table_12_6_Data$GROUP <- as.factor(Table_12_6_Data$GROUP)
   - levels(Table_12_6_Data$GROUP) <- list("Strength"=1, "Control"=0)

6. Generate a correlation matrix for the three dependent variables (variables 2, 3 and 4 in the data frame):

   - cor(Table_12_6_Data[,2:4])

7. To find the mean of SITUPS, PULLUPS, and VERTJUMP by each group:

   - tapply(X=Table_12_6_Data$SITUPS,INDEX=list(Table_12_6_Data$GROUP), FUN=mean)

- tapply(X=Table_12_6_Data$PULLUPS,INDEX=list(Table_12_6_Data$GROUP), FUN=mean)
- tapply(X=Table_12_6_Data$VERTJUMP,INDEX=list(Table_12_6_Data$GROUP), FUN=mean)

8. To find the standard deviation of SITUPS, PULLUPS, and VERTJUMP by each group:

- tapply(X=Table_12_6_Data$SITUPS,INDEX=list(Table_12_6_Data$GROUP), FUN=sd)
- tapply(X=Table_12_6_Data$PULLUPS,INDEX=list(Table_12_6_Data$GROUP), FUN=sd)
- tapply(X=Table_12_6_Data$VERTJUMP,INDEX=list(Table_12_6_Data$GROUP), FUN=sd)

9. Conduct a MANOVA on these data. First, we combine the three dependent variables into one "outcome" object. Next, we specify the contrast to give us the relevant SS that we are interested in. Then we conduct the MANOVA and request the output in

- outcome <- cbind(Table_12_6_Data$SITUPS,Table_12_6_Data$PULLUPS,Table_12_6_Data$VERTJUMP)
- options(contrasts = c("contr.helmert", "contr.poly"))
- manova.model <- manova(outcome ~ GROUP,data=Table_12_6_Data)
- summary(manova.model, intercept=TRUE)
- summary(manova.model, intercept=TRUE, test="Wilks")
- summary(manova.model, intercept=TRUE, test="Hotelling")
- summary(manova.model, intercept=TRUE, test="Roy")
- Anova(manova.model,type=3)

10. Finally, explore the univariate results:

- summary.aov(manova.model,intercept = TRUE)

Highland and Run all of the RStudio commands you have entered

1. Why did you decide to conduct a MANOVA?
2. What is your null hypothesis?
3. What is your alternative hypothesis?
4. Describe the correlation results.
5. What are the dependent variable means and standard deviations by group?
6. Interpret the MANOVA results.

7. Present and interpret the univariate results.
8. How would you explain the pull-up results that you obtained?
9. What other independent variables might you add to the model if you conducted future research in this area?

REFERENCES

Church, T. S., Blair, S. N., Cocreham, S., Johannsen, N., Johnson, W., . . . Earnest, C. P. (2010). Effects of aerobic and resistance training on hemoglobin A1c levels in patients with type 2 diabetes: A randomized controlled trial. *JAMA*, *304*, 2253–2262.

Morrow, J. R. Jr., & Frankiewicz, R. G. (1979). Strategies for the analysis of repeated and multiple measures designs. *Research Quarterly*, *50*, 297–304.

Physical Activity Guidelines Advisory Committee. (2018). *Physical activity guidelines advisory committee scientific report*. Washington, DC: U.S. Department of Health and Human Services.

Schutz, R. W., & Gessaroli, M. E. (1987). The analysis of repeated measures designs involving multiple dependent variables. *Research Quarterly for Exercise and Sport*, *58*, 132–149.

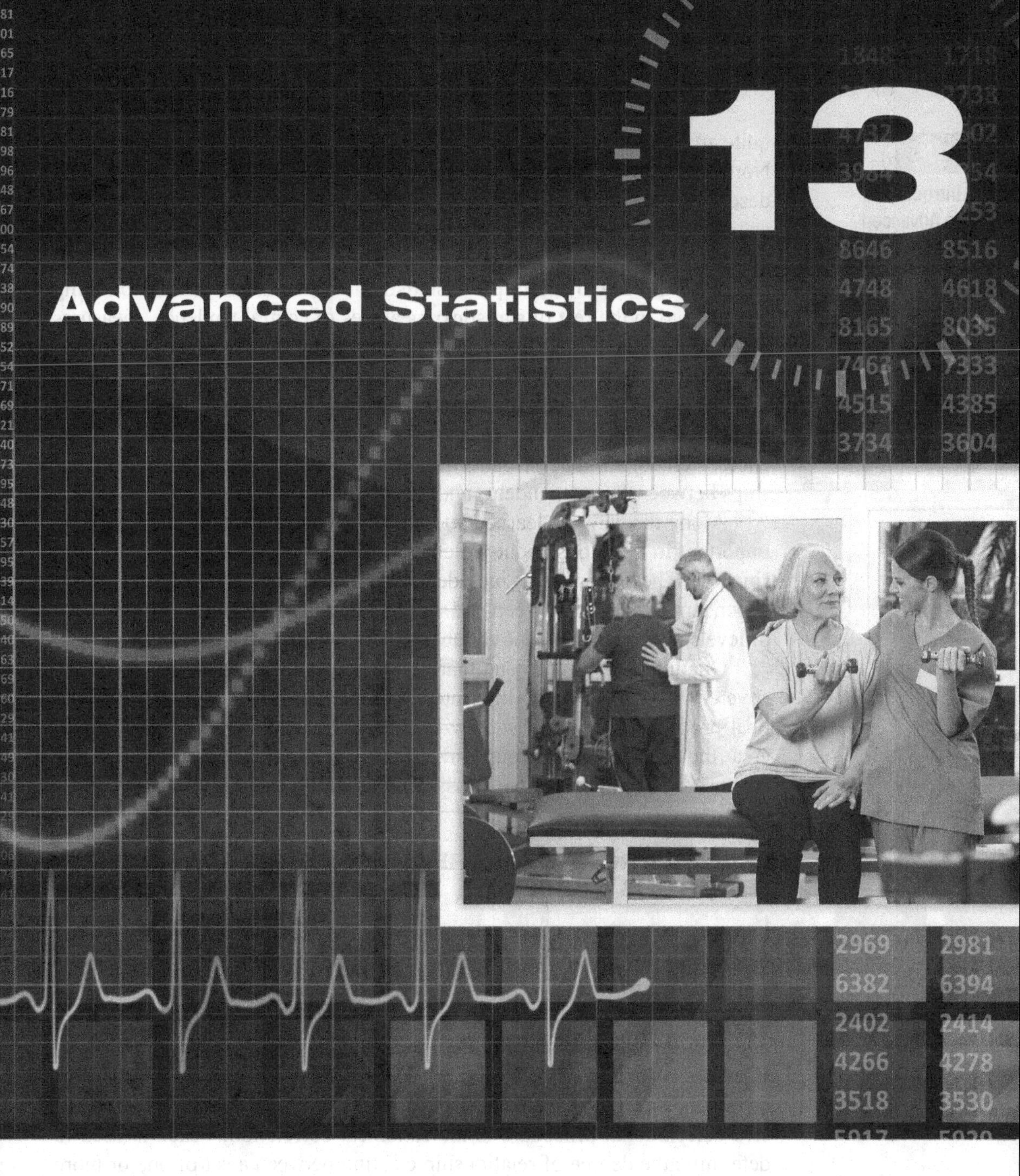

13 Advanced Statistics

INTRODUCTION

In this chapter, we will summarize much of what we have presented thus far, to provide a basis for describing and explaining some advanced statistical procedures. For these advanced procedures, we will not concern ourselves so much with the calculations involved but, rather, with the situations where it would be appropriate to use a particular procedure, the types and numbers of variables each procedure includes, and the results the procedure yield.

We could organize the study of statistical tests in many ways. In fact, all of these tests are connected to one another. For example, the *t*-test can be understood from a correlational point of view, and $t^2 = F$. It is actually

V.13.1 **Introduction to Advanced Statistics**

quite artificial to think of statistical tests as belonging to discrete categories. Nonetheless, we can group the statistics that we have discussed so far as descriptive statistics, correlational and regression statistics, and inferential statistics used primarily to determine group differences (i.e., *t*-test, ANOVA, and MANOVA).

We started with descriptive statistics because these are the building blocks used in the rest of statistics: measures of central tendency, measures of variability, the normal curve, and so forth. Next, we presented some basic correlational statistics, because not only are they descriptive (e.g., the degree of association between two variables), but they are a bridge that we needed to proceed to the branch of inferential statistics that examines group differences. Remember, that ultimately all inferential statistics examine relationships between *X* (independent) and *Y* (dependent) variables.

Along the way, we learned about many descriptors of variables that are important in decisions about which statistical test is appropriate in a given situation. These descriptors include levels of measurement, independent and dependent variables, between and within, random and fixed, and the number of levels of the independent variable(s).

Our intent now is to summarize the statistical tests we have already covered in terms of these descriptors and then to move to advanced statistical procedures. We will start with statistics in the correlation/regression area, then move to the group differences category, and finish with a few advanced techniques that we classify as "other".

CORRELATIONAL AND REGRESSION STATISTICS (REVIEW)

In a real sense, many of the relational statistical procedures, processes, and methods that we have presented throughout this book can be thought of as some portion of what is generally referred to as canonical correlation or the canonical model. In all correlational techniques, the researcher is relating one or more independent variables, *X* (predictor), to one or more dependent variables, *Y* (predicted). The number and the nature of the *X* and *Y* variables dictate the specific statistical analytical procedure to be used. The all-inclusive correlational procedure is the canonical correlation, which is used to determine the degree of relationship existing between a set of one or more *X* variables and another set of one or more *Y* variables. Leading up to this ultimate procedure are all sorts of correlational and regression techniques. The choice of procedure depends on the number and nature of the variables involved, as well as the nature of the problem to be solved. However, the methods, the strategy, the logic, and the interpretation are typically very similar for all of these procedures. The researcher develops a null hypothesis stating that there is no relationship between the *X* and *Y* variables. The researcher then collects data and conducts analyses to determine the probability of the results obtained *when the null hypothesis is true*. If the probability is considerably low (generally, less than .05), then the researcher rejects the null hypothesis in favor of an alternative hypothesis that suggests a relationship does exist between *X* and *Y*.

Let's turn our attention to a review of the correlation and regression techniques presented to this point. We will then provide a summary of some advanced topics.

Pearson Product-Moment Correlation Coefficient

The PPM correlation coefficient is a descriptive statistic. It describes the degree of relationship (i.e., direction and magnitude) existing between two continuously measured variables. The test statistic, r, ranges from −1.00 (perfect negative correlation) to 0.00 (no correlation) to +1.00 (perfect positive correlation). The two variables are designated X and Y at this point, but neither is generally identified as being the independent or dependent variable. X and Y become predictor (independent) and predicted (dependent) in other techniques. See Chapter 5 for complete information about this statistic and its relatives, used when variables with non-continuous levels of measurement are involved.

Bivariate Regression

If some amount of correlation is present between two variables, then it is possible to predict an outcome on one variable (Y) from knowledge of the other variable (X). It is also possible to determine the accuracy of that prediction. The variable for which scores are predicted is labeled the dependent variable (Y), and the predictor variable is called the independent variable (X). The larger the absolute value of the correlation between the two variables, the more accurate the prediction. Bivariate regression results in a regression equation of the form $\hat{Y} = bX + c$, where $\hat{Y}$ = the predicted score on variable Y, b = a coefficient that is multiplied by X (the value of the independent variable), and c = a constant. Recall that b represents the slope of the regression line, and c is the Y-intercept. To estimate the accuracy of the prediction, we use a statistic called the standard error of estimate (also called the standard error of prediction or simply the standard error). As the correlation approaches 1.00 in absolute value, the amount of error in prediction approaches zero. See Chapter 5 for complete information about these statistics.

Multiple Regression

Multiple regression is an extension of bivariate regression in which two or more independent (X, predictor) variables are combined to create a regression equation that predicts the value of a dependent variable (Y). By using what is called the least squares solution, we can perform multiple regression to find the linear combination of the independent variables that yields the highest possible correlation. This correlation is called the multiple correlation, R. Independent variables that correlate highly with the dependent variable but do not correlate highly with each other will generally result in higher multiple correlations. Figure 13.1 represents multiple regression with two predictor variables. In this figure, the square represents the total variance in the dependent variable (Y), and the two shaded areas represent

FIGURE 13.1

Representation of multiple regression with two predictor variables.

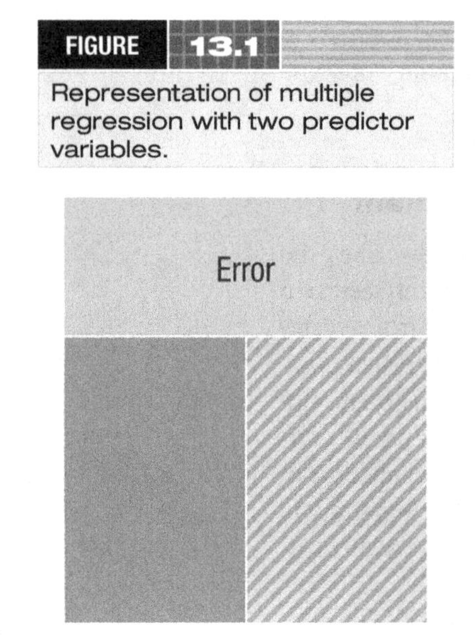

different independent variables *(X)*. The "error" portion is the amount of variance in the dependent variable that is not accounted for by the two independent variables. The intent of multiple regression is to continue to add independent variables (*X*s) to the model to help explain the variation in the dependent variable, thus reducing the error in prediction.

Multiple regression requires that several assumptions be met. Some are issues of design and the selection of independent variables, and others have to do with conditions necessary to minimize prediction errors. The researcher also must make decisions about the order in which the independent variables are added to the regression equation.

In general, a multiple regression equation takes the form of

$$\hat{Y} = b_1X_1 + b_2X_2 + b_3X_3 + \ . \ . \ . \ + b_kX_k + c \quad \text{Eq.13.1}$$

Once again, we refer you to Chapter 5 for additional information about this procedure.

ADVANCED CORRELATIONAL AND REGRESSION STATISTICS

Logistic Regression

The main purpose of logistic regression is to use information from several independent variables (which may be categorically or continuously measured) to predict the value of a dependent variable that is dichotomously measured. In other words, the goal is to predict membership in one of two groups. Typically, the researcher uses logistic regression to predict the membership of a subject in dichotomous groups such as alive/dead or diseased/non-diseased. Rather than predict the specific value for each subject, logistic regression produces the odds that the individual belongs to one of the two groups as a function of change in the predictor variables. In logistic regression, the *X* variables are often "exposure" variables. These exposure variables may be continuous (e.g., hours of physical activity per week) or categorical (e.g., smoker/nonsmoker). Logistic regression allows the researcher to identify the change in odds (called an odds ratio) of the subject's being in one or the other category of the *Y* variable when a change occurs in the exposure *(X)* variable.

Discriminant Analysis

Discriminant analysis (also called discriminant function analysis) can be considered an extension of logistic regression. In logistic regression, the categorical dependent variable consists of only two groups, and the results from the logistic regression are odds ratios. In discriminant analysis, the categorical dependent variable has two or more categories. The independent variables may be continuous or dichotomous. Discriminant analysis can help us identify how each of the independent variables contributes toward classifying an individual into one of the categories of the dependent variable. With discriminant analysis, the researcher develops a combination of the independent *X* variables that best predicts each subject's group membership in the categories represented by the *Y* variable. An important outcome of discriminant analysis is the percentage of observations that are correctly classified. Any discriminant analysis should be cross-validated with another sample to confirm its accuracy. Once a model has been developed, then the researcher can measure subsequent individuals on the predictor variables *(X)* and predict the group *(Y)* into which the individuals are most likely to be categorized.

For example, consider a track and field coach who takes a variety of physical measures on potential team members. He then groups the candidates as sprinters, distance runners, and throwers. With the data collected, he develops a model that predicts group membership from the performance and anthropometric variables. For subsequent individuals who might join the track team, the coach could enter their data into the model to determine which group they most resemble. Obviously, no prediction model is without error, so the coach must be careful when interpreting the results.

Path Analysis

Path analysis, related to regression analysis, involves two or more continuous independent *(X)* variables and a single continuously measured outcome *(Y)* variable. The researcher's intent is to systematically review the various bivariate correlations among the *X* variables as they form a "path" toward predicting the outcome variable. At each step in the path, the variables are adjusted and accounted for, and statements can be made about the relationships between specific predictor variables and the outcome variable. Path analysis helps the researcher identify the direct and indirect effects of *X* variables on the *Y* variable. Its purpose is to identify the relationships among the predictor variables and outcomes when intervening variables are considered. For example, Konopack and McAuley (2012) developed a path analytic model with physical activity and quality of life. Path analysis is a special case of confirmatory factor analysis (which is presented later in this chapter). Figure 13.2 illustrates a path model with various paths provided. Note that the predictor variables have direct paths to En2, as well as indirect paths through En1.

FIGURE 13.2

Path analysis example.

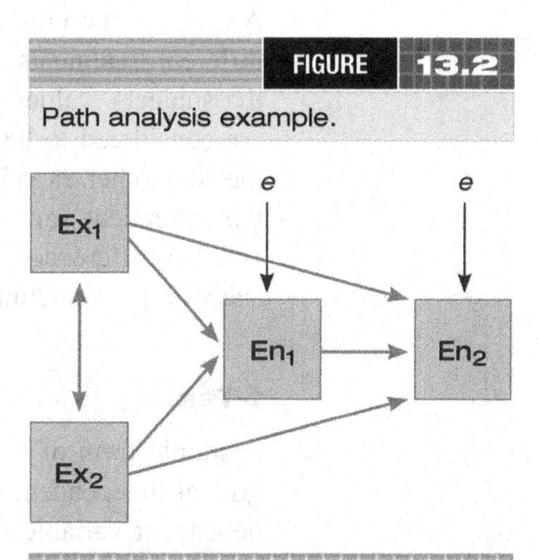

Canonical Correlation

Canonical correlation, mentioned at the beginning of this section, is related to the canonical model.

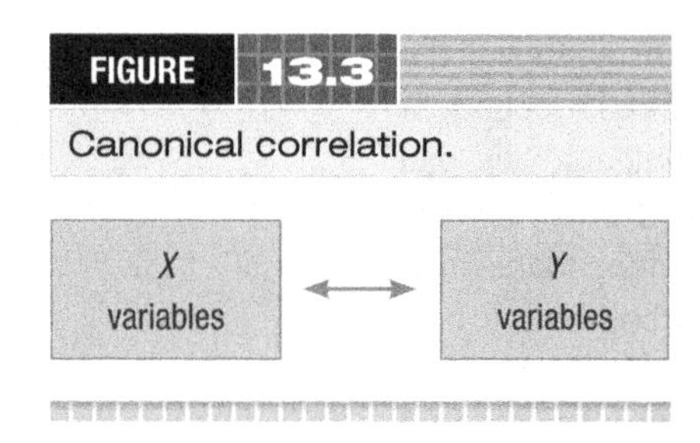

FIGURE 13.3 Canonical correlation.

In fact, canonical correlation can be viewed as the all-encompassing model for inference and hypothesis testing. With canonical correlation, one can have any number of continuous *Y* variables and any number of continuous *X* variables. Figure 13.3 illustrates the general nature of canonical correlation, where the researcher's interest lies in determining the relationship between a "set" of *X* variables and a "set" of *Y* variables. Of course, the "set" could consist of one or more than one variable. If the "sets" are made up of one continuous *X* variable and one continuous *Y* variable, then this is simply the Pearson product-moment correlation. The researcher identifies whether the *X* variables are related to the *Y* variables, and, if so, which *X* variables are most closely related to which *Y* variables. If we change the *X* variable to a categorical variable, we have the *t*-test. Make the *X* variable continuous and the *Y* variable a dichotomy, and we have logistic regression. These examples illustrate that, although the variables might have different characteristics and exist in varying numbers, we can still think of all of these situations as cases of the general canonical model. Chung, Zhao, Liu, and Quach (2017) used canonical correlation to explore the relationship between 7 functional fitness variables (Xs) and 8 health-related quality of life variables (Ys) in older adults. They concluded functional fitness had a significant influence on quality of life.

V.13.2 **Correlational and Regression Statistics**

Summary of Uses

Table 13.1 identifies the appropriate correlational procedure according to (1) the objective, (2) the type and number of independent variables, and (3) the type and number of dependent variables of interest.

MEAN DIFFERENCE STATISTICS (REVIEW)

A very common research design is to expose different groups of subjects to different treatments and then determine whether the exposure *(X)* changes the subjects' values on some other variable *(Y)*. The different treatments are considered to be levels of the independent variable. The variable that the researcher examines to determine whether changes occurred is called the dependent variable *(Y)*. The most common metric used to determine whether differences occur is the mean of the various groups, although other statistics are sometimes used, such as medians or correlation coefficients.

t-Test

A simple form of the type of analysis described above includes one categorical independent variable *(X)* with only two levels and one continuous dependent variable *(Y)*. After the subjects are exposed to one of the two

Objective, type and number of independent and dependent variables, and appropriate statistical test.

TABLE 13.1

CORRELATION AND REGRESSION STATISTICS

OBJECTIVE	TYPE AND NUMBER OF INDEPENDENT VARIABLES	TYPE AND NUMBER OF DEPENDENT VARIABLES	STATISTICAL TEST
To determine the relationship between two variables	Ordinal, 1	Ordinal, 1	Spearman rank order correlation (see Ch. 14)
To determine the relationship between two variables	Continuous, 1	Continuous, 1	Pearson product-moment correlation coefficient
To predict one variable *(Y)* from another *(X)*	Continuous, 1	Continuous, 1	Bivariate regression
To predict one variable *(Y)* from two or more other variables (*X*s)	Continuous, 2 or more	Continuous, 1	Multiple regression
To predict the subject's odds of belonging to one of two categories	Categorical, continuous, or combination, 1 or more	Dichotomous, 1	Logistic regression
To predict that the subject belongs to one of two or more categories	Continuous or dichotomous, 1 or more	Categorical, 1	Discriminant analysis
To determine causal correlations among variables in a theoretical model determined a priori	Continuous, 2 or more	Continuous, 1	Path analysis
To determine the relationship between two sets of variables	Continuous, 1 or more	Continuous, 1 or more	Canonical correlation

treatments *(X)*, the researcher compares the means of the dependent variable *(Y)* from the two groups to determine whether they

- are equal;
- are different, but with a difference that can reasonably be ascribed to chance; or
- are different, with a difference that is too large to be attributable to chance.

The two groups in this design may be composed in one of two ways. Either each group can be made up of different sets of subjects, or the same set of subjects can be measured on two separate occasions.

The appropriate statistical test for this design is called the *t*-test. When the subjects are divided into two different sets, we use an independent *t*-test, and when the same subjects are measured on two occasions (a repeated-measures or within-groups scenario) we use the correlated *t*-test (also called the paired *t*-test). You are encouraged to review Chapter 7 for a complete discussion of these statistical tests.

One-Way ANOVA

The one-way ANOVA is akin to the *t*-test (to be specific, the *t*-test is a special case of one-way ANOVA) and is appropriate when the research design includes one categorically measured independent variable *(X)* and one continuous dependent variable *(Y)*. The difference between the *t*-test and one-way ANOVA is that one-way ANOVA accommodates any number of levels of the independent variable *(X)*. As with the *t*-test, the subjects may be assigned to the treatment groups in two possible ways, for either a between-groups or a within-groups (repeated measures) design. Chapter 9 discusses the one-way between-groups ANOVA, and Chapter 10 covers the one-way repeated-measures (within groups) ANOVA.

Factorial ANOVA

When a research design involves more than one categorically measured independent variable (each with any number of levels) and one continuously measured dependent variable, the appropriate statistical test is a factorial ANOVA. Many new considerations arise with this statistical procedure, including interaction, possible combinations of between-groups and within-groups independent variables, and new assumptions required for the use of this technique. These and other considerations are found in Chapter 11.

One-Way and Factorial MANOVA

The next logical step is to consider research designs with one categorically measured independent variable and two or more continuously measured dependent variables (one-way MANOVA). Designs with multiple categorically measured independent variables and multiple continuously measured dependent variables are factorial MANOVAs. The complex issues associated with MANOVA tests are discussed in Chapter 12.

We will now proceed to describe some advanced statistics in the area of testing group mean differences that are typically used in special situations. The understanding of these techniques is predicated on knowledge of the concepts and techniques thus far presented.

ADVANCED MEAN DIFFERENCES STATISTICS

One-Way Analysis of Covariance (ANCOVA)

Analysis of covariance is essentially a marriage of ANOVA and regression. One-way ANCOVA is a statistical procedure that includes two or more independent *(X)* variables, of which at least one is continuously measured and one is categorically measured, and a single continuously measured dependent *(Y)* variable. With ANCOVA, the researcher first seeks to adjust the *Y* variable for known effects associated with the continuous independent *(X)* variable(s). This *X* variable is called the *covariate* or, sometimes, the *concomitant* variable. The researcher then conducts ANOVA with the adjusted *Y* variable and the categorical independent variable (the one that the researcher is interested in manipulating). The adjustment that occurs generally alters the subjects' *Y* values as if they were all equal on the covariate. Although this adjustment requires that one degree of freedom be subtracted from the error term (which reduces power), this cost is almost always worthwhile, as long as the covariate is correlated to some degree with the dependent variable. Essentially, the variation in the *Y* variable associated with the covariate variable is extracted, and then ANOVA is conducted on the "residualized *Y*" variable.

An example might be helpful. Assume that a researcher is interested in whether physical activity differs as a function of gender. However, the researcher knows that a person's amount of physical activity is also related to age. Therefore, the researcher decides to conduct an ANCOVA with self-reported physical activity as the continuous dependent variable, Gender as the categorical independent variable, and Age as the continuously measured covariate variable. In other words, the researcher wants to "control" or "adjust" for the participants' ages. Thus, the self-reported physical activity value is adjusted for the subject's age, and then the residualized self-reported physical activity, which is unrelated to age, serves as the dependent variable. The researcher can now discuss the ANCOVA results regarding whether self-reported physical activity behaviors differ between males and females, after the effects of age have been removed. Dias, Ferreira, Resende, and Pereira-Baldon (2018) used ANCOVA to evaluate the effectiveness of pelvic floor exercises for pregnant women. The covariate was the pretest strength measure. They concluded the pelvic floor exercises did not change the pelvic floor strength after adjusting for pretest strength. Although this procedure can be useful, especially when total random sampling is not possible, the researcher must take care with the interpretation of the results. It is possible to obtain a result for a world that doesn't exist. For example, it would not make much sense to have initial body weight as a covariate for a study of various diets for mice and elephants.

Factorial ANCOVA

An ANCOVA design can include more than one categorically measured independent variable *(X)*. For example, in the physical activity study in the previous paragraph, the researcher might want to include race/ethnicity as an additional independent variable. The analysis would then become a factorial (two-way) ANCOVA.

It is also possible, both in the one-way and in the factorial ANCOVA, to include more than one continuously measured covariate. Perhaps the researcher wishes to adjust self-reported physical activity for the distance to the nearest park (measured in city blocks), as well as for age. This would result in the loss of two degrees of freedom in the error term. Note that factorial ANCOVA involves a single continuous dependent variable *(Y)* but, potentially, multiple categorically measured independent variables (*X*s) and multiple continuously measured covariate variables (*X*s) can be included in the ANCOVA analysis.

One-Way and Factorial MANCOVA

MANCOVA is an extension of ANCOVA that accommodates multiple continuously measured dependent *(Y)* variables. The independent *(X)* variables are categorical, and the covariates are continuously measured. A number of independent variables (at least one categorical and one continuous) can be

V.13.3
Mean Difference Statistics

used, but multiple continuously measured dependent variables will always be involved in MANCOVA. As an illustration, Hsu and colleagues (2011) used a variety of statistical techniques, including ANCOVA, the PPM correlation, and logistic regression when they investigated the relationships among physical activity, sedentary behaviors, and metabolic syndrome in Latino and African-American youth. They concluded that relationships existed between physical activity and sedentary behaviors and health outcomes. In another use of factorial ANCOVA, Lauer, Jackson, Martin, and Morrow (2017) used MANCOVA adjusting for age and gender when investigating the relationship between six health markers and various levels of physical activity based upon national physical activity guidelines. Results suggested that achieving at least 500 METmin/week was associated with an enhanced health profile.

Summary of Uses

Table 13.2 summarizes the various tests of mean differences that we have presented and helps you identify the appropriate statistical test to conduct based on the number and nature of the independent *(X)* and dependent *(Y)* variables.

OTHER ADVANCED STATISTICAL PROCEDURES

Factor Analysis (Exploratory)

Factor analysis is a bit different from the inferential procedures we have presented so far. Nevertheless, it is an important and widely used statistical procedure. Factor analysis is sometimes referred to as "exploratory" factor analysis, because the researcher is interested in learning what "factors" underlie the many variables that have been measured. Factor analysis is also known as a data-reduction procedure.

Suppose that a researcher wants to investigate the broad topic of motor performance variables. She reviews the literature and identifies 30 tests believed to assess various aspects of motor performance. For example, she might identify tests covering such aspects of motor behavior as speed, strength, endurance, reaction time, balance, flexibility, and kinesthetic perception. She designs a study in which each subject completes all 30 tests. It would be rather difficult for her to discuss the entire "package" of 30 variables. It would also be difficult to discuss all 30 tests one at a time. Thus, the researcher could examine the correlations among all of the tests (there would be 435 of them) and eliminate those that are highly correlated (because they measure the same thing). However, looking at and interpreting a 30 x 30 correlation matrix would itself be rather difficult.

Factor analytic procedures would help this researcher reduce the number of observed variables (30, in this example) to some number, less than 30, of "unobservable" but theoretically existing variables, called *factors*. A factor analytic procedure called principal components would identify 30 factors that would account for all the variance in the original 30 variables. This might not appear to be much of a data reduction. However, the 30

TABLE 13.2 Objective, type and number of independent, dependent, and covariate variables, and appropriate statistical test for mean differences.

MEAN DIFFERENCE STATISTICS

OBJECTIVE	TYPE AND NUMBER OF INDEPENDENT VARIABLES	TYPE AND NUMBER OF DEPENDENT VARIABLES	TYPE AND NUMBER OF COVARIATE VARIABLES	STATISTICAL TEST
To determine whether group means are significantly different	Dichotomous, 1	Continuous, 1	None	*t*-test
To determine whether group means are significantly different	Categorical, 1, with 2 or more levels	Continuous, 1	None	One-way ANOVA
To determine whether group means are significantly different	Categorical, 2 or more	Continuous, 1	None	Factorial ANOVA
To determine whether a linear combination of dependent variables results in significant mean differences	Categorical, 1, with 2 or more levels	Continuous, 2 or more	None	One-way MANOVA
To determine whether a linear combination of dependent variables results in significant mean differences	Categorical, 2 or more	Continuous, 2 or more	None	Factorial MANOVA
To determine whether group means are significantly different after the effect of a related variable is removed from the dependent variable	Categorical, 1	Continuous, 1	Continuous, 1 or more	One-way ANCOVA
To determine whether group means are significantly different after the effect of a related variable is removed from the dependent variable	Categorical, 2 or more	Continuous, 1	Continuous, 1 or more	Factorial ANCOVA
To determine whether a linear combination of dependent variables results in significant mean differences	Categorical, 1	Continuous, 2 or more	Continuous, 1 or more	One-way MANCOVA
To determine whether a linear combination of dependent variables results in significant mean differences	Categorical, 2 or more	Continuous, 2 or more	Continuous, 1 or more	Factorial MANCOVA

factors would be ordered from the one that accounts for the most variance to the one that accounts for the least variance. Associated with each factor would be a value called an *eigenvalue*, calculated by dividing the amount of variance that the factor explains by the average amount of variance that each original variable would be expected to explain, if they were all equally effective in explaining the total amount of variance in the whole data set. In our example, this denominator would be 100/30 = 3.33%. Usually, the first three to five factors will explain a large amount of the variance.

Let's imagine that, in our example, the first five factors explained 35%, 30%, 25%, 3%, and 2.5% of the variance, respectively. The remaining 25 factors would continue to decline in the amount of the unique variance they explain. The eigenvalues associated with the first five factors would be 10.5 (i.e., 35/3.33), 9.0, 7.5, 0.9, and 0.75, respectively. Commonly, a researcher includes only those factors with eigenvalues greater than 1.0 in the solution. Because our solution includes three factors with eigenvalues greater than 1.0, it would be called a *three-factor* solution.

Other factoring procedures follow this step, and their purpose is usually to use iterative rotations of the data to make the remaining factors as orthogonal (uncorrelated) to each other as possible. These rotational procedures do not change the amount of variance explained by the factors, but they help to clarify the factors. An alternative to uncorrelated factors is an "oblique" solution, where the resulting factors are actually correlated.

Regardless of the method of factor determination, the question then becomes, "What are these factors?" This is where the expertise of the researcher becomes important. One of the results of a factor analysis is a table indicating the amount of correlation between each of the original 30 variables and each newly identified factor. These correlations are called *factor loadings*. By examining these factor loadings and noticing which of the original variables correlate at a high level and at a low level with the factors, the researcher decides on a name for each factor. For example, after examining the factor structure in our illustration, the researcher might conclude that the first factor is strength, the second is associated with speed of movement, and the third is associated with flexibility. These three factors summarize the 30 variables and are more easily discussed than the full set of 30 variables. Consider how much easier it will be for the researcher to conduct, for example, a MANOVA on three dependent variables rather than on the original 30 variables. An example of a study using factor analysis is provided by Furzer, Bebich-Philip, Wright, Reid, and Thorton (2017). To examine the construct the Resistance Training Skills Battery for Children, they conducted a factor analysis on strength components, age, and lean muscle mass which resulted in a single factor. Their results provided support for the construct validity of the Resistance Training Skills Battery for Children.

Confirmatory Factor Analysis (Structural Equation Modeling)

Confirmatory factor analysis (CFA), also called structural equation modeling (abbreviated SEM, but not to be confused with the standard error of the mean or standard error of measurement) can serve to explore as well as confirm the relationships among variables. With CFA, the researcher develops a hypothesis about the relationships among variables, essentially building a hypothetical model of the way the variables might be causally related. The researcher then tests this theoretical model with the collected data to see whether the hypothetical relationships are confirmed. Special software (e.g., LISREL, EQS, AMOS, Mplus) is employed for CFA/SEM. The researcher makes decisions about how well the data fit the hypothesized, theoretical model and can draw causal inferences based on the obtained relationships and parameters. For example, with 970 Finnish children aged 9–15 years, Syvaoja et al. (2018) studied the association between physical activity, sedentary behavior, and academic achievement while investigating whether aerobic fitness, obesity, and bedtime mediate these relationships. They concluded that participating in physical activity, avoiding excessive screen time, and going to bed earlier may benefit academic achievement.

Cluster Analysis

Cluster analysis results in observations being assigned to groups with similar characteristics. In a sense, cluster analysis can be thought of as "data mining". The researcher has obtained a large number of variables on subjects and wants to learn whether any characteristics exist within the variables that cause the observations to "cluster" together in similar groups. A variety of clustering algorithms are available, which result in different clustering criteria. The bottom line for all of them is that subjects with similar characteristics (values) on the *X* variables cluster together, because they are similar to one another but different from the members of other clusters. For example, Priego Quesada, Kerr, Bertucci, and Carpes (2018) used cluster analytic procedures to categorize amateur cyclists into homogeneous groups. Four cycling activity measures yielded 5 distinct groups (competitive road, recreational road, competitive mountain bike, recreational mountain bike, and competitive triathlon).

Meta-Analysis

Researchers use meta-analytic procedures when they want to quantitatively combine the results from a large number of studies to draw conclusions about what the studies suggest overall. Historically, a researcher might qualitatively review a large number of studies about the relationship between variables *X* and *Y*. Some of these studies might have suggested a relationship, and others might not. For example, many studies have found significant positive effects of stretching before participating in physical activity, but many other studies have found no significant positive effects. How should researchers and the general population interpret these contradictory and perhaps confusing results? Meta-analysis permits researchers to combine the results of many different studies, quantitatively. Meta-analysis procedures involve the use of effect sizes, which can result from several different statistics (essentially, an effect size can be a z-score or an R^2). The effect sizes become values of the dependent variables of interest, permitting a systemic, quantitative interpretation of the relationship between *X* and *Y* across a large number of studies that are theoretically related. In effect, with this quantification of studies, the studies become the subjects and serve as data points with appropriate statistical techniques. Payne and Morrow (1993), and Payne, Morrow, Johnson, and Dalton (1997) provide illustrations of employing meta-analytic techniques to summarize studies on the effect of aerobic and strength training in children and youth. Numerous

V.13.4
Other Advanced Statistics

meta-analytic studies have been conducted on the effects of various exercise modalities on blood pressure, including those of Igarashi and Nogami (2018) and Casonatto, Goessler, Cornelissen, Cardoso, and Polito (2016)

Summary of Uses

Table 13.3 summarizes the various advanced statistical procedures that have been presented. This table can help you identify the appropriate procedure to conduct based on your objective, the data input variables, and the desired result.

SUMMARY

In this chapter, we discussed correlational and regression statistics, mean difference statistics, and other advanced statistical procedures. In each case, we began with the simplest model and moved to advanced procedures. We provided tables summarizing the conditions under which various statistical procedures are appropriate.

TABLE 13.3 Objective, data input variables, desired result, and statistical procedures for advanced statistics.

OBJECTIVE	DATA INPUT VARIABLES	DESIRED RESULT	STATISTICAL TEST
Data reduction	Many continuous variables	Some "small" number of factors	Exploratory factor analysis
To test to see how well the data fit a hypothesized model	Many continuous variables	Model explaining relationships among variables	Structural equation modeling (confirmatory factor analysis) or cluster analysis
To combine results from related studies quantitatively to draw overall conclusions	Effect sizes from a large number of related studies	To use various statistical tests to determine whether means differ significantly	Meta-analysis

Review Questions and Practice Problems

PART A (ANSWERS PROVIDED IN APPENDIX C)

1. ANCOVA is a combination of which two statistical techniques?
   A. Multiple regression and ANOVA
   B. Parametric and nonparametric statistics
   C. Hypothesis testing and sampling
   D. Reliability and validity

2. What would be the *poorest* choice for a covariate in ANCOVA?
   A. A pretest on the dependent variable
   B. A variable highly related to the dependent variable
   C. A variable moderately related to the dependent variable
   D. A variable unrelated to the dependent variable

3. The cost of using ANCOVA is a loss of one degree of freedom in the error term for each covariate that is used. This could lead to an increase in what?
   A. Interpretation problems
   B. Power
   C. Type I errors
   D. Type II errors
4. A factor having which of the following eigenvalues would probably be retained in a factor analysis?
   A. 0.00
   B. 0.50
   C. 0.99
   D. 1.62
5. Which of the following describes a logistic regression?
   A. Independent variable is continuous and dependent variable is continuous
   B. Independent variable is categorical and dependent variable is continuous
   C. Independent variable is continuous and dependent variable is categorical
   D. Independent variable is categorical and dependent variables are categorical
6. Which statistical procedure would be appropriate to classify people as very fit, somewhat fit, and unfit, if data is available on their strength, VO_2max, muscular endurance, and flexibility?
   A. ANOVA
   B. Discriminant analysis
   C. MANOVA
   D. Factor analysis

PART B

For each of the items listed in Column 1, select the most appropriate statistical procedure in Column 2. Record your answer on the line preceding the item number. Each statistical procedure in Column 2 may be used once, more than once, or not at all.

Column 1	Column 2
_____ 7. Adjusted sum of squares	a. ANCOVA
_____ 8. Bar graph	b. Dependent *t*-test
_____ 9. Coefficient alpha	c. Discriminant analysis
_____ 10. Data reduction	d. Factor analysis
_____ 11. Difference between two independent group means	e. Frequency distribution
_____ 12. Eigenvalues	f. Histogram
_____ 13. Factor A, factor B, interaction	g. Independent *t*-test
_____ 14. Four levels of one independent variable, one dependent variable	h. MANOVA
_____ 15. Interval size	i. Meta-analysis
_____ 16. Intraclass reliability	j. Multiple regression
_____ 17. Mathematical correction for initial differences	k. One-way ANOVA
	l. Pearson *r*
	m. Repeated-measures ANOVA
	n. Three-way ANOVA
	o. Two-way ANOVA

_____ 18. Multiple dependent variables
_____ 19. Nominally measured dependent variable
_____ 20. Wilks' lambda
_____ 21. Predicting one variable from several others
_____ 22. Posttest/pretest design
_____ 23. $R_{0.1234}$
_____ 24. Relationship between X and Y
_____ 25. Seven F-tests
_____ 26. Tabulation of score counts

COMPUTER ACTIVITIES

ANCOVA

TABLE 13.4

Data set.

DRUG 1		DRUG 2	
Before	After	Before	After
5	20	7	19
10	23	12	26
12	30	27	33
9	25	24	35
23	34	18	30
21	40	22	31
14	27	26	34
18	38	21	28
6	34	14	23
13	31	9	22

Twenty male volunteers who are overweight were randomly placed into two groups of 10 subjects each. The purpose of the study was to determine whether a difference existed in the mood of these individuals after they ingested one of two different diet drugs. The concern was that these drugs might cause mood changes as a side-effect. The subjects' mood was measured with a five-item checklist, which was administered just prior to administration of the drug and again two hours later. The higher the score on the checklist, the higher the subject's positive mood. Table 13.4 gives the data collected.

Using SPSS

Examine the data, and then enter the data into SPSS with three columns: Drug, Before, and After. First, run a one-way ANOVA, using the "Before" scores as the dependent variable, to see whether the two groups differed on mood scores before ingesting the drug, even though they were assigned to the groups randomly.

1. What is the overall mood mean for all of the subjects?
2. What are the Before mood means for the two groups?
3. Are the two means significantly different at $\alpha = .05$?

What the researcher really wants to know is whether there is a difference in mood after administration of the drug, so now run a one-way ANOVA, using the "After" scores as the dependent variable.

4. What is the overall mood mean for all subjects?
5. What happened to the overall average mood of the subjects after administration of the drugs?
6. What are the After mood means for the two groups?
7. Are the two mood means significantly different at a = .05?
8. How many degrees of freedom does the error term have in this ANOVA?
9. On the basis of these two ANOVAs, is it reasonable to conclude that either drug results in an increase in positive mood and that the two drugs do not differ significantly in the level of improvement in positive mood?

Being a perceptive statistics student and a statistical wizard, you realize that the two groups in this study are somewhat different (although not statistically significantly) on the initial measurement of mood. You decide that an analysis of covariance (ANCOVA) is in order.

Use the following commands to conduct an ANCOVA with SPSS:

- Go to **Analyze→General Linear Model→Univariate.**
- Enter "Drug" in the Fixed Factor box, "After" in the Dependent Variable box, and "Before" in the Covariate box.
- Click on **Options** and move (OVERALL) and Drug into the Display means for: box. Check Descriptive statistics.
- Click **Continue** and **OK.**

Use the **Tests of Between-Subjects Effects** box to answer the following questions:

10. In effect, what does the ANCOVA do to each subject's "After" score?
11. What is the F-ratio for the drugs?
12. Is it significant at $\alpha = .05$?
13. How many degrees of freedom does the error term have in the ANCOVA?
14. Why did the degrees of freedom change from the one-way ANOVA?

Go to the **Estimated Marginal Means Grand Mean** box.

15. What is the grand mean?
16. How does this compare to your answer to question 4?
17. Why didn't the grand mean change?

Go to box 2 under **Estimated Marginal Means for "Drug"**.

18. What are the mood means for the two groups?
19. How do they compare to your answer to question 6?
20. What has happened to the adjusted average after mood score for the subjects taking drug 1?
21. What has happened to the adjusted average after mood score for the subjects taking drug 2?

22. Why were these adjustments made?
23. On the basis of the ANCOVA and the information available to you, which drug should the company manufacture?

Using R

Use the following R commands to obtain the RStudio output necessary to answer questions 1 through 23 above. Note the format and labeling for the RStudio output is different than the SPSS output (e.g., there is no reference to "boxes" in the RStudio output.)

1. Open RStudio.
2. Install any necessary R packages.

   For Chapter 13, we will need to install the "emmeans" and the "car" package to generate the estimated marginal means (if these packages are not already installed).

   In the lower right pane of the RStudio, click on the "Packages" tab and then click on the "Install" button.

   In the "Packages" field, type in "emmeans" and click "Install".

   1. Repeat for the "car" package

   [Note that from this point forward, the packages "emmeans" and "car" will be installed, and you do not have to install them again.]

3. Set up your RStudio workspace.

   In the RStudio menu bar, click on **File → New File → R Script**.

   On the first line of the R Script (upper left pane of RStudio), type #Chapter 13.

   On the second line of the R Script, type "library(emmeans)".

   On the third line of the R Script, type "library(car)".

   When you have done that, you should see the following in your new R Script:

   - #Chapter 13
   - library(emmeans)
   - library(car)

   Highlight these three lines that you've typed and click **Run**.

   [Note that the second line and third line is where we tell R that we will require the functions that are included in the "emmeans" and "car" packages.]

Conducting the R analysis

4. Import the data from Table 13.4.

5. Before starting our analysis, we can label DRUG (1="Drug 1", 1="Drug 2") by first instructing R to treat DRUG as a factor and then assigning the appropriate level.
   - Table_13_4_Data$DRUG <- as.factor(Table_13_4_Data$DRUG)
   - levels(Table_13_4_Data$DRUG) <- list("DRUG1"=1,"DRUG 2"=2)

6. To generate the mean for all BEFORE mood mean values and the BEFORE mood mean values by DRUG, type the following:
   - #BEFORE
   - mean(Table_13_4_Data$BEFORE)
   - tapply(X=Table_13_4_Data$BEFORE,INDEX=list(Table_13_4_Data$DRUG), FUN=mean)

7. To conduct a one-way ANOVA on the BEFORE mood mean values type the following:
   - before.aov <- aov(BEFORE ~ DRUG, data=Table_13_4_Data)
   - summary(before.aov)

8. To generate the mean for all AFTER mood mean values and the AFTER mood mean values by DRUG, type the following:
   - #AFTER
   - mean(Table_13_4_Data$AFTER)
   - tapply(X=Table_13_4_Data$AFTER, INDEX=list(Table_13_4_Data$DRUG), FUN=mean)

9. To conduct a one-way ANOVA on the AFTER mood mean values type the following:
   - after.aov <- aov(AFTER ~ DRUG, data=Table_13_4_Data)
   - summary(after.aov)

10. To run an ANCOVA on these data, we can type the following:
    - #ANCOVA
    - ancova <- aov(AFTER ~ DRUG + BEFORE, data=Table_13_4_Data)
    - summary(ancova, type = 3)

11. To display the grand mean and the estimated marginal means for DRUG, we can type the following:
    - model.tables(ancova,"means", cterms="DRUG")
    - emmeans(ancova, ~DRUG)

Highlight and **Run** all of the RStudio commands you have entered.

REFERENCES

Casonatto, J., Goessler, K. F., Cornelissen, V. A., Cardoso, J. R., & Polito, M. D. (2016). The blood pressure-lowering effect of a single bout of resistance exercise: A systematic review and meta-analysis of randomized controlled trials. *European Journal of Preventive Cardiology*, *23*(16), 1700–1714.

Chung, P. K., Zhao, Y., Liu, J. D., & Quach, B. (2017). A canonical correlation analysis on the relationship between functional fitness and health-related quality of life in older adults. *Archives of Gerontology and Geriatrics*, *68*, 44–48.

Dias, N. T., Ferreira, L. R., Resende, A. P. M., & Pereira-Baldon, V. S. (2018). A pilates exercise program with pelvic floor muscle contration: Is it effective for pregnant women? A randomized controlled study. *Neurourology and Urodynamics*, *37*, 379–384.

Furzer, B. J., Bebich-Philip, M. D., Wright, K. E., Reid, S. L., & Thorton, A. L. (2017). Reliability and validity of the adapted resistance training skills battery for children. *Journal of Science and Medicine in Sport*, *21*, 822–827. doi:10.1016/jsams.2017.12.010.

Hsu, Y. W., Belcher, B. R., Ventura, E. E., Byrd-Williams, C. E., Weigensberg, M. J., Davis, J. N., . . . Spruijt-Metz, D. (2011, December). Physical activity, sedentary behavior, and the metabolic syndrome in minority youth. *Medicine & Science in Sport and Exercise*, *43*(12), 2307–2313.

Igarashi, Y., & Nogami, Y. (2018). The effect of regular aquatic exercise on blood pressure: A meta-analysis of randomized controlled trials. *European Journal of Preventive Cardiology*, *25*(2), 190–199. doi:10.1177/2047487317731164.

Konopack, J. F., & McAuley, E. (2012, May). Efficacy mediated effects of spirituality and physical activity on quality of life: A path analysis. *Health and Quality of Life Outcomes*, *29*(10), 57. doi:10.1186/1477-7525-10-57.

Lauer, E. E., Jackson, A. W., Martin, S. B., & Morrow, J. R. Jr. (2017). Meeting USDHHS physical activity guidelines and health outcomes. *International Journal of Exercise Science*, *10*, 121–127.

Payne, V. G., & Morrow, J. R. Jr. (1993). The effects of physical training on children's VO2max: A meta-analysis. *Research Quarterly for Exercise and Sport*, *64*, 305–313.

Payne, V. G., Morrow, J. R. Jr., Johnson, L., & Dalton, S. N. (1997). Resistance training in children and youth: A meta-analysis. *Research Quarterly for Exercise and Sport*, *68*, 80–88.

Priego Quesada, J. I., Kerr, Z. Y., Bertucci, W. M., & Carpes, F. P. (2018). The categorization of amateur cyclists as research participants: Findings from an observational study. *Journal of Sports Science*, *25*, 1–7.

Syvaoja, H. J., Kankaanpaa, A., Kallio, J., Hakonen, H., Kulmala, J., Hillman, C. H., . . . Tammelin, T. H. (2018). The relation of physical activity, sedentary behaviors, and academic achievement is mediated by fitness and bedtime. *Journal of Physical Activity and Health*, *15*, 135–143.

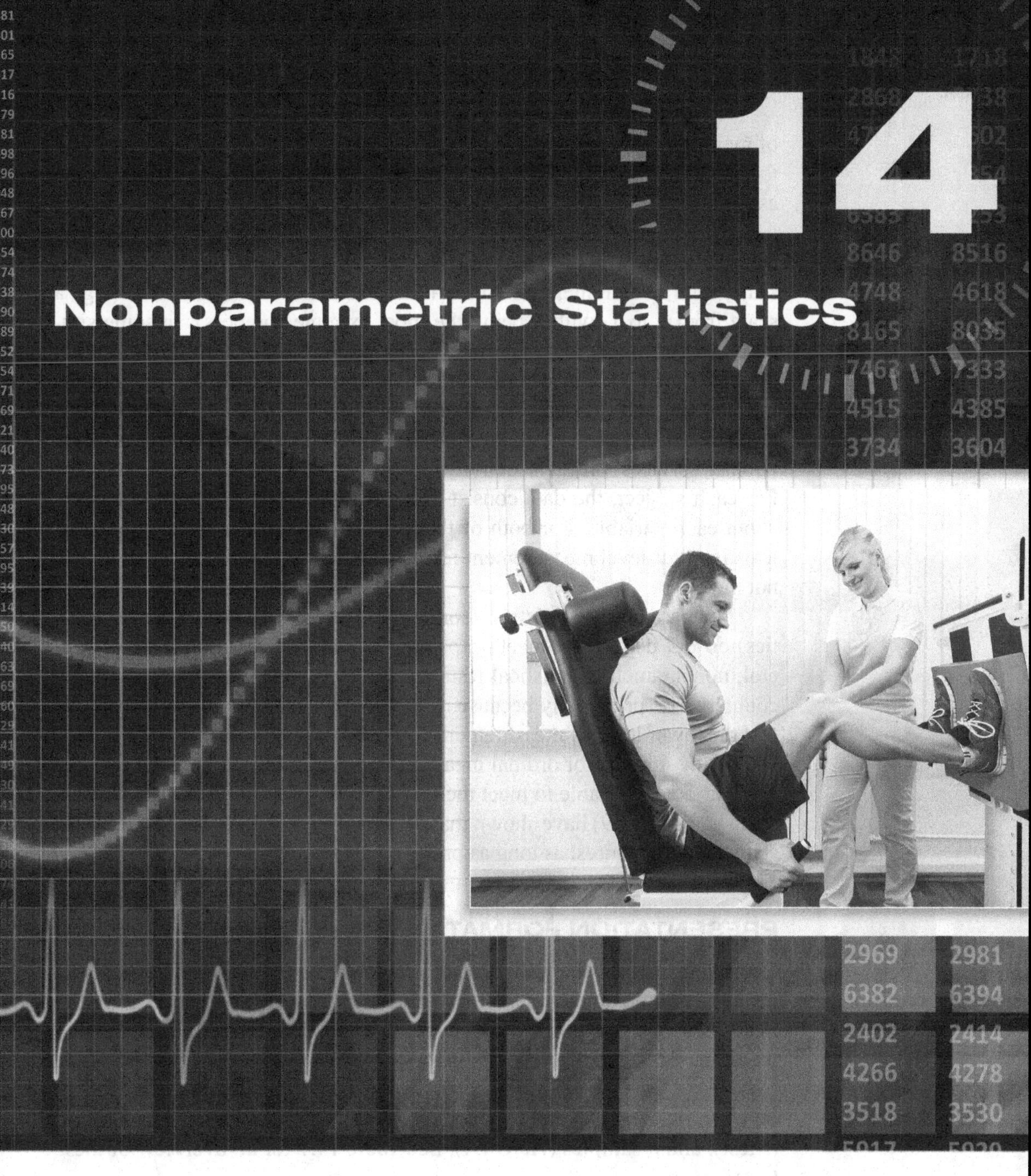

14 Nonparametric Statistics

INTRODUCTION

For the most part, the statistical tests we have examined to this point are called *parametric statistics*. We call them parametric because of the nature of the distributions that are assumed with procedures such as correlation, the *t*-test, and ANOVA. Their accuracy depends on the validity of certain assumptions about the population from which the samples have been drawn and about the samples themselves. In most cases, we listed these assumptions for each of the tests we presented. The most common of these assumptions are normality, equal variances, and at least the interval level of measurement for the dependent variable. As we have indicated, many of the statistical procedures we have presented are robust to violations of the assumptions.

Occasionally, it is not possible to meet these assumptions. For example, the measurement of muscle soreness often is limited to the use of an ordinal scale, such as extreme, quite sore, moderate, not too sore, and none (e.g., a scale from 5 to 1, with 5 being the most extreme soreness). In some situations, the available data are nominal in nature. For example, a researcher might be interested in determining whether an association exists between tobacco use and the achievement of 150 minutes of moderate to vigorous physical activity weekly, as suggested by the U.S. Department of Health and Human Services' Physical Activity Guidelines Advisory Committee (2018). For each subject, the data consist of an answer of yes or no to a question about each variable. For both of these two variables, the assumption of at least interval-level measurement for parametric statistical tests is obviously not met.

Nonparametric statistics (sometimes called distribution-free statistics) do not depend on meeting parametric statistics assumptions. In general, nonparametric statistical tests are not as powerful as their parametric counterparts (principally because of the diminished amount of information that the lower level of measurement conveys), but they are extremely useful when only nominal or ordinal measurement is possible. Where the concern is about not being able to meet the normality assumption, Thomas, Nelson, and Thomas (1999) have shown that very little power, if any, is lost for most statistical procedures, as long as ordinal data are available.

PRESENTATION FORMAT

The intent of this chapter is to introduce a representative sample of the available nonparametric statistical tests. Many other nonparametric statistical tests exist, but we will not present all of them here. In this chapter, we focus on nonparametric statistics that are analogous to the following parametric tests: correlation between two variables (PPM) and differences between independent and dependent groups with one sample, two samples (t-test), and n samples (ANOVA). See Table 14.1 for an overview of these tests.

In discussing the tests, we will follow the same order as the presentation of parametric statistical tests in the earlier chapters. Tests examining the amount of association or correlation between variables are followed by tests to determine whether differences exist between groups, starting with the one-sample case, then tests to determine whether differences exist between two samples, and finally the n-sample case. For each test, we present the situation for which the test is appropriate, the equations involved, and a brief example, either fictitious or drawn from the literature, that illustrates the calculation of the nonparametric statistic involved.

Introduction to Nonparametric Statistics

TABLE 14.1 Summary of nonparametric tests presented in this chapter and their parametric equivalents.

Parametric Test	Nonparametric Equivalent	
	Level of Measurement	
↓	*Nominal*	*Ordinal*
Tests of Association		
Pearson's *r*	Phi coefficient and kappa coefficient of agreement	Spearman rank order correlation coefficient and Kendall coefficient of concordance
Tests of Significant Differences		
One-sample case	Binomial test and chi-square test of goodness of fit	
Two-sample case		
Independent *t*-test	Chi-square for two independent samples	Median test and Mann–Whitney *U* test
Dependent *t*-test		Sign test and Wilcoxon signed-rank test
N-sample case		
Independent groups ANOVA	Chi-square for *n*-independent samples	Kruskal–Wallis test for independent samples
Repeated-measures ANOVA	Cochran's *Q* test	Friedman two-way ANOVA by ranks test

NONPARAMETRIC STATISTICAL TESTS OF ASSOCIATION

Phi Coefficient (r_Φ)

This test is analogous to the parametric Pearson product-moment correlation coefficient. Like the Pearson correlation, the limits for the phi coefficient are −1.00 to +1.00. Essentially, the phi coefficient *is* the Pearson correlation between variables that are dichotomous in nature (i.e., scored 0 or 1). To examine the data, we generally create in a 2 x 2 contingency table, as illustrated in Table 14.2.

TABLE 14.2 Generic 2 x 2 contingency table for calculating r_Φ.

	VARIABLE *X*		
VARIABLE *Y*	0	1	TOTALS
0	*a*	*b*	*a* + *b*
1	*c*	*d*	*c* + *d*
Totals	*a* + *c*	*b* + *d*	*N*

Equation

The equation for the phi coefficient is

$$r_\Phi = \frac{ad - bc}{\sqrt{(a+b)(c+d)(a+c)(b+d)}} \qquad \text{Eq. 14.1}$$

where

r_Φ = the phi coefficient and

a, *b*, *c*, and *d* = the frequencies in the cells of the data table.

Example

For a random sample of 105 adults, researchers wanted to determine whether their body mass index was above or below 25, along with their answer to

TABLE 14.3

Example data in a 2 x 2 contingency table.

	VARIABLE X (BMI)		
VARIABLE Y	Above 25 (0)	Below 25 (1)	Row total
Elevator (0)	38	12	50
Stairs (1)	28	27	55
Column total	66	39	105

the question, "When an option is available, do you normally take the stairs or the elevator?" Table 14.3 gives the resulting data.

For this example,

$$r_\Phi = \frac{(38)(27)-(12)(28)}{\sqrt{(50)(55)(66)(39)}} = .259$$

A positive association exists between the two variables, indicating that the frequencies classified as 1s for both variables tend to associate with each other, and the frequencies classified as 0s tend to go together. Inspecting the data, we find that subjects with a BMI below 25 tend to take the stairs more often (69%) than those with a BMI over 25 (42%).

By using a statistic (X^2) that is related to a nonparametric statistic called chi-square (χ^2), described later in this chapter, we can determine whether or not the value of r_Φ is significantly different from zero. That is, we can determine whether the obtained value of r_Φ is greater than the highest value that could be reasonably be obtained by chance at a selected confidence level when the two variables are not actually related. The equation for calculating X^2 is

$$X^2 = \frac{N\left(|ad-bd| - \frac{N}{2}\right)^2}{(a+b)(c+d)(a+c)(b+d)} \quad \text{Eq. 14.2}$$

where

X^2 = a statistic that is distributed as chi-square (χ^2) with 1 *df*, and

a, b, c, and d = the frequencies in the cells of the data table. For this example,

$$X^2 = \frac{105(|\ (38)(27) - (12)(28)\ | - \frac{105}{2})^2}{(50)(55)(66)(39)}$$

$$= \frac{105(|\ 1026 - 336\ | - 52.5)^2}{7078500} = \frac{42672656}{7078500} = 6.03$$

If the desired level of α is .05, the critical value of χ^2 with 1 *df* is 3.84 (see Appendix A.6). Since 6.03 is greater than 3.84, the conclusion is to reject the null hypothesis and declare that the r_Φ of .259 is statistically significant. In other words, the probability of obtaining an r_Φ of .259 by chance from a population where the two variables are not related is less than 5%.

Kappa Coefficient of Agreement

Situation

Like the phi coefficient, we use the kappa coefficient of agreement to evaluate the association between two nominally measured variables. The most common use of kappa is in assessing inter-rater agreement. With human performance data, researchers most often use kappa to assess the reliability and validity of criterion-referenced measures. For example, suppose a researcher uses the FITNESSGRAM PACER to determine whether subjects achieve the Healthy Fitness Zone on consecutive Mondays. This researcher can examine the consistency (reliability) of subjects' achieving (or not achieving) the Healthy Fitness Zone with the kappa coefficient of agreement. See Table 14.4 for an example of how data would be set up to calculate test reliability.

TABLE 14.4

Generic 2 x 2 contingency table for calculating kappa for reliability.

	RETEST		
TEST	0 (fail)	1 (pass)	Row totals
0 (fail)	a	b	$a+b$
1 (pass)	c	d	$c+d$
Column totals	$a+c$	$b+d$	N

Equation

The equation for the kappa coefficient of agreement is

$$K = \frac{P - P_c}{1 - P_c} \quad \text{Eq. 14.3}$$

where

K = the kappa coefficient of agreement,
P = the proportion of agreement, and
P_c = the proportion of agreement attributable to chance.

Other statistics might be useful in this situation, but many of these do not remove the amount of agreement that is attributable to chance. A virtue of the kappa statistic is that it gives a conservative estimate of reliability or validity, because it does remove the chance agreements.

Example

The data in Table 14.5 resulted from an investigation of the reliability of a 12-minute run for distance test administered to a group of 300 subjects on two different occasions.

We must first calculate percent agreement (P) with equation 14.4:

$$P = \frac{(a + d)}{N} \quad \text{Eq. 14.4}$$

where

P = percent agreement,
a = the frequency in cell a,
d = the frequency in cell d, and
N = the number of subjects.

Hence,

$$P = \frac{(60 + 187)}{300} = .823$$

TABLE 14.5 Example data format for calculating kappa for reliability.

		DAY 2		
		Did not achieve criterion	Did achieve criterion	Total
DAY 1	Did not achieve criterion	60	15	75
	Did achieve criterion	38	187	225
	Total	98	202	300

Given the distribution of scores, the expected values of the "non-agreement" cells (b and c) represent the chance agreements. We calculate these expected values by multiplying the appropriate margin values for the respective row and column totals together and dividing by N^2. The equation is

$$P_c = \frac{(a+c)(a+b)}{N^2} + \frac{(b+d)(c+d)}{N^2} \quad \text{Eq. 14.5}$$

Hence,

$$P_c = \frac{(98)(75)}{(300)^2} + \frac{(202)(225)}{(300)^2} = \frac{7350}{90000} + \frac{45450}{90000} = .587$$

We now have the values necessary to calculate kappa. We substitute these values in equation 14.3 and find:

$$K = \frac{.823 - .587}{1 - .587} = \frac{.236}{.413} = .571$$

Thus, the percent of agreement of .823 decreases to .571 when the chance agreements are removed. The phi coefficient for these data is .58, which is similar to the slightly more conservative kappa coefficient.

V.14.2
SPSS Phi and Kappa

V.14.2R
RStudio Phi and Kappa

To illustrate the use of kappa to assess validity, imagine that the data in Table 14.5 represent the number of subjects who do or do not meet the criterion for the 12-minute run for distance, as one variable, and a treadmill test to measure VO_2max, as the other variable. In this case kappa would now represent a validity coefficient. Higher values of kappa indicate stronger relationships between variables. Viera and Garrett (2005) suggest that kappa values less than .20 indicate poor agreement, those between .41 and .60 are moderate, and those above .60 are substantial.

Spearman Rank Order Correlation Coefficient

Situation

When we wish to determine the amount of association between two variables that are both measured on an ordinal scale, the Spearman rank order correlation coefficient (ρ or r_s) is the appropriate statistic. If the data include no ties, ρ results in the same value as the Pearson product-moment correlation coefficient *(r)* between the ranks. Although equations are available to correct for ties, the corrected and uncorrected estimates of ρ, even when ties are present, will differ by a very slight amount.

Equation

The equation for the Spearman rank order correlation coefficient is

$$\rho = 1 - \frac{6\sigma d^2}{N(N^2 - 1)} \qquad \text{Eq. 14.6}$$

where

ρ = the Spearman rank order correlation coefficient (also r_s),
σd^2 = the sum of the squared differences between each subject's two ranks, and
N = the number of subjects.

Example

To gather evidence for the validity of a battery of tests to measure volleyball playing ability among female high school students, a researcher administered the test battery to a random sample of 20 subjects and summed their scores in such a way as to produce a ranking of the subjects from 1 to 20. In addition, expert volleyball coaches created a second ranking of the same subjects by observing the subjects while they played volleyball. The resulting data are displayed in Table 14.6

Example data format for calculating ρ. TABLE 14.6

PLAYER	BATTERY RANK	PLAYING RANK	DIFFERENCE IN RANKS	DIFFERENCE IN RANKS SQUARED
1	2	7	–5	25
2	7	8	–1	1
3	12	16	–4	16
4	5	3	2	4
5	17	17	0	0
6	1	4	–3	9
7	14	9	5	25
8	19	13	6	36
9	9	5	4	16
10	13	10	3	9
11	18	20	–2	4
12	8	15	–7	49
13	10	12	–2	4
14	20	14	6	36
15	15	11	4	16
16	11	19	–8	64
17	3	1	2	4
18	4	6	–2	4
19	16	18	–2	4
20	6	2	4	16
				$\Sigma d_2 = 342$

When we substitute these values in equation 14.6, we find

$$\rho = 1 - \frac{6(342)}{20(400-1)} = 1 - \frac{2052}{7980} = 1 - .26 = .743$$

We can interpret ρ^2 like the coefficient of determination, r^2. Here, $r_s^2 = (.743)^2 = .552$, indicating that the test results account for about 55% of the variance in the judges' ratings. Thus, ρ offers some evidence of validity for the test battery.

Testing the Statistical Significance of ρ

If N is 20 or greater, we can examine the statistical significance of ρ by using the following equation:

$$z = \rho\sqrt{N-1} \qquad \text{Eq. 14.7}$$

where

z = a z-score,
ρ = the obtained value of ρ, and
N = the number of subjects.

V.14.3
SPSS Spearman Rank Correlation

V.14.3R
RStudio Spearman Rank Correlation

For the data above, this equation results in a z-score of 3.24. We can conclude that it would be extremely unlikely to obtain a value of .743 for ρ by chance from a population in which the two variables are not correlated. Recall that a z-score of 3.24 is in the extreme tail of the z-distribution, and we reject the null hypothesis with $\alpha = .05$ when the z-score exceeds 1.96.

For situations with Ns less than 20, tables are available to determine whether or not the ρ value obtained is statistically significant. Appendix A.7 provides one such table, presenting the critical values of ρ.

Kendall Coefficient of Concordance (W)

Situation

We use the Spearman rank order correlation coefficient to assess the degree of relationship between two ordinally measured variables. To determine the degree of relationship among more than two sets of rankings, we can use the Kendall coefficient of concordance. For example, judges' scores determine the results of many athletic events (e.g., springboard diving, figure skating, and gymnastics), and it may be of interest to determine whether a particular set of judges' scores results in a consistent ranking of the participants.

Data Arrangement

To calculate W, we arrange the data in a two-way table, with the rows containing the rankings of the judges and the columns representing the participants. For each column, we obtain the sum of the ranks (R_i) and the mean of the ranks $(\bar{R}_i)$. Then, we calculate W with equation 14.8.

Equation

The equation for the Kendall coefficient of concordance is

$$W = \frac{\sigma(\bar{R}_i - \bar{R})^2}{\frac{N(N^2 - 1)}{12}}$$ Eq. 14.8

where

W = the Kendall coefficient of concordance,
$\bar{R}_i \cdot$ = the mean of the rankings of column i,
$\bar{R}$ = the mean of all of the rankings, and
N = the number of participants.

Example

The data in Table 14.7 represent the rankings of three judges (numbered 1, 2, and 3) scoring six participants (A through F) in a gymnastics event.

We first calculate $\bar{R}$ by using the following equation:

$$\bar{R} = \frac{\textit{number of participants} + 1}{2}$$

$$= \frac{6+1}{2} = 3.5$$

Next, W is calculated, as follows:

$$W = \frac{\sigma((2-3.5)^2 + (5.67-3.5)^2 + (1.33-3.5)^2 + (5.33-3.5)^2 + (3.33-3.5)^2 + (3.33-3.5)^2)}{\frac{6(35)}{12}}$$

$$= \frac{\sigma(2.25 + 4.71 + 4.71 + 3.35 + 0.029 + 0.029)}{\frac{6(35)}{12}}$$

$$= \frac{15.078}{17.5} = .862$$

The range of the W statistic is from 0 to +1.0. The value of 1.0 signifies perfect agreement among the judges (i.e., the judges all ranked the participants in the exact same order), and the value of 0 indicates a complete lack of

TABLE 14.7 Example data format for calculating the Kendall coefficient of concordance.

	PARTICIPANT					
Judge	**A**	**B**	**C**	**D**	**E**	**F**
1	1	5	2	6	4	3
2	3	6	1	5	2	4
3	2	6	1	5	4	3
R_i	6	17	4	16	10	10
$\bar{R}_i$	2.0	5.67	1.33	5.33	3.33	3.33

agreement among the judges. In this example, we conclude that the amount of agreement among this set of judges is reasonably high.

Issues

If the data include a small number of tied rankings, the effect on the value of W is to reduce it, but only minimally. Correction equations are available for cases where the number of tied rankings is fairly large.

We can test the statistical significance of W by using tables when N is 7 or less or, when N is greater than 7, an equation resulting in a statistic that is distributed as a chi-square. These tests are beyond the scope of this textbook, but they may be found in Siegel and Castellan (1988). For the example presented above, the value of .862 for W is statistically significant with an alpha level of .01.

NONPARAMETRIC TESTS OF SIGNIFICANT DIFFERENCES

The remaining nonparametric statistical tests we present are analogous to parametric statistical tests that indicate whether significant differences exist between a sample statistic and a population parameter or between two or more samples taken from the same or identical populations. We present the procedures in the same order as the parametric tests presented in earlier chapters: tests appropriate for one sample, for two independent samples, for two related samples, for n independent samples, and for n related samples. Generally, within each category we present them in order from the least to the most powerful.

One-Sample Case

As in parametric statistics, the situation sometimes arises where one sample is randomly drawn from a population, and we wish to determine the probability that the sample could have come from a population with known or hypothesized parameters, such as the mean or expected frequencies.

Binomial Test

The most fundamental one-sample case occurs when the population sampled is dichotomous. That is, it consists of only two possibilities (e.g., male/female, smoker/nonsmoker). To determine the probability that the sample came from a population with known or hypothesized values for the percentages of individuals in each of the two categories of the dichotomy, we use the binomial test.

Data Format

A subject from a dichotomous population can have only two possible values, and it is customary to label these outcomes as 1 and 0. The probability of obtaining a 1, labeled p, is the same as the proportion of 1s in the population. The probability of obtaining a 0 is $1 - p$, and this probability is labeled q.

Equation

The equation for the binomial test is

$$P = \binom{N}{k} p^k q^{N-k} \quad \text{Eq. 14.9}$$

where

P = the probability of obtaining k objects in one category and $N - k$ in the other category,
p = the proportion of 1s in the population,
q = the proportion of 0s in the population,
k = the number of 1s in the sample,
N = the sample size, and

$$\binom{N}{k} = \frac{N!}{(k!(N-k)!)}$$

Where $N!$ = N factorial (e.g., when $N = 5$, $N! = 5 * 4 * 3 * 2 * 1$).

Example

Assume that the proportion of individuals who smoke tobacco in a certain population is 25%. If we randomly select five subjects from that population, what is the probability that two of them will be smokers?

In this situation, $N = 5$, $k = 2$, $p = .25$ and $q = .75$. Hence,

$$P = \binom{N}{k} p^k\, q^{N-k}$$

$$= \frac{5!}{2!(5-2)!} * (.25)^2 * (.75)^3 = \frac{(5)(4)(3)(2)(1)}{(2)(1)(3)(2)(1)} *(.25)^2*(.75$$

$$= (10)*(.25)^2*(.75)^3 = (10)(.0625)(.422) = 0.26$$

Thus, 26% of the samples of five individuals from a population made up of 25% smokers and 75% nonsmokers are expected to contain two smokers and three nonsmokers.

Chi-Square (χ^2) Goodness of Fit

The χ^2 goodness of fit is one of the most useful and most frequently used nonparametric statistical tests. Its primary function is to compare obtained frequencies in a sample with hypothesized expected frequencies in a population. The expected frequencies in the population may be available because the character of the population is known, or they may be hypothesized to represent a particular theory. Unlike the binomial test, chi-square can accommodate more than two categories.

If we know the proportion of observations in various categories in the population, then we can use chi-square to determine the probability that a sample could reasonably have been drawn from that population. For example, if from a large national study researchers have determined the percentages of individuals with BMIs in the normal, overweight, obese, and extremely obese categories, we could measure a random sample of individuals from a particular community, classify the observations into these categories, and use chi-square to determine whether the sample percentages differ significantly from the national data.

When the proportion of subjects in the population is not known, we can use chi-square to determine whether the distribution of subjects in the sample either conforms to a researcher's hypothesis about how they would be distributed or, more commonly, how it differs from being equally distributed among all possible categories. For example, if we wanted to determine which of four different treatments to relieve muscle soreness is the most effective, we might treat each of four groups of subjects with one of the four modalities. Then, we would classify all of the subjects into the categories of good results, fair results, and poor results. The null hypothesis in this case would state that equal proportions of subjects in each of the treatment groups would be classified into each of the three results categories. Chi-square would be the appropriate statistical test to determine whether the distribution of subjects conformed to the null hypothesis (i.e., no significant differences in results categories among the modalities) or one or more of the treatment groups differed from this hypothesized outcome.

Equation

The equation for the chi-square test is

$$\chi^2 = \frac{\sigma_{i=1}^{N}(O_i - E_i)^2}{E_i} \qquad \text{Eq. 14.10}$$

where

χ^2 = the chi-square statistic,
O_i = the observed frequency in category i,
E_i = the expected frequency in category i, and
N = the number of categories.

Example

To determine whether the percentage of the citizens of a particular city smoke at the same rate as the citizens of the United States (say, 25%), we select 100 random subjects and ask whether they smoke or not. Fourteen answer yes, and 86 answer no. The data are presented in Table 14.8.

TABLE 14.8 Example data format for calculating chi-square for goodness of fit.

	SAMPLE	POPULATION	TOTAL
Smoke (%)	14	25	39
Do not smoke (%)	86	75	161
Total	100	100	200

To determine the expected frequency for each cell if the null hypothesis is true (i.e., no difference exists between the percentage of smokers in the population and the sample), we multiply the cell margins for each cell and divide the resulting product by *N*. For example, the expected frequency for the smokers in the sample cell is (39 * 100)/200 = 19.5. We do this operation for each cell in the table and then use equation 14.10 to calculate the value of χ^2:

$$\begin{aligned} \text{cell a} &= (39 * 100)/200 = 19.5, \\ \text{cell b} &= (39 * 100)/200 = 19.5, \\ \text{cell c} &= (161 * 100)/200 = 80.5, \text{ and} \\ \text{cell d} &= (161 * 100)/200 = 80.5. \end{aligned}$$

Hence,

$$\begin{aligned} \chi^2 &= \frac{(14-19.5)^2}{19.5} + \frac{(25-19.5)^2}{19.5} + \frac{(86-80.5)^2}{80.5} + \frac{(75-80.5)^2}{80.5} \\ &= 1.55 + 1.55 + .38 + .38 = 3.86 \end{aligned}$$

We find the *df* for χ^2 by multiplying the number of rows in the data table, less one (i.e., R − 1), by the number of columns in the data table, less one (i.e., *C* − 1). In the example, (R − 1) * (C − 1) = (2 − 1) * (2 − 1) = 1. With $\alpha = .05$ and $df = 1$, the critical value for χ^2 is 3.84 (see Appendix A.6). Since the obtained value of 3.86 exceeds the critical value of 3.84, we reject the null hypothesis, concluding that a significant difference exists between the percentage of smokers in the city and in the population. Examination of the data reveals that a significantly smaller proportion of people in the sample city smoke than in the population.

Nonparametric Statistical Tests for Two Independent Samples

The following statistical tests are analogous to the parametric independent groups *t*-test, which assumes interval-level measurement and that the means for each group come from independent random samples from normally distributed populations having equal variances. When these assumptions cannot be met, one of the following nonparametric tests may be appropriate. We present them in order of increasing power.

Chi-Square Test for Two Independent Samples

Chi-square is a very useful test, appropriate when the data represent one variable with two levels (i.e., the two independent groups) and the dependent variable is categorical with two or more levels. We arrange the data in a 2 x *k* contingency table, as shown in Table 14.9.

In Table 14.9, F_{11} is the frequency of subjects in group 1 who belong to the first category of the variable of interest, G_1 is the total number of subjects in group 1, F_1 is the total number of subjects in category 1 of the variable of interest, and *N* is the total number of subjects.

TABLE 14.9 Generic data table for chi-square test for two independent samples.

	CATEGORY OF DEPENDENT VARIABLE					TOTAL
	1	2	3	...	k	
GROUPS						
1	F_{11}	F_{12}	F_{13}	...	F_{1k}	G_1
2	F_{21}	F_{22}	F_{23}	...	F_{2k}	G_2
			...			
Totals	F_1	F_2	F_3	...	Fk	N

Equation

The equation for the chi-square test for two independent samples is

$$\chi^2 = \sigma_{i=1}^{G}\sigma_{k=1}^{F}\frac{(F_{ik} - E_{ik})^2}{E_{ik}} \quad \text{Eq. 14.11}$$

where

χ^2 = the chi-square statistic,
F_{ik} = the frequency of subjects in group i who belong to the kth variable of interest,
G = the number of subjects in a group, and
E_{ik} = the expected frequency of subjects in group i who belong to the kth variable of interest.

We obtain the expected frequencies in the same way as explained in the example for the previously presented chi-square for goodness of fit.

Example

Researchers used the chi-square statistic to determine whether the attitudes of males and females differed on an item embedded in a questionnaire assessing feelings about physical activity. The item was, "I like physical activity because it makes me sweat". The possible responses were strongly agree (SA), agree (A), disagree (D), and strongly disagree (SD). The resulting data are presented in Table 14.10.

TABLE 14.10 Data from item on questionnaire assessing attitudes toward physical activity.

	RESPONSE				
	SA	A	D	SD	Total
Males	15	10	10	5	40
Females	10	10	20	20	60
Total	25	20	30	25	100

Once again, to determine the expected frequency for each cell if the null hypothesis is true (i.e., no relationship exists between the responses and gender), we multiply the cell margins for each cell and divide the resulting product by N. For example, the expected frequency for the males in the strongly agree cell is $(40 * 25)/100 = 10$. We carry out this operation for each cell in the table and then use equation 14.11 to calculate the value of χ^2.

$$\chi^2 = \frac{(15-10)^2}{10} + \frac{(10-8)^2}{8} + \frac{(10-12)^2}{12} + \frac{(5-10)^2}{10}$$
$$+ \frac{(10-15)^2}{15} + \frac{(10-12)^2}{12} + \frac{(20-18)^2}{18} + \frac{(20-15)^2}{15}$$
$$\chi^2 = 2.5 + 0.5 + 0.33 + 2.5 + 1.67 + 0.33 + 0.22 + 1.67 = 9.72$$

We compare this value of χ^2 to the critical value found in Appendix A.6 with the alpha level selected and the appropriate degrees of freedom. The degrees of freedom are calculated in the same way as previously indicated, by multiplying the number of rows in the table, less one, by the number of columns in the table, less one $[(R-1) * (C-1)]$. Thus, in this example, the degrees of freedom are $(2-1) * (4-1) = 3$, and the critical value for χ^2 is 7.81 (from Appendix A.6). Our conclusion is to reject the null hypothesis. By inspecting the data, we find that males and females differ significantly on their response to this item, with males more likely to agree and females more likely to disagree with it.

V.14.4 **SPSS Chi-Square Analysis for Two Groups**

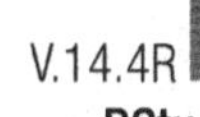

V.14.4R **RStudio Chi-Square Analysis for Two Groups**

The Median Test

Whereas the chi-square test is applicable when the data available are in the form of frequencies in mutually exclusive categories, the median test can help us determine whether median differences exist between independent groups when the data are ordinal. It uses the chi-square statistic to do so.

Situation

We first measure the subjects in two independent groups on a variable of interest. Then, we arrange *all* of the measurements in numerical (chronological) order and find the median. To determine which score in the ordered data set is the median, we first divide (N + 1)/2. The resulting value indicates the position of the score in the data set that is the median. We then note the number of subjects that fall into each of four possible categories: Group 1 above the median, Group 1 below the median, Group 2 above the median, and Group 2 below the median. We place these frequencies in a 2 x 2 contingency table and calculate the chi-square statistic (see equation 14.11).

Example

An investigator asked 50 students in an introductory university statistics class whether they have ever taken an algebra course. The researcher then determined the median score on the initial examination for the 50 students in the class and classified each student as being either above or below the median. Because these students did not represent a random sample from a defined population, the investigator was uncomfortable about meeting the assumptions for the independent groups *t*-test to determine whether a significant mean difference existed between the groups. Instead, she used the median test. The contingency table is presented in Table 14.11.

Use equation 14.10 to verify that the chi-square value for these data is 5.26, which results in rejection of the null hypothesis at the .05 level of significance with 1 *df*. The investigator would conclude that a positive

TABLE 14.11 Example data format for calculating the median test.

	DID NOT TAKE ALGEBRA	DID TAKE ALGEBRA	TOTAL
Below median	17	10	27
Above median	7	16	23
Total	24	26	50

relationship exists between having taken an algebra course and scoring above the median on the examination.

Mann-Whitney U Test of Medians

The Mann–Whitney U test is another nonparametric version of the independent groups *t*-test. It is appropriate for use when you have ordinal data or when you are unable to meet other *t*-test assumptions. For example, you might have interval- or ratio-level data, but because the sample sizes are small, you have doubts about meeting the normality assumption. In a situation like this, you can use the Mann–Whitney U test to determine whether the medians (not means) of the two independent samples are significantly different.

Procedure

We draw two random samples from the same population or one random sample from each of two populations that are assumed to be equal. The samples do not have to be of equal size, but they do have to be independent. We then rank all of the subjects, regardless of group membership, on the variable of interest. We then re-sort the subjects back into their original groups, carrying their ranks with them.

Equations

The equations for the Mann–Whitney U test are

$$U_1 = N_1N_2 + \frac{(N_1(N_1 + 1))}{2} - \sigma R_1 \text{ and} \quad \text{Eq. 14.12}$$

$$U_2 = N_1N_2 - U_1 \quad \text{Eq. 14.13}$$

where

U_1 and U_2 = the Mann–Whitney values,
N_1 = the number of subjects in group 1,
N_2 = the number of subjects in group 2, and
σR_1 = the sum of the ranks of the subjects in group 1 (it doesn't matter which group is designated as group 1).

Example

A researcher classified 28 students as participating in intramurals or not participating in intramurals and ranked all 28 (from high to low) on the basis of their grade point averages. The resulting data are presented in Table 14.12.

For these data,

$$U_1 = (13)(15) + \frac{(13)(14)}{2} - 156.5$$
$$= 195 + \frac{182}{2} - 156.5$$
$$= 195 + 91 - 156.5 = 129.5$$

and

$$U_2 = (13)(15) - 129.5$$
$$= 195 - 129.5 = 65.5$$

When the sample sizes are small ($N_1 + N_2 < 20$), there are tables available to determine whether to retain or reject the null hypothesis. These tables are not presented in this textbook but can be found in Siegel and Castellan (1988). When $N_1 + N_2 \geq 20$, we can use the following equation, which results in a z-score:

$$z = \frac{U - \frac{N_1 N_2}{2}}{\sqrt{N_1 N_2 \frac{(N_1 + N_2 + 1)}{12}}} \quad \text{Eq. 14.14}$$

where

z = z-score,
U = either U_1 or U_2 (they will result in the same absolute value but with opposite signs),
N_1 = the number of subjects in group 1, and
N_2 = the number of subjects in group 2.

For the example above,

$$z_1 = \frac{129.5 - 97.5}{21.71} = \frac{32}{21.71} = 1.47$$

or

$$z_2 = \frac{65.5 - 97.5}{21.71} = \frac{-32}{21.71} = -1.47$$

Since the critical z-value in Appendix A.1 for significance at an alpha of .05 is 1.96 (two-tailed test), we conclude that the median GPAs for the two groups are not significantly different.

TABLE 14.12

Example data format for calculating the Mann–Whitney U test.

RANK ON GPA	
INTRAMURALS	NO INTRAMURALS
1	2
4.5	3
6	4.5
7	9
8	10
11	15
12	16
13.5	18
13.5	19
17	23
20	24
21	25
22	26
	27
	28
$N = 13$	$N = 15$
$\Sigma R = 156.5$	$\Sigma R = 249.5$

Nonparametric Statistical Tests for Two Related Samples

The following two tests are analogous to the dependent-samples t-test. Recall that for samples to be dependent or related, either the subjects are

TABLE 14.13 Generic format for calculating the sign test.

Note: In Table 14.13, S_{11} is a score for subject 1, and S_{12} is either a second score for subject 1 or the score for the subject matched with subject 1. The third column is simply the sign (not the value) obtained when the value in Column 2 is subtracted from the value in Column 1.

	COLUMN 1	COLUMN 2	SIGN OF DIFFERENCE
Subjects			
1	S_{11}	S_{12}	– or +
2	S_{21}	S_{22}	– or +
3	S_{31}	S_{32}	– or +
		. . .	
N	S_{N1}	S_{N2}	– or +

matched on important variables, or, more commonly, the subjects serve as their own controls by being measured twice (e.g., pre- and post-testing).

The Sign Test

The sign test is appropriate when the measurement process is so inexact that, although it is possible to decide which of two measurements reflect "more of the variable", it is not possible to determine "how much more" with any certainty (i.e., ordinal data). The generic data format for the sign test is shown in Table 14.13.

If the null hypothesis is true, we would expect to find an approximately equal number of plus signs (+) and minus signs (−). If the number of plus signs and the number of minus signs are very different, at some point it becomes unreasonable to believe that chance is responsible for the disparity, and rejection of the null hypothesis is warranted. For example, if the scores in column 1 were recorded before the administration of a very effective treatment and the scores in column 2 followed the treatment, evidence of the change made by the treatment would be revealed by a substantial number of negative signs versus positive signs in the right-hand column.

For small samples ($N < 30$), we calculate the probability of various distributions of negative and positive signs by using the binomial expansion. Appendix A.8 provides these probabilities. For large samples ($N \geq 30$), we can use an equation to convert the binomial distribution to an approximation of the normal distribution. The equation is:

$$z = \frac{2X \pm 1 - N}{\sqrt{N}}$$ Eq. 14.15

where

z = a z-score,
X = the frequency of the most frequent sign, and
N = the number of pairs of scores.

Small-Sample Example

An investigator administered to 10 subjects a 20-item questionnaire designed to obtain a score reflecting the subjects' attitudes toward physical activity (higher scores indicate a more positive attitude). The subjects then viewed a film about the negative effects of obesity, and the questionnaire was re-administered. The resulting data are presented in Table 14.14.

The results are 7 negatives, 2 positives, and 1 no change. By using Appendix A.8 (N = 9, k = 2), we find the probability of obtaining 7 or more of one sign out of 9 possibilities (the one "no change" is omitted) is .090 (one-tailed test). If the level of significance is .05, the conclusion is to retain the null hypothesis. If this had been a two-tailed test, the value from Appendix A.8 would have been doubled (.18), which would also result in retaining the null hypothesis.

TABLE 14.14

Data for calculating the sign test.

SUBJECT	BEFORE FILM	AFTER FILM	SIGN OF DIFFERENCE
1	15	17	–
2	12	14	–
3	7	9	–
4	14	13	+
5	9	9	0
6	10	12	–
7	15	18	–
8	8	11	–
9	16	17	–
10	13	12	+

Note: If X is less than $N/2$, then +1 is used in the equation, if X is more than $N/2$, then –1 is used.

Large-Sample Example

From the same experiment as above but with a larger sample ($N = 100$), the following results are obtained:

negative signs = 61,

positive signs = 34, and

no change = 5.

Since 61 is larger than $N/2$ (50), the equation uses -1 instead of +1 in the numerator to calculate z is:

$$z = \frac{2 * 61 - 1 - 95}{\sqrt{100}} = \frac{122 - 96}{10} = \frac{26}{10} = 2.6$$

From Appendix A.1, the probability of obtaining a z-score of 2.6 is .0047. If this is a one-tailed test, the conclusion is to reject the null hypothesis at the .05 alpha level. If this is a two-tailed test, the probability is .0094, and the null hypothesis would still be rejected at the .05 alpha level.

Wilcoxon Signed-Rank Test

If we can determine the amount as well as the direction of the difference between the matched subjects or the repeated measures of the same subjects, the Wilcoxon signed-rank test is appropriate. Because of the additional measurement information, it is more powerful than the sign test. The data format is similar to that of the sign test, with the addition of one column. See Table 14.15.

Small-Sample Example ($N < 16$)

Let's return to the data in Table 14.14 presented for the sign test and calculate the Wilcoxon signed-rank test statistic.

There is no equation involved with the Wilcoxon signed-rank test for small samples, but it is necessary to calculate the sum of the negative

TABLE 14.15 Generic data format for the Wilcoxon signed-rank test.

SUBJECT	COLUMN 1 FIRST SCORE	COLUMN 2 SECOND SCORE	COLUMN 3 DIFFERENCE *(D)*	COLUMN 4 RANK OF DIFFERENCE (WITH SIGN) *R*
1	S_{11}	S_{12}	$\pm D_1$	$\pm R_1$
2	S_{21}	S_{22}	$\pm D_2$	$\pm R_2$
3	S_{31}	S_{32}	$\pm D_3$	$\pm R_3$
		. . .		
N	S_{N1}	S_{N2}	$\pm D_N$	$\pm R_N$

Note: In Table 14.15, S_{11} is the first score for subject 1, and S_{12} is either a second score for subject 1 or the score for the subject matched with subject 1, *D* is the difference and the sign obtained when the value in column 2 is subtracted from the value in column 1, and $\pm R$ is the rank of the absolute values in column 3, with the sign from column 3 carried to column 4.

and positive ranks. To do this, imagine that we add another column to Table 14.14. In this new column we include, not only the sign of the difference, but also the amount of the difference. Going from subject 1 down to subject 10 the column would read −2, −2, −2, +1, 0, −2, −3, −3, −1, +1. The next step is to put these values in chronological order (disregarding the signs and the one value of 0). This result is 1, 1, 1, 2, 2, 2, 2, 3, 3. The next step is to rank these values. The three 1's would all receive the rank of 2, the four 2's would each receive the rank of 5.5 and the two 3's, the rank of 8.5. When the signs are reattached to ranks, the result is 2, −2, 2, −5.5, −5.5 −5.5, −5.5, −8.5, −8.5. The sum of the negative ranks is −41.0, and the positive ranks sum to 4. Note that the absolute value of the two sums added together will always equal $N * (N + 1)/2$. In our example, $N = 9$, because subject 5 is omitted, as there was no difference in the two measurements. Therefore, 9 * (10)/2 = 45, and 41 + 4 = 45.

The next step is to consult Appendix A.9 to determine the probability of this outcome. The highest sum (in this case 41) and *N* (9 in this case) are all that are required to use Appendix A.9. Verify that the one-tailed probability of this event is .0137 and the two-tailed probability is .027 (.0137 * 2). In either situation, the conclusion would be to reject the null hypothesis (at $\alpha = .05$) and conclude that the film had a significant influence on the subjects' attitudes. Notice that the conclusion for the sign test did not permit rejection of the null hypothesis. This demonstrates the increased power of the Wilcoxon signed-rank test.

Large-Sample Example

In a motor learning experiment, the investigator measured the accuracy with which 25 subjects could estimate the distance they moved a lever along a sliding scale with their nondominant hand before and after a training session in which they practiced the task with their dominant hand. The resulting data were as follows:

number of subjects who improved = 18 (sum of their ranks = 210),
number of subjects whose scores decreased = 5 (sum of their ranks = 66), and
number of subjects whose scores did not change = 2.

When the sample size is at least 16, we can use an equation to convert the larger sum of ranks into a *z*-score so that we can use the normal distribution to determine the probability of the event. The equation is:

$$z = \frac{S - \frac{N(N+1)}{4}}{\sqrt{\frac{N(N+1)(2N+1)}{24}}} \quad \text{Eq. 14.16}$$

where

z = a *z*-score,
S = the larger sum of ranks, and
N = the number of pairs of scores (remember that each difference of zero results in reducing N by 1). Thus, N = 23.

Substituting the experimental data results in

$$z = \frac{210 - \frac{23(24)}{4}}{\sqrt{\frac{23(24)(2*23+1)}{24}}} = \frac{210 - \frac{552}{4}}{\sqrt{\frac{23*24*47}{24}}} = \frac{210-138}{\sqrt{\frac{25{,}944}{24}}}$$
$$= \frac{210-138}{\sqrt{1{,}081}} = \frac{210-138}{32.87} = 2.19$$

Because 2.19 is larger than the critical *z*-value (1.96) when $\alpha = .05$, the researcher rejects the null hypothesis and concludes that the training with the dominant hand transferred to improve the performances with the non-dominant hand.

Nonparametric Tests of Significant Differences for n Independent Samples

The following two nonparametric tests are analogous to the independent groups ANOVA and are used when the assumptions for ANOVA (samples are drawn from normal distributions having equal variances and the measurements are of at least the interval level) cannot be met. The first test (chi-square) is most appropriate when the data are in the form of frequencies in discrete categories (nominal), and the second test (Kruskal–Wallis) is useful when the data are ordinal.

Chi-square Test for n Independent Samples

The chi-square test for *n* independent samples is an extension of the chi-square test for goodness of fit and the chi-square test for two independent samples presented earlier in the chapter. Instead of just two levels of the independent variable, with this chi-square test we can have two or more levels. The data are arranged in a contingency table that has as many columns as there are levels of the independent variable and as many rows as there are categories of the dependent variable. The equation to calculate the chi-square value is the same as presented earlier:

$$\chi^2 = \sigma_{i=1}^{G}\sigma_{k=1}^{F}\frac{(F_{ik} - E_{ik})^2}{E_{ik}} \qquad \text{(repeated)}$$

Eq. 14.11

Example

Researchers devised an experiment to determine the effectiveness of four different modalities for the treatment of chondromalacia (damage to the cartilage under the kneecap). Eighty subjects suffering from this malady were randomly assigned to one of four treatments: (1) complete rest, (2) moderate exercise, (3) anti-inflammatory medication, and (4) both exercise and medication. The researchers then administered a questionnaire to assess the level of the subjects' pain, and each subject was placed in one of three categories: no pain, some pain, and much pain (the dependent variable). The results are displayed in Table 14.16.

The critical value for chi-square is a function of the alpha level and the degrees of freedom, which is (Columns −1) * (Rows −1). In the current example, $df = (4 - 1)(3 - 1) = 6$. If $\alpha = .05$ and $df = 6$, the critical value for chi-square is 12.59 (see Appendix A.6). Because the obtained chi-square value is 6.91, the researchers retain the null hypothesis, indicating that no significant difference exists among the four modalities in alleviating the chondromalacia pain.

A significant chi-square value, like a significant *F*-ratio in an ANOVA, indicates that significant differences exist among the groups, but, again like the *F*-ratio, it does not identify which differences are significant. For the

TABLE 14.16 Data for calculating chi-square for *n* independent samples.

Determine the expected value for each cell and then use Equation 14.11 to verify that the chi-square value for these data is 6.91.

	REST	EXERCISE	MEDICATION	EXERCISE & MEDICATION	TOTAL
PAIN CATEGORY					
None	5	3	2	7	17
Some	10	15	14	11	50
Much	5	2	4	2	13
Total	20	20	20	20	80

ANOVA, we presented several multiple comparison techniques to determine which differences were significant. With chi-square, one can use a technique that involves partitioning the data into as many 2 x 2 tables as there are degrees of freedom to determine which differences are significant. Most computer programs that calculate chi-square include this possibility (See Siegel & Castellan, 1988).

V.14.5 **SPSS Chi-Square Analysis for k Groups**

V.14.5R  **RStudio Chi-Square Analysis for k Groups**

Kruskal–Wallis Test for n Independent Samples

The Kruskal–Wallis test for *n* independent samples is analogous to the one-way independent groups ANOVA, except that it is appropriate for use when the dependent variable is measured on an ordinal level rather than an interval or ratio level. The data are placed in a table that has as many columns as there are levels of the independent variable, and the *overall ranks* of the subjects are listed in each column. The numbers of subjects in each group do not have to be equal.

To construct the table, all the subjects (regardless of what group they are in) are ranked on the dependent variable. Then the subjects' ranks are listed in the column to which they belong.

Equation

The equation to calculate the Kruskal–Wallis *(H)* value is:

$$H = \frac{12}{N(N+1)} \sum_{j=1}^{k} N_j(\bar{R}_j - \bar{R})^2 \qquad \text{Eq. 14.17}$$

where

H = Kruskal–Wallis value,
N = the total number of subjects,
k = the number of levels of the independent variable,
$\bar{R}_f$ = the mean rank of the *j*th group, and
$\bar{R}$ = the mean of all the ranks (this is calculated by $(N + 1)/2$).

As is common with many nonparametric statistical tests, when the samples are small, we must consult a table to compare the obtained value with a critical value to determine whether the null hypothesis should be retained or rejected. With large samples, the sampling distribution of the obtained value often approximates either a chi-square or a normal distribution.

For the Kruskal–Wallis test, small samples are considered to be those that have two or three levels of the independent variable and five or fewer subjects in each group. For large samples (those with more than three groups and more than five subjects per group), the sampling distribution of H approximates the chi-square distribution with $df = k - 1$, where k is the number of levels of the independent variable.

Small-Sample Example

The students in several sections of a physical conditioning class take a physical fitness test at the end of the course. An investigator used the

TABLE 14.17 Data for calculating the Kruskal–Wallis test.

	INSTRUCTORS		
	BILL	WILL	PHIL
SECTION			
A	250	267	232
B	235	271	285
C	249	282	248
D			265

When the scores are changed to ranks, the following data are generated.

	INSTRUCTORS		
	BILL	WILL	PHIL
SECTION			
A	5	7	1
B	2	8	10
C	4	9	3
D			6
ΣR	11	24	20
$\bar{R}j$	3.67	8	5.0

$\bar{R} = (N + 1) / 2 = 11 / 2 = 5.5$

Kruskal–Wallis test to determine whether significant differences existed between course instructors. The data in Table 14.17 are the average fitness scores for the students in each section for each instructor.

For these data,

$$H = \frac{12}{(10)(11)}(3 * (3.67 - 5.5)^2 + 3 * (8.00 - 5.5)^2 + 4 * \left(5.0 - 5.5)^2\right)$$

$$= \frac{12}{110}(10.05 + 18.75 + 1)$$

$$= .109 * 28.8 = 3.25$$

To use Appendix A.10 for determining if 3.25 is statistically significant, put the ns in chronological order (i.e., 4, 3, 3) and locate the column for $\alpha = .05$. The value of 3.25 is less than the critical value of 5.73, thus the null hypothesis is retained. Recall that nonparametric tests are low in power, especially when sample sizes are small. We conclude that no statistically significant difference exists among the median scores of the classes for the three instructors. Remember that the data analyzed are ranks.

Large-Sample Example

Researchers classified 40 sixth-grade children in a physical education class as being underweight, normal, overweight, or obese by using BMI as a criterion. The students completed a questionnaire designed to assess self-esteem (with high scores representing high self-esteem), and *all* of the resulting scores were ranked, with the ranks listed under each BMI category. In Table 14.18 the resulting data are presented.

From the data in Table 14.18,

$$H = \frac{12}{(40)(41)}(7(20.5 - 20.5)^2 + 18(26.75 - 20.5)^2 + 10(16.1 - 20.5)^2 + 5\left(6.8 - 20.5)^2\right)$$

$$= \frac{12}{1640}(0 + 703.13 + 193.6 + 938.45)$$

$$= (.007317)(1835.18) = 13.43$$

V.14.6
SPSS Kruskal–Wallis

V.14.6R
RStudio Kruskal–Wallis

When we compare the calculated H value of 13.43 to the critical value of chi-square from Appendix A.6 ($df = 4 - 1 = 3$ and $\alpha = .05$) of 7.81, we decide to reject the null hypothesis and conclude that this distribution of ranks would occur less than 5 times out of 100 by chance. (Actually, it would occur less than 1 time out of 100 by chance, since the critical value for chi-square at $\alpha = .01$ is 11.34.)

Data (ranks) for calculating large-sample Kruskal–Wallis test. TABLE 14.18

	UNDER WEIGHT	NORMAL WEIGHT	OVERWEIGHT	OBESE
	10	11	2	1
	12.5	14	4	3
	20	15	7	5
	22	16	8	6
	24	17	9	19
	27	18	12.5	
	28	21	25	
		23	26	
		29	30.5	
		30.5	37	
		33		
		33		
		33		
		35		
		36		
		38		
		39		
		40		
ΣR	143.5	481.5	161	34
$\bar{R}_J$	20.5	26.75	16.1	6.8
$\bar{R} = (N + 1) / 2 = (40 + 1) / 2 = 20.5$				

Issues

When ties in ranks are present, we can use an equation to correct the value of H, but this calculation typically reduces the obtained value of H very minimally, and, unless the obtained value and the critical value are extremely close, the correction is not necessary. A multiple comparison technique also exists to follow up a significant finding (see Siegel & Castellan, 1988).

Nonparametric Tests for Differences Between n Related Samples

As with the tests for two related samples, when we have n related samples, we can either match the subjects in the levels of the independent variable or measure the same subjects repeatedly. The difference with this test is that we can have more than two levels of the independent variable. This is analogous to the repeated-measures ANOVA presented in Chapter 10. The first test presented (Cochran's Q) is appropriate when the level of measurement of the dependent variable is nominal, and the second test (Friedman two-way ANOVA by ranks test) requires ordinal measurement for the dependent variable.

The Cochran's Q Test

For Cochran's Q the final form of the data is a dichotomy. Examples might include items on a written test scored correct or incorrect, made a goal or didn't make the goal, pass/fail, or yes/no. These typically take the form of 1 for the "positive" category and 0 for the "negative" category.

Equation

The equation for Cochran's Q is

$$Q = \frac{(k-1)\left[k(\sigma_{j=1}^{k} p_j^2) - (\sigma_{j=1}^{k} p_j)^2\right]}{k\sigma_{i=1}^{N} T_i - \sigma_{i=1}^{N} T_i^2} \quad \text{Eq. 14.18}$$

where

Q = the Cochran's Q value,
k = the number of levels of the independent variable,
P_j = the total number of 1s in the jth column,
Ti = the total number of 1s in the ith row, and
N = the number of sets of subjects.

Note: The sampling distribution of the Q statistic takes the form of chi-square with $k - 1$ degrees of freedom.

Example

Thirty subjects, all having BMI values between 25 and 29 (overweight), are divided into triads based on initial body weight. In each of the triads, one person uses diet plan 1, one person uses diet plan 2, and the third person uses diet plan 3. The goal for all subjects is to lose 5% of their body weight over two months. The results are presented in Table 14.19.

Substituting the experimental data into Equation 14.18 results in

$$Q = \frac{(3-1) * (3 * [7^2 + 8^2 + 2^2] - 17^2)}{3 * (17) - 33}$$

$$= \frac{(2) * (3 * [117] - 289)}{51 - 33}$$

$$= \frac{(2) * (351 - 289)}{18}$$

$$= \frac{(2) * (62)}{18}$$

$$= \frac{124}{18} = 6.89$$

The critical value for chi-square with $df = k - 1 = 2$ (number of levels of the independent variable −1) and $\alpha = .05$ is 5.99. Because the calculated

Data format for the calculation of the Cochran's Q test (1 = reached goal, 0 = did not reach goal). TABLE 14.19

TRIAD	PLAN 1	PLAN 2	PLAN 3	T_i	T_i^2
1	1	0	0	1	1
2	1	1	0	2	4
3	0	1	0	1	1
4	1	0	1	2	4
5	1	1	0	2	4
6	1	1	0	2	4
7	0	1	0	1	1
8	1	1	1	3	9
9	1	1	0	2	4
10	0	1	0	1	1
Totals	7	8	2	17	33

value for Q of 6.89 exceeds the critical value, the conclusion is to reject the null hypothesis. Inspection of the data suggests that diet plan 3 is not as effective as the other two plans.

The Friedman Two-Way ANOVA by Ranks Test

The Friedman two-way ANOVA by ranks test is appropriate with either matched subjects or a repeated-measures design where it is possible to rank the subjects' performances among the k levels of the independent variable. The data are arranged in a two-way table having a column for each level of the independent variable and a row for each set of matched subjects (or for the same subject measured k times).

Equation

The equation for the Friedman two-way ANOVA by ranks test is

$$T = \left[\frac{12}{Nk(k+1)} \sigma_{j=1}^{k} R_j^2 \right] - 3N(k+1) \quad \text{Eq. 14.19}$$

where
T = the Friedman statistic,
N = the number of rows,
k = the number of columns, and
R_j = the sum of the ranks of the jth column.

Small-sample example

Unless N is quite large, a special table is required to determine the critical value for T, when k is five or less. See Appendix A.11.

Five rats from each of four different litters were randomly assigned to one of five different exercise protocols for six weeks and then tested for lactic acid levels following a maximal exercise test. The resulting data are displayed in Table 14.20.

To calculate the Friedman T value, we convert these results in each row to ranks, as shown in Table 14.21.

For the experimental data,

$$T = \left(\frac{12}{4 * 5 * 6}\right) * \sigma(6^2 + 11^2 + 17^2 + 15^2 + 11^2) - 3 * 4 * 6$$

$$= \left(\frac{12}{120}\right) * \sigma(36 + 121 + 289 + 225 + 121) - 72$$

$$= (.1 * 792) - 72$$

$$= 79.2 - 72 = 7.2$$

The critical value for T when $N = 4$ and $k = 5$ ($\alpha = .05$) is 8.80 (see Appendix A.11). Because 7.20 is less than this critical value, the decision is to retain the null hypothesis and conclude that the various exercise protocols did not differ significantly in lactic acid production in the rats.

Large-Sample Example

When N and k are both large, the sampling distribution of T approximates the chi-square distribution with $k - 1$ degrees of freedom.

TABLE 14.20 Results of lactic acid test in rats after a maximal exercise test (mmol/l).

	EXERCISE PROTOCOL				
Litter	A	B	C	D	E
1	2.15	3.41	2.07	1.98	3.24
2	3.12	2.03	2.57	1.84	2.71
3	3.02	2.74	1.89	2.45	2.36
4	2.39	2.17	1.90	2.33	2.28

TABLE 14.21 Data format for calculating the Friedman two-way ANOVA by ranks test.

	EXERCISE PROTOCOL				
Litter	A	B	C	D	E
1	3	1	4	5	2
2	1	4	3	5	2
3	1	2	5	3	4
4	1	4	5	2	3
Totals *(Rj)*	6	11	17	15	11

In a test of athletic performance, the 24 members of a basketball team practiced free throws for a substantial number of practices and then each attempted 15 free throws under three different distracting conditions (quiet, moderate noise, and very noisy). Four players were randomly assigned to each of the six possible orders in which the conditions could be presented (ABC, ACB, . . . CBA). The number of free throws made was converted into ranks for each player. The results (which have already been converted to ranks across the three conditions with the highest rank being 1 in each row) are displayed in Table 14.22.

TABLE 14.22 Data format for calculating a large-sample Friedman two-way ANOVA by ranks test.

	CONDITIONS		
SUBJECTS	QUIET	MODERATE NOISE	VERY NOISY
1	2	3	1
2	1	3	2
3	2	3	1
4	1	3	2
5	1	2	3
6	2	3	1
7	1	3	2
8	1	2	3
9	2	3	1
10	2	3	1
11	1	3	2
12	2	1	3
13	1	3	2
14	2	3	1
15	2	3	1
16	2	3	1
17	2	3	1
18	1	3	2
19	2	1	3
20	1	3	2
21	2	1	3
22	1	3	2
23	1	3	2
24	2	1	3
Totals	37	62	45

Hence,

$$T = \left(\frac{12}{24 * 3 * 4}\right) * \sigma(37^2 + 62^2 + 45^2) - 3 * 24 * 4$$
$$= \left(\frac{12}{288}\right) * \sigma(1369 + 3844 + 2025) - 288$$
$$= (.04167 * 7238) - 288$$
$$= 301.61 - 288 = 13.61$$

Since the critical value for chi-square ($\alpha = .05$, $df = 2$) is 5.99 and the observed value for T is 13.61, the null hypothesis is rejected. The test reveals that at least one significant difference exists among the distracting conditions.

Issues

As with many nonparametric tests, we can correct the calculated T statistic when ties are present. Unless the number of ties is great, this correction seldom changes the conclusion. A multiple comparison technique exists to determine which differences are significant for the Friedman two-way ANOVA by ranks test (see Siegel & Castellan, 1988).

SUMMARY

When the assumptions of parametric statistical tests cannot be met, it may still be possible to determine whether significant correlations or significant differences exist between and among groups. The branch of statistics employed to do this is called nonparametric statistics. Although nonparametric tests are typically less powerful than their parametric counterparts, these statistics are useful in certain situations. In this chapter, we presented nonparametric statistics that are analogous to parametric tests for the correlation between two variables (PPM) and for differences between independent and dependent groups [one-sample, two samples (*t*-test), and *n*-samples (ANOVA)]. We presented illustrations of these tests, often including both small- and large-sample examples. The nonparametric tests presented for the above categories were listed in order of increasing power, and often tests were provided involving both nominal and ordinal levels of measurement for the dependent variable.

Review Questions and Practice Problems

PART A (ANSWERS ARE PROVIDED IN APPENDIX C)

1. Nonparametric statistics are sometimes called distribution-free statistics. This is because they do not require which assumption?

   A. At least interval measurement

   B. Homoscedasticity

   C. Independence

   D. Normality

2. If a student received a score of 33 on a 100-item multiple-choice test with each question having four possible responses, did he do better than chance at $\alpha = .05$?

3. If 356 individuals were categorized on smoking and social-economic status as shown in Table 14.23, do these data reveal a significant relationship between these two variables at α = .05?

TABLE 14.23 Data set.

	SOCIAL-ECONOMIC STATUS		
SMOKING	HIGH	MIDDLE	LOW
Current	51	22	43
Former	92	21	28
Never	68	9	22

PART B

4. Two teaching assistants (TAs) graded the writing assignments of 12 students and arranged them in rank order. What is the correlation between the two TAs' ranks, if the ranks were distributed as in Table 14.24?

TABLE 14.24 Data set.

STUDENT	RANK OF TA 1	RANK OF TA 2
1	2.5	5
2	2.5	2
3	9	8
4	5.5	7
5	12	12
6	7.5	11
7	1	2
8	10	6
9	4	2
10	5.5	4
11	7.5	10
12	11	9

5. A researcher measured 15 individuals (7 wrestlers and 8 non-wrestlers) on overall flexibility and ranked these measurements, with the results given in Table 14.25. Is there a difference in the flexibility of the wrestlers and the non-wrestlers at α = .05?

TABLE 14.25 Data set.

WRESTLERS	NON-WRESTLERS
5.5	7.5
1.5	9
7.5	10
1.5	15
3	11
4	14
5.5	13
	12

6. Students ranked 14 faculty members, and the scores for three different types of teachers are given in Table 14.26. Higher scores represent better student evaluations. Do the teachers' scores differ significantly at α = .05? *Hint:* Remember to change the scores to ranks.

TABLE 14.26 Data set.

TALL PHYSICS TEACHERS	LOUD KINESIOLOGY TEACHERS	FUNNY STATISTICS TEACHERS
96	68	115
128	124	149
83	132	166
101	109	147
61	135	

7. The resting heart rates of 20 subjects were obtained before training, after one month of training, and after two months of training. The heart rates were then ranked across time for each subject (1 = low, 3 = high), with the results given in Table 14.27. Is there a training effect at $\alpha = .05$?

TABLE 14.27 Data set.

SUBJECT	BEFORE TRAINING	AFTER ONE MONTH	AFTER TWO MONTHS
1	3	2	1
2	3	2	1
3	3	1	2
4	2	3	1
5	3	2	1
6	3	2	1
7	3	2	1
8	3	1	2
9	2	3	1
10	3	2	1
11	3	2	1
12	3	2	1
13	3	2	1
14	3	2	1
15	3	2	1
16	3	1	2
17	3	2	1
18	2	3	1
19	3	1.5	1.5
20	3	2	1

COMPUTER ACTIVITIES
Nonparametric Statistics

ACTIVITY 1 Spearman Rank Order Correlation Coefficient

A National Football League team is trying to draft the best players. Following the Combines Camp, the head coach and the chief assistant rank the top 15 players so they can compare their rankings prior to draft day. Table 14.28 presents their ranks.

1. What is the Pearson product-moment correlation coefficient between the two rankings?
2. What is the Spearman rank correlation coefficient (ρ) between the two coaches' rankings?
3. Why are the correlations the same?
4. The team owner brings the coaches to her office and asks whom they should draft. Briefly describe the discussion that will take place in the owner's office.

TABLE 14.28 Data set.

PLAYER	HEADCOACH	CHIEFASSISTANT
1	12.5	6
2	2.5	3.5
3	5	2
4	9	11
5	15	7
6	14	13
7	7	9
8	1	1
9	4	5
10	7	10
11	7	3.5
12	12.5	15
13	2.5	8
14	11	14
15	10	12

Using SPSS

Enter the data into SPSS and answer questions 1–3.

- Go to **Analyze→Correlate→Bivariate**
- Put the head coach's and the chief assistant's ranks in the Variables window.
- Check the boxes for Pearson and Spearman Correlation Coefficients.
- Click **OK.**

Using R

1. Open RStudio.
2. Install any necessary R packages.

   For Chapter 14, we will need to install the "DescTools" package to conduct the multiple comparisons tests.

   In the lower right pane of the RStudio, click on the "Packages" tab and then click on the "Install" button.

   In the "Packages" field, type in "DescTools" and click "Install".

   [Note that from this point forward, the package "DescTools" will be installed, and you do not have to install it again.]

3. Set up your RStudio workspace.

   In the RStudio menu bar, click on **File → New File → R Script**

Enter the following two lines in your R Script:

- #Chapter 14
- library(DescTools)

Highlight these two lines that you've typed and click **Run**.

[Note that the second line is where we tell R that we will require the functions that are included in the "DescTools" package.]

Import data from Table 14.28 into R.

4. To generate the Pearson and Spearman Correlation Coefficients, type the following:

- cor(Table_14_28_Data$HEADCOACH,Table_14_28_Data$CHIEF ASSISTANT,method="pearson")
- cor(Table_14_28_Data$HEADCOACH,Table_14_28_Data$CHIEF ASSISTANT,method="spearman")

Highlight these two lines and click **Run**.

ACTIVITY 2 Nonparametric Statistics for Assessing Association

The middle school basketball coaches are excited that 60 girls signed up to try out for the three team levels (A, B, and C). Both coaches evaluated all of the girls with a six-item skills/drills activity. The coaches' first cut was an attempt to pick 20 girls for the A team. After observing the girls' tryouts, each coach listed their 20 girls. The results of their picks are listed in Table 14.29. Use SPSS or R to answer questions 5–14. The coaches are interested in testing whether significant agreement exists in their picks at the .05 level.

5. Compare your 2 x 2 contingency table from SPSS or RStudio with Table 14.29 in the textbook. If it does not match, go back to your entered data and correct it.
6. What is the null hypothesis being tested in this example?
7. What are the degrees of freedom for this chi-square test?
8. What is the critical value for chi-square with this number of degrees of freedom (alpha = .05)? (See Appendix A.6).
9. What is the value of the calculated chi-square?
10. Is this statistically significant?
11. On what percentage of the players did the coaches agree?
12. What is the phi coefficient for their picks?
13. What is the kappa coefficient for their picks?
14. Describe the discussion that the coaches had following the first tryout.

TABLE 14.29 Data set.

		COACH2		
		No (*not* picked = 0)	Yes (picked = 1)	
COACH1	No (*not* picked = 0)	28	12	40
	Yes (picked = 1)	12	8	20
		40	20	60

Using SPSS

Enter the data for the 60 girls into SPSS. The variable names are COACH1 and COACH2. Indicate whether each girl was picked (1) or not (0). There will be 28 (0,0), 12 (0,1), 12 (1,0), and 8 (1,1) entries. Alternatively, you can load Table_14_29_Data from the textbook website.

- Use **Analyze→Descriptive Statistics→Crosstabs.**
- Enter COACH1 in the rows and enter COACH2 in the columns.
- Click on **Statistics**.
- Check the boxes for (1) Chi-square, (2) Correlations, (3) Phi and Cramer's V, and (4) Kappa.
- Click **Continue**.
- Click **OK**.

Using R

Remember to enter and run "library(DescTools)" if you are returning to RStudio after having closed the program.

For this example, you will read in the data from Table 14.29 directly into a data matrix with statement #1 below.

1. Enter the data into R by typing the following:
   - Table_14_29<-matrix(c(28,12,12,8),2,2,byrow=TRUE)
   - rownames(Table_14_29) <- c("NO", "YES")
   - colnames(Table_14_29) <- c("NO", "YES")
2. To calculate the Chi-square for these data, type the following:
   - chisq.test(Table_14_29,correct=FALSE)
3. To calculate Pearson's phi, Cramer's V as well as Cohen's kappa for these data, type the following:
   - Phi(Table_14_29)
   - CramerV(Table_14_29)
   - CohenKappa(Table_14_29)

Highlight and Run the RScript commands you have entered.

ACTIVITY 3 Kruskal–Wallis Test

A life coach is interested in the physical self-confidence that children gain as a result of participating in physical activities. The life coach wants to test the hypothesis that engagement in increasingly demanding physical activities results in increased physical self-confidence. He identifies 40 children as illustrated here:

Those not in PE, intramurals, or school athletic activities	$n = 10$
Those in PE only (not intramurals or school athletic activities)	$n = 10$
Those in intramurals only (not PE or school athletic activities)	$n = 10$
Those in school athletic activities only	$n = 10$

The life coach administers the Logan Kinley Physical Self-Confidence Inventory to all 40 children and ranks all of the scores from high (1 = most confident) to low (40 = least confident). The results are reflected in Table 14.30.

15. What is the life coach's null hypothesis?
16. The alternative hypothesis is stated in the scenario. Which one do you test?
17. What are the mean ranks for the four activity levels?
18. What are the degrees of freedom associated with the test statistic?
19. Recall that the Kruskal–Wallis H statistic with large samples is distributed as a chi-square. What is the critical value for the Kruskal–Wallis test at $\alpha = .05$? See Appendix A.6.
20. What is your decision about the null hypothesis?
21. What does the life coach conclude?

TABLE 14.30 Data set.

NO PE	PE ONLY	INTRAMURALS	SCHOOL ATHLETICS
35	32	4	26
36	22	3	16
37	39	13	15
40	19	34	25
38	10	9	8
31	33	30	7
28	24	29	6
27	17	12	11
20	14	21	2
23	5	18	1

Using SPSS

Enter the data into SPSS. There will be two variables (Activity Level and Rank). For Activity Level, use the following codes: 1 = No PE, 2 = PE; 3 = Intramurals; and 4 = Athletics. Enter the ranks from Table 14.30 for each of the 40 children based on their activity level.

- Use **Analyze→Nonparametric Tests→Legacy Dialogs→K Independent Samples**.
- Move Rank into the Test Variable List window.
- Move Activity Level into the Grouping Variable window.
- Click on Define Range.
- Enter the minimum range (1) and maximum range (4) for the grouping variable, and click **Continue**.
- Check the Kruskal–Wallis *H* box under Test Type and click **OK**.

Using R

Remember to enter and run "library(DescTools)" if you are returning to RStudio after having closed the program.

1. Import data from Table 14.30 into R and to ensure that R treats your ACTIVITY variable as a factor, type the following:
   - Table_14_30_Data$ACTIVITYLEVEL <- as.factor(Table_14_30_Data$ACTIVITYLEVEL)
2. To calculate the mean rank for each activity level, type the following:
   - tapply(X=Table_14_30_Data$RANK, INDEX=list(Table_14_30_Data$ACTIVITYLEVEL), FUN=mean)
3. To run the Kruskal–Wallis nonparametric test, type the following:
   - kruskal.test(RANK ~ ACTIVITYLEVEL,data=Table_14_30_Data)

Highlight and Run the RScript commands you have entered.

REFERENCES

Physical Activity Guidelines Advisory Committee. (2018). *Physical activity guidelines advisory committee scientific report*. Washington, DC: U.S. Department of Health and Human Services.

Siegel, S., & Castellan, N. J. (1988). *Nonparametric statistics for the behavioral sciences* (2nd ed.). New York: McGraw-Hill.

Thomas, J. R., Nelson, J. K., & Thomas, K. T. (1999). A generalized rank-order method for nonparametric analysis of data from exercise science: A tutorial. *Research Quarterly for Exercise and Sport*, *70*(1), 11–23.

Viera, A. J., & Garrett, J. M. (2005). Understanding interobserver agreement: The kappa statistic. *Family Medicine*, *37*, 360–363.

Appendices

Statistical Tables

APPENDIX A.1 The *z*-table.

Normal-Curve Areas

z	0.00	0.01	0.02	0.03	0.04	0.05	0.06	0.07	0.08	0.09
0.0	00.00	00.40	00.80	01.20	01.60	01.99	02.39	02.79	03.19	03.59
0.1	03.98	04.38	04.78	05.17	05.57	05.96	06.36	06.75	07.14	07.53
0.2	07.93	08.32	08.71	09.10	09.48	09.87	10.26	10.64	11.03	11.41
0.3	11.79	12.17	12.55	12.95	13.31	13.68	14.06	14.43	14.80	15.17
0.4	15.54	15.91	16.28	16.64	17.00	17.36	17.72	18.08	18.44	18.79
0.5	19.15	19.50	19.85	20.19	20.54	20.88	21.23	21.57	21.90	22.24
0.6	22.57	22.91	23.24	23.57	23.89	24.22	24.54	24.86	25.17	25.49
0.7	25.80	26.11	26.42	26.73	27.04	27.34	27.64	27.94	28.23	28.52
0.8	28.81	29.10	29.39	29.67	29.95	30.23	30.51	30.78	31.06	31.33
0.9	31.59	31.86	32.12	32.38	32.64	32.90	33.15	33.40	33.65	33.89
1.0	34.13	34.38	34.61	34.85	35.08	35.31	35.54	35.77	35.99	36.21
1.1	36.43	36.65	36.86	37.08	37.29	37.49	37.70	37.90	38.10	38.30
1.2	38.49	38.69	38.88	39.07	39.25	39.44	39.62	39.80	39.97	40.15
1.3	40.32	40.49	40.60	40.82	40.99	41.15	41.31	41.47	41.62	41.77
1.4	41.92	42.07	42.22	42.36	42.51	42.65	42.79	42.92	43.06	43.19
1.5	43.32	43.45	43.57	43.70	43.83	43.94	44.06	44.18	44.29	44.41
1.6	44.52	44.63	44.74	44.84	44.95	45.05	45.15	45.25	45.35	45.45
1.7	45.54	45.64	45.73	45.82	45.91	45.99	46.08	46.16	46.25	46.33
1.8	46.41	46.49	46.54	46.64	46.71	46.78	46.86	46.93	46.99	47.06
1.9	47.13	47.19	47.26	47.32	47.38	47.44	47.50	47.56	47.61	47.67
2.0	47.72	47.78	47.83	47.88	47.93	47.98	48.03	48.08	48.12	48.17
2.1	48.21	48.26	48.30	48.34	48.38	48.42	48.46	48.50	48.54	48.57
2.2	48.61	48.64	48.68	48.71	48.75	48.78	48.81	48.84	48.87	48.90
2.3	48.93	48.96	48.98	49.01	49.04	49.06	49.09	49.11	49.13	49.16
2.4	49.18	49.20	49.22	49.25	49.27	49.29	49.31	49.32	49.34	49.36
2.5	49.38	49.40	49.41	49.43	49.45	49.46	49.48	49.49	49.51	49.52
2.6	49.53	49.55	49.56	49.57	49.59	49.60	49.61	49.62	49.63	49.64
2.7	49.65	49.66	49.67	49.68	49.69	49.70	49.71	49.72	49.73	49.74
2.8	49.74	49.75	49.76	49.77	49.77	49.78	49.79	49.79	49.80	49.81
2.9	49.81	49.82	49.82	49.83	49.84	49.84	49.85	49.85	49.86	49.86
3.5	49.98									
4.0	49.997									
5.0	49.99997									

Source: Based on Lindquist 1942.

Statistical significance of the correlation coefficient (df = N − 2). **APPENDIX A.2**

Values of the Correlation Coefficient *(r)*

df	.10	.05	.01
1	.9877	.9969	.9999
2	.900	.950	.990
3	.805	.878	.959
4	.729	.811	.917
5	.669	.754	.875
6	.621	.707	.834
7	.582	.666	.798
8	.549	.632	.765
9	.521	.602	.735
10	.497	.576	.708
11	.476	.553	.684
12	.457	.532	.661
13	.441	.514	.641
14	.426	.497	.623
15	.412	.482	.606
16	.400	.468	.590
17	.389	.456	.575
18	.378	.444	.561
19	.369	.433	.549
20	.360	.423	.537
25	.323	.381	.487
30	.296	.349	.449
35	.275	.325	.418
40	.257	.304	.393
45	.243	.288	.372
50	.231	.273	.354
60	.211	.250	.325
70	.195	.232	.302
80	.183	.217	.283
90	.173	.205	.267
∞	.164	.195	.254

Source: From *Biometrika Tables for Statisticians* (Vol. 1) (3rd ed.) by E.S. Pearson and H.O. Hartley (Eds.), 1966, London: Biometrika Trustees. Copyright 1966 by Biometrika Trustees. Reprinted by permission of Oxford University Press.

APPENDIX A.3 Values for the *t*-distributions.

	Two-tailed test				One-tailed test		
df	.10	.05	.01	*df*	.10	.05	.01
1	6.314	12.706	63.657	1	3.078	6.314	31.821
2	2.920	4.303	9.925	2	1.886	2.920	6.965
3	2.353	3.182	5.841	3	1.638	2.353	4.541
4	2.132	2.776	4.604	4	1.533	2.132	3.747
5	2.015	2.571	4.032	5	1.476	2.015	3.365
6	1.943	2.447	3.707	6	1.440	1.943	3.143
7	1.895	2.365	3.499	7	1.415	1.895	2.998
8	1.860	2.306	3.355	8	1.397	1.860	2.896
9	1.833	2.262	3.250	9	1.383	1.833	2.821
10	1.812	2.228	3.169	10	1.372	1.812	2.764
11	1.796	2.201	3.106	11	1.363	1.796	2.718
12	1.782	2.179	3.055	12	1.356	1.782	2.681
13	1.771	2.160	3.012	13	1.350	1.771	2.650
14	1.761	2.145	2.977	14	1.345	1.761	2.624
15	1.753	2.131	2.947	15	1.341	1.752	2.602
16	1.746	2.120	2.921	16	1.337	1.746	2.583
17	1.740	2.110	2.898	17	1.333	1.740	2.567
18	1.734	2.101	2.878	18	1.330	1.734	2.552

Values for the t-distributions, continued. APPENDIX A.3

	Two-tailed test				One-tailed test		
df	.10	.05	.01	*df*	.10	.05	.01
19	1.729	2.093	2.861	19	1.328	1.729	2.539
20	1.725	2.086	2.845	20	1.325	1.725	2.528
21	1.721	2.080	2.831	21	1.323	1.721	2.518
22	1.717	2.074	2.819	22	1.321	1.717	2.508
23	1.714	2.069	2.807	23	1.319	1.714	2.500
24	1.711	2.064	2.797	24	1.318	1.711	2.499
25	1.708	2.060	2.787	25	1.316	1.708	2.485
26	1.706	2.056	2.779	26	1.315	1.706	2.479
27	1.703	2.052	2.771	27	1.314	1.703	2.473
28	1.701	2.048	2.763	28	1.313	1.701	2.467
29	1.699	2.045	2.756	29	1.311	1.699	2.462
30	1.697	2.042	2.750	30	1.310	1.697	2.457
40	1.684	2.021	2.704	40	1.303	1.684	2.423
60	1.671	2.000	2.660	60	1.296	1.671	2.390
120	1.658	1.980	2.617	120	1.289	1.658	2.385
∞	1.645	1.960	2.576	∞	1.282	1.645	2.326

Source: From *Biometrika Tables for Statisticians* (Vol. 1) (3rd ed.) by E.S. Pearson and H.O. Hartley (Eds.), 1966, London: Biometrika Trustees. Copyright 1966 by Biometrika Trustees. Reprinted by permission of Oxford University Press.

APPENDIX A.4 The F-distribution.

Outside column is *df* for the denominator; adjacent column is alpha; remaining columns have the *df* for the numerator in the first row.

df	α	1	2	3	4	5	6	7	8	9	10	12	15
1	.25	5.828	7.500	8.200	8.581	8.820	8.983	9.102	9.192	9.263	9.320	9.406	9.493
	.10	39.86	49.50	53.59	55.83	57.24	58.20	58.91	59.44	59.86	60.19	60.71	61.22
	.05	161.4	199.5	215.7	224.6	230.2	234.0	236.8	238.9	240.5	241.9	243.9	245.9
	.025	647.8	799.5	864.2	899.6	921.8	937.1	948.2	956.6	963.3	968.6	976.7	984.9
	.01	4052	4999	5404	5624	5764	5859	5928	5981	6022	6056	6107	6157
2	.25	2.571	3.000	3.153	3.232	3.280	3.312	3.335	3.353	3.366	3.377	3.393	3.410
	.10	8.526	9.000	9.162	9.243	9.293	9.326	9.349	9.367	9.381	9.392	9.408	9.425
	.05	18.51	19.00	19.16	19.25	19.30	19.33	19.35	19.37	19.38	19.40	19.41	19.43
	.025	38.51	39.00	39.17	39.25	39.30	39.33	39.36	39.37	39.39	39.40	39.41	39.43
	.010	98.50	99.00	99.16	99.25	99.30	99.33	99.36	99.38	99.39	99.40	99.42	99.43
	.001	998.4	998.8	999.3	999.3	999.3	999.3	999.3	999.3	999.3	999.3	999.3	999.3
3	.25	2.024	2.280	2.356	2.390	2.409	2.422	2.430	2.436	2.441	2.445	2.450	2.455
	.10	5.538	5.462	5.391	5.343	5.309	5.285	5.266	5.252	5.240	5.230	5.216	5.200
	.05	10.13	9.552	9.227	9.117	9.013	8.941	8.887	8.845	8.812	8.785	8.745	8.703
	.025	17.44	16.04	15.44	15.10	14.88	14.73	14.62	14.54	14.47	14.42	14.34	14.25
	.010	34.12	30.82	29.46	28.71	28.24	27.93	27.67	27.49	27.34	27.23	27.05	26.87
	.001	167.1	148.5	141.1	137.1	134.6	132.8	131.6	130.6	129.9	129.2	128.3	127.4
4	.25	1.807	2.000	2.047	2.064	2.072	2.077	2.079	2.080	2.081	2.082	2.083	2.083
	.10	4.545	4.325	4.191	4.107	4.051	4.010	3.979	3.955	3.936	3.920	3.896	3.870
	.05	7.709	6.944	6.591	6.388	6.256	6.163	6.094	6.041	5.999	5.964	5.912	5.858
	.025	12.22	10.65	9.979	9.604	9.364	9.197	9.074	8.980	8.905	8.844	8.751	8.657
	.010	21.20	18.00	16.69	15.98	15.52	15.21	14.98	14.80	14.66	14.55	14.37	14.20
	.001	74.13	61.25	56.17	53.43	51.72	50.52	49.65	49.00	48.47	48.05	47.41	46.76
5	.25	1.692	1.853	1.884	1.893	1.895	1.894	1.894	1.892	1.891	1.890	1.888	1.885
	.10	4.060	3.780	3.619	3.520	3.453	3.405	3.368	3.339	3.316	3.297	3.268	3.238
	.05	6.608	5.786	5.409	5.192	5.050	4.950	4.876	4.818	4.772	4.735	4.678	4.619
	.025	10.01	8.434	7.764	7.388	7.146	6.978	6.853	6.757	6.681	6.619	6.525	6.428
	.010	16.26	13.27	12.06	11.39	10.97	10.67	10.46	10.29	10.16	10.05	9.888	9.722
	.001	47.18	37.12	33.20	31.08	29.75	28.83	28.17	27.65	27.24	26.91	26.42	25.91
6	.25	1.621	1.762	1.784	1.787	1.785	1.782	1.779	1.776	1.773	1.771	1.767	1.762
	.10	3.776	3.463	3.289	3.181	3.108	3.055	3.014	2.983	2.958	2.937	2.905	2.871
	.05	5.987	5.143	4.757	4.534	4.387	4.284	4.207	4.147	4.099	4.060	4.000	3.938
	.025	8.813	7.260	6.599	6.227	5.988	5.820	5.695	5.600	5.523	5.461	5.366	5.269
	.010	13.75	10.92	9.780	9.148	8.746	8.466	8.260	8.102	7.976	7.874	7.718	7.559
	.001	35.51	27.00	23.71	21.92	20.80	20.03	19.46	19.03	18.69	18.41	17.99	17.56
7	.25	1.573	1.701	1.717	1.716	1.711	1.706	1.701	1.697	1.693	1.690	1.684	1.678
	.10	3.589	3.257	3.074	2.961	2.883	2.827	2.785	2.752	2.725	2.703	2.668	2.632
	.05	5.591	4.737	4.347	4.120	3.972	3.866	3.787	3.726	3.677	3.637	3.575	3.511
	.025	8.073	6.542	5.890	5.523	5.285	5.119	4.995	4.899	4.823	4.761	4.666	4.568
	.010	12.25	9.547	8.451	7.847	7.460	7.191	6.993	6.840	6.719	6.620	6.469	6.314
	.001	29.25	21.69	18.77	17.20	16.21	15.52	15.02	14.63	14.63	14.08	13.71	13.32

The *F*-distribution, continued. APPENDIX A.4

Outside column is *df* for the denominator; adjacent column is alpha; remaining columns have the *df* for the numerator in the first row.

20	25	30	40	50	100	120	200	500	1000	∞	α	*df*
9.581	9.634	9.670	9.714	9.741	9.795	9.804	9.822	9.838	9.844	9.849	.25	*1*
61.74	62.05	62.26	62.53	62.69	63.01	63.06	63.17	63.26	63.30	63.33	.10	
248.0	249.3	250.1	251.1	251.8	253.0	253.3	253.7	254.1	254.2	254.3	.05	
993.1	998.1	1001	1006	1008	1013	1014	1016	1017	1018	1018	.025	
6209	6240	6260	6286	6302	6334	6340	6350	6360	6363	6366	.010	
3.426	3.436	3.443	3.451	3.46	3.47	3.47	3.47	3.47	3.48	3.48	.25	*2*
9.441	9.451	9.458	9.466	9.47	9.48	9.48	9.49	9.49	9.49	9.49	.10	
19.45	19.46	19.46	19.47	19.5	19.5	19.5	19.5	19.5	19.5	19.5	.05	
39.45	39.46	39.46	39.47	39.5	39.5	39.5	39.5	39.5	39.5	39.5	.025	
99.45	99.46	99.47	99.48	99.5	99.5	99.5	99.5	99.5	99.5	99.5	.010	
999.3	999.3	999.3	999.3	999.3	999.3	999.3	999.3	999.3	999.3	999.3	.001	
2.460	2.463	2.465	2.467	2.469	2.471	2.472	2.473	2.474	2.474	2.474	.25	*3*
5.184	5.175	5.168	5.160	5.155	5.144	5.143	5.139	5.136	5.135	5.134	.10	
8.660	8.634	8.617	8.594	8.581	8.554	8.549	8.540	8.532	8.529	8.526	.05	
14.17	14.12	14.08	14.04	14.01	13.96	13.95	13.93	13.91	13.91	13.90	.025	
26.69	26.58	26.50	26.41	26.35	26.24	26.22	26.18	26.15	26.14	26.13	.010	
126.4	125.8	125.4	125.0	124.7	124.1	124.0	123.7	123.6	123.5	123.5	.001	
2.083	2.083	2.082	2.082	2.082	2.081	2.081	2.081	2.081	2.081	2.081	.25	*4*
3.844	3.828	3.817	3.804	3.795	3.778	3.775	3.769	3.764	3.762	3.761	.10	
5.803	5.769	5.746	5.717	5.699	5.664	5.658	5.646	5.635	5.632	5.628	.05	
8.560	8.501	8.461	8.411	8.381	8.319	8.309	8.288	8.270	8.264	8.257	.025	
14.02	13.91	13.84	13.75	13.69	13.58	13.56	13.52	13.49	13.47	13.46	.010	
46.10	45.69	45.43	45.08	44.88	44.47	44.40	44.27	44.14	44.09	44.05	.001	
1.882	1.880	1.878	1.876	1.875	1.872	1.872	1.871	1.870	1.870	1.869	.25	*5*
3.207	3.187	3.174	3.157	3.147	3.126	3.123	3.116	3.109	3.107	3.105	.10	
4.558	4.521	4.496	4.464	4.444	4.405	4.398	4.385	4.373	4.369	4.365	.05	
6.329	6.288	6.227	6.175	6.144	6.080	6.069	6.048	6.028	6.022	6.015	.025	
9.553	9.449	9.379	9.291	9.238	9.130	9.112	9.075	9.042	9.032	9.020	.010	
25.39	25.08	24.87	24.60	24.44	24.11	24.06	23.95	23.85	23.82	23.79	.001	
1.757	1.753	1.751	1.748	1.746	1.741	1.741	1.739	1.738	1.737	1.737	.25	*6*
2.836	2.815	2.800	2.781	2.770	2.746	2.742	2.734	2.727	2.725	2.722	.10	
3.874	3.835	3.808	3.774	3.754	3.712	3.705	3.690	3.678	3.673	3.669	.05	
5.168	5.107	5.065	5.012	4.980	4.915	4.904	4.882	4.862	4.856	4.849	.025	
7.396	7.296	7.229	7.143	7.091	6.987	6.969	6.934	6.901	6.891	6.880	.010	
17.12	16.85	16.67	16.44	16.31	16.03	15.98	15.89	15.80	15.77	15.75	.001	
1.671	1.667	1.663	1.659	1.657	1.651	1.650	1.648	1.646	1.646	1.645	.25	*7*
2.595	2.571	2.555	2.535	2.523	2.497	2.493	2.484	2.476	2.473	2.471	.10	
3.445	3.404	3.376	3.340	3.319	3.275	3.267	3.252	3.239	3.234	3.230	.05	
4.467	4.405	4.362	4.309	4.276	4.210	4.199	4.176	4.156	4.149	4.142	.025	
6.155	6.058	5.992	5.908	5.858	5.755	5.737	5.702	5.671	5.660	5.650	.010	
12.93	12.69	12.53	12.33	12.20	11.95	11.91	11.82	11.75	11.72	11.70	.001	

APPENDIX A.4 The *F*-distribution, continued.

Outside column is *df* for the denominator; adjacent column is alpha; remaining columns have the *df* for the numerator in the first row.

df	α	1	2	3	4	5	6	7	8	9	10	12	15
8	.25	1.538	1.657	1.668	1.664	1.658	1.651	1.645	1.640	1.635	1.631	1.624	1.617
	.10	3.458	3.113	2.924	2.806	2.726	2.668	2.624	2.589	2.561	2.538	2.502	2.464
	.05	5.318	4.459	4.066	3.838	3.688	3.581	3.500	3.438	3.388	3.347	3.284	3.218
	.025	7.571	6.059	5.416	5.053	4.817	4.652	4.529	4.433	4.357	4.295	4.200	4.101
	.010	11.26	8.649	7.591	7.006	6.632	6.371	6.178	6.029	5.911	5.814	5.667	5.515
	.001	25.41	18.49	15.83	14.39	13.48	12.86	12.40	12.05	12.05	11.54	11.19	10.84
9	.25	1.512	1.624	1.632	1.625	1.617	1.609	1.602	1.596	1.591	1.586	1.579	1.570
	.10	3.360	3.006	2.813	2.693	2.611	2.551	2.505	2.469	2.440	2.416	2.379	2.340
	.05	5.117	4.256	3.863	3.633	3.482	3.374	3.293	3.230	3.179	3.137	3.073	3.006
	.025	7.209	5.715	5.078	4.718	4.484	4.320	4.197	4.102	4.026	3.964	3.868	3.769
	.010	10.56	8.022	6.992	6.422	6.057	5.802	5.613	5.467	5.351	5.257	5.111	4.962
	.001	22.86	16.39	13.90	12.56	11.71	11.13	10.70	10.37	10.11	9.894	9.570	9.239
10	.25	1.491	1.598	1.603	1.595	1.585	1.576	1.569	1.562	1.556	1.551	1.543	1.534
	.10	3.285	2.924	2.728	2.605	2.522	2.461	2.414	2.377	2.347	2.323	2.284	2.244
	.05	4.965	4.103	3.708	3.478	3.326	3.217	3.135	3.072	3.020	2.978	2.913	2.845
	.025	6.937	5.456	4.826	4.468	4.236	4.072	3.950	3.855	3.779	3.717	3.621	3.522
	.010	10.04	7.559	6.552	5.944	5.636	5.386	5.200	5.057	4.942	4.849	4.706	4.558
	.001	21.04	14.90	12.55	11.28	10.48	9.926	9.517	9.204	8.956	8.754	8.446	8.129
11	.25	1.475	1.577	1.580	1.570	1.560	1.550	1.542	1.535	1.528	1.523	1.514	1.504
	.10	3.225	2.860	2.660	2.536	2.451	2.389	2.342	2.304	2.274	2.248	2.209	2.167
	.05	4.844	3.982	3.587	3.357	3.204	3.095	3.012	2.948	2.896	2.854	2.788	2.719
	.025	6.724	5.256	4.630	4.275	4.044	3.881	3.759	3.644	3.588	3.526	3.430	3.330
	.010	9.646	7.206	6.217	5.668	5.316	5.069	4.886	4.744	4.632	4.539	4.397	4.251
	.001	19.69	13.81	11.56	10.35	9.579	9.047	8.655	8.355	8.116	7.923	7.625	7.321
12	.25	1.461	1.560	1.561	1.550	1.539	1.529	1.520	1.512	1.505	1.500	1.490	1.480
	.10	3.177	2.807	2.606	2.480	2.394	2.331	2.283	2.245	2.214	2.188	2.147	2.105
	.05	4.747	3.885	3.490	3.259	3.106	2.996	2.913	2.849	2.796	2.753	2.687	2.617
	.025	6.554	5.096	4.474	4.121	3.891	3.728	3.607	3.512	3.436	3.374	3.277	3.177
	.010	9.330	6.927	5.953	5.412	5.064	4.821	4.640	4.499	4.388	4.296	4.155	4.010
	.001	18.64	12.97	10.80	9.633	8.892	8.378	8.001	7.711	7.480	7.292	7.005	6.709
13	.25	1.450	1.545	1.545	1.534	1.521	1.511	1.501	1.493	1.486	1.480	1.470	1.459
	.10	3.136	2.763	2.560	2.434	2.347	2.283	2.234	2.195	2.164	2.138	2.097	2.053
	.05	4.667	3.806	3.411	3.179	3.025	2.915	2.832	2.767	2.714	2.671	2.604	2.533
	.025	6.414	4.965	4.347	3.996	3.767	3.604	3.483	3.388	3.312	3.250	3.153	3.053
	.010	9.074	6.701	5.739	5.205	4.862	4.620	4.441	4.302	4.191	4.100	3.960	3.815
	.001	17.82	12.31	10.21	9.073	8.355	7.856	7.489	7.206	6.982	6.799	6.519	6.231
14	.25	1.440	1.533	1.532	1.519	1.507	1.495	1.485	1.477	1.470	1.463	1.453	1.441
	.10	3.102	2.726	2.522	2.395	2.307	2.243	2.193	2.154	2.122	2.095	2.054	2.010
	.05	4.600	3.739	3.344	3.112	2.958	2.848	2.764	2.699	2.646	2.602	2.534	2.463
	.025	6.298	4.857	4.242	3.892	3.663	3.501	3.380	3.285	3.209	3.147	3.050	2.949
	.010	8.862	6.515	5.564	5.035	4.695	4.456	4.278	4.140	4.030	3.939	3.800	3.656
	.001	17.82	11.78	9.730	8.622	7.922	7.436	7.078	6.802	6.583	6.404	6.130	5.848

The *F*-distribution, continued. APPENDIX A.4

Outside column is *df* for the denominator; adjacent column is alpha; remaining columns have the *df* for the numerator in the first row.

20	25	30	40	50	100	120	200	500	1000	∞	α	df
1.609	1.603	1.600	1.595	1.591	1.585	1.584	1.581	1.579	1.578	1.578	.25	**8**
2.425	2.400	2.383	2.361	2.348	2.321	2.316	2.307	2.298	2.295	2.293	.10	
3.150	3.108	3.079	3.043	3.020	2.975	2.967	2.951	2.937	2.932	2.928	.05	
3.999	3.937	3.894	3.840	3.807	3.739	3.728	3.705	3.684	3.677	3.670	.025	
5.359	5.263	5.198	5.116	5.065	4.963	4.946	4.911	4.880	4.869	4.859	.010	
10.48	10.26	10.11	9.919	9.804	9.572	9.532	9.453	9.382	9.358	9.333	.001	
1.561	1.555	1.551	1.545	1.541	1.534	1.533	1.530	1.527	1.527	1.526	.25	**9**
2.298	2.272	2.255	2.232	2.218	2.189	2.184	2.174	2.165	2.162	2.159	.10	
2.936	2.893	2.864	2.826	2.803	2.756	2.748	2.731	2.717	2.712	2.707	.05	
3.667	3.604	3.560	3.505	3.472	3.403	3.392	3.368	3.347	3.340	3.333	.025	
4.808	4.713	4.649	4.567	4.517	4.415	4.398	4.363	4.332	4.321	4.311	.010	
8.898	8.689	8.547	8.368	8.260	8.038	8.002	7.926	7.858	7.836	7.813	.001	
1.523	1.517	1.512	1.506	1.502	1.493	1.492	1.489	1.486	1.485	1.484	.25	**10**
2.201	2.174	2.155	2.132	2.117	2.087	2.082	2.071	2.062	2.059	2.055	.10	
2.774	2.730	2.700	2.661	2.637	2.588	2.580	2.563	2.548	2.543	2.538	.05	
3.419	3.355	3.311	3.255	3.221	3.152	3.140	3.116	3.094	3.087	3.080	.025	
4.405	4.311	4.247	4.165	4.115	4.014	3.996	3.962	3.930	3.920	3.909	.010	
7.803	7.604	7.469	7.297	7.192	6.980	6.944	6.872	6.807	6.785	6.762	.001	
1.493	1.486	1.481	1.474	1.469	1.460	1.459	1.455	1.452	1.451	1.450	.25	**11**
2.123	2.095	2.076	2.052	2.036	2.005	2.000	1.989	1.979	1.975	1.972	.10	
2.646	2.601	2.570	2.531	2.507	2.457	2.448	2.431	2.415	2.410	2.404	.05	
3.226	3.162	3.118	3.061	3.027	2.956	2.944	2.920	2.898	2.890	2.883	.025	
4.099	4.005	3.941	3.860	3.810	3.708	3.690	3.656	3.624	3.613	3.602	.010	
7.008	6.815	6.884	6.517	6.416	6.210	6.175	6.105	6.041	6.020	5.998	.001	
1.468	1.460	1.454	1.447	1.443	1.433	1.431	1.428	1.424	1.423	1.422	.25	**12**
2.060	2.031	2.011	1.986	1.970	1.938	1.932	1.921	1.911	1.097	1.904	.10	
2.544	2.498	2.466	2.426	2.401	2.350	2.341	2.323	2.307	2.302	2.296	.05	
3.073	3.008	2.963	2.906	2.871	2.800	2.787	2.763	2.740	2.733	2.275	.025	
3.858	3.765	3.701	3.619	3.569	3.467	3.449	3.414	3.382	3.372	3.361	.010	
6.405	6.217	6.090	5.928	5.829	5.627	5.593	5.524	5.462	5.441	5.420	.001	
1.447	1.438	1.432	1.425	1.420	1.409	1.408	1.404	1.400	1.399	1.398	.25	**13**
2.007	1.978	1.958	1.931	1.915	1.882	1.876	1.864	1.853	1.850	1.846	.10	
2.459	2.412	2.380	2.339	2.314	2.261	2.252	2.234	2.218	2.212	2.206	.05	
2.948	2.882	2.837	2.780	2.744	2.671	2.659	2.634	2.611	2.603	2.595	.025	
3.665	3.571	3.507	3.425	3.375	3.272	3.255	3.219	3.187	3.176	3.165	.010	
5.934	5.751	5.626	5.467	5.370	5.172	5.138	5.070	5.009	4.988	4.967	.001	
1.428	1.420	1.414	1.405	1.400	1.389	1.387	1.383	1.380	1.378	1.377	.25	**14**
1.962	1.933	1.912	1.885	1.869	1.834	1.828	1.816	1.805	1.801	1.797	.10	
2.388	2.341	2.308	2.266	2.241	2.187	2.178	2.159	2.142	2.136	2.131	.05	
2.844	2.778	2.732	2.674	2.638	2.565	2.552	2.526	2.503	2.495	2.487	.025	
3.505	3.412	3.348	3.266	3.215	3.112	3.094	3.059	3.026	3.015	3.004	.010	
5.557	5.377	5.254	5.098	5.002	4.807	4.773	4.707	4.645	4.625	4.604	.001	

APPENDIX A.4 The *F*-distribution, continued.

Outside column is *df* for the denominator; adjacent column is alpha; remaining columns have the *df* for the numerator in the first row.

df	α	1	2	3	4	5	6	7	8	9	10	12	15
15	.25	1.523	1.523	1.520	1.507	1.494	1.482	1.472	1.463	1.456	1.449	1.438	1.426
	.10	2.695	2.695	2.490	2.361	2.273	2.208	2.158	2.119	2.086	2.059	2.017	1.972
	.05	3.682	3.682	3.287	3.056	2.901	2.790	2.707	2.641	2.588	2.544	2.475	2.403
	.025	4.765	4.765	4.153	3.804	3.576	3.415	3.293	3.199	31.23	3.060	2.963	2.862
	.010	6.359	6.359	5.417	4.893	4.556	4.318	4.142	4.004	3.895	3.805	3.666	3.522
	.001	11.34	11.34	9.335	8.253	7.567	7.091	6.741	6.471	6.256	6.081	5.812	5.535
16	.25	1.514	1.514	1.510	1.497	1.483	1.471	1.460	1.451	1.443	1.437	1.426	1.413
	.10	2.668	2.668	2.462	2.33	2.244	2.178	2.128	2.088	2.055	2.028	1.985	1.940
	.05	3.634	3.634	3.239	3.007	2.852	2.741	2.657	2.591	2.538	2.494	2.425	2.352
	.025	4.687	4.687	4.077	3.729	3.502	3.341	3.219	3.125	3.049	2.986	2.889	2.788
	.010	6.226	6.226	5.292	4.773	4.437	4.202	4.026	3.890	3.780	3.691	3.553	3.409
	.001	10.97	10.97	9.006	7.944	7.272	6.805	6.460	6.195	5.984	5.812	5.547	5.275
17	.25	1.419	1.506	1.502	1.487	1.473	1.460	1.450	1.441	1.433	1.426	1.414	1.401
	.10	3.026	2.645	2.437	2.308	2.218	2.152	2.102	2.061	2.028	2.001	1.958	1.912
	.05	4.451	3.592	3.197	2.965	2.810	2.699	2.614	2.548	2.494	2.450	2.381	2.308
	.025	6.042	4.619	4.011	3.665	3.438	3.277	3.156	3.061	2.985	2.922	2.825	2.723
	.010	8.400	6.112	5.185	4.669	4.336	4.101	3.927	3.791	3.682	3.593	3.455	3.312
	.001	15.72	10.66	8.727	7.683	7.022	6.562	6.224	5.962	5.754	5.584	5.324	5.055
18	.25	1.413	1.499	1.494	1.479	1.464	1.452	1.441	1.431	1.423	1.416	1.404	1.391
	.10	3.007	2.624	2.416	2.286	2.196	2.130	2.079	2.038	2.005	1.977	1.933	1.887
	.05	4.414	3.555	3.160	2.928	2.773	2.661	2.577	2.510	2.456	2.412	2.342	2.269
	.025	5.978	4.560	3.954	3.608	3.382	3.221	3.100	3.005	2.929	2.866	2.769	2.667
	.010	8.285	6.013	5.092	4.579	4.248	4.015	3.841	3.705	3.597	3.508	3.371	3.227
	.001	15.38	10.39	8.487	7.460	6.808	6.355	6.021	5.763	5.557	5.390	5.132	4.866
19	.25	1.408	1.493	1.487	1.472	1.457	1.444	1.432	1.423	1.414	1.407	1.395	1.382
	.10	2.990	2.606	2.397	2.266	2.176	2.109	2.058	2.017	1.984	1.956	1.912	1.865
	.05	4.381	3.522	3.127	2.895	2.740	2.628	2.544	2.477	2.423	2.378	2.308	2.234
	.025	5.922	4.508	3.903	3.559	3.333	3.172	3.051	2.956	2.880	2.817	2.720	2.617
	.010	8.185	5.926	5.010	4.500	4.171	3.939	3.765	3.631	3.523	3.434	3.297	3.153
	.001	15.08	10.16	8.280	7.265	6.622	6.175	5.845	5.591	5.387	5.22	4.967	4.703
20	.25	1.404	1.487	1.481	1.465	1.450	1.437	1.425	1.415	1.407	1.399	1.387	1.374
	.10	2.975	2.589	2.380	2.249	2.158	2.091	2.040	1.999	1.965	1.937	1.892	1.845
	.05	4.351	3.493	3.098	2.866	2.711	2.599	2.514	2.447	2.393	2.348	2.278	2.203
	.025	5.871	4.461	3.859	3.515	3.289	3.128	3.007	2.913	2.837	2.774	2.676	2.573
	.010	8.096	5.849	4.938	4.431	4.103	3.871	3.699	3.564	3.457	3.368	3.231	3.088
	.001	14.82	9.953	8.098	7.096	6.461	6.019	5.692	5.440	5.239	5.075	4.823	4.562
22	.25	1.396	1.477	1.470	1.454	1.438	1.424	1.413	1.402	1.394	1.386	1.374	1.359
	.10	2.949	2.561	2.351	2.219	2.128	2.060	2.008	1.967	1.933	1.904	1.859	1.811
	.05	4.301	3.443	3.049	2.817	2.661	2.549	2.464	2.397	2.342	2.297	2.226	2.151
	.025	5.786	4.383	3.783	3.440	3.215	3.055	2.934	2.839	2.763	2.700	2.602	2.498
	.010	7.945	5.719	4.817	4.313	3.988	3.758	3.587	3.453	3.346	3.258	3.121	2.978
	.001	14.38	9.612	7.796	6.814	6.191	5.758	5.437	5.190	4.993	4.832	4.583	4.326

The *F*-distribution, continued. APPENDIX A.4

Outside column is *df* for the denominator; adjacent column is alpha; remaining columns have the *df* for the numerator in the first row.

20	25	30	40	50	100	120	200	500	1000	∞	α	*df*
1.413	1.404	1.397	1.389	1.383	1.372	1.370	1.366	1.362	1.360	1.359	.25	***15***
1.924	1.894	1.873	1.845	1.828	1.793	1.787	1.774	1.763	1.759	1.755	.10	
2.328	2.280	2.247	2.204	2.178	2.123	2.114	2.095	2.078	2.072	2.066	.05	
2.756	2.689	2.644	2.585	2.549	2.474	2.461	2.435	2.411	1.403	2.395	.025	
3.372	3.278	3.214	3.132	3.081	2.977	2.959	2.923	2.891	2.880	2.868	.010	
5.249	5.071	4.950	4.796	4.702	4.508	4.475	4.408	4.348	4.327	4.307	.001	
1.399	1.390	1.383	1.374	1.369	1.356	1.354	1.350	1.346	1.345	1.343	.25	***16***
1.891	1.860	1.839	1.811	1.793	1.757	1.751	1.738	1.726	1.722	1.718	.10	
2.276	2.227	2.194	2.151	2.124	2.068	2.059	2.039	2.022	2.016	2.010	.05	
2.681	2.614	2.568	2.509	2.472	2.396	2.383	2.357	2.333	2.324	2.316	.025	
3.259	3.165	3.101	3.018	2.967	2.863	2.845	2.808	2.775	2.764	2.753	.010	
4.992	4.817	4.697	4.545	4.451	4.259	4.226	4.160	4.100	4.080	4.059	.001	
1.387	1.377	1.370	1.361	1.355	1.343	1.341	1.336	1.332	1.330	1.329	.25	***17***
1.862	1.831	1.809	1.781	1.763	1.726	1.719	1.706	1.694	1.690	1.686	.10	
2.230	2.181	2.148	2.104	2.077	2.020	2.011	1.991	1.973	1.967	1.960	.05	
2.616	2.548	2.502	2.442	2.405	2.329	2.315	2.289	2.264	2.256	2.247	.025	
3.162	3.068	3.003	2.920	2.869	2.764	2.746	2.709	2.676	2.664	2.653	.010	
4.775	4.602	4.484	4.332	4.240	4.049	4.016	3.950	3.890	3.870	3.849	.001	
1.376	1.366	1.359	1.350	1.344	1.331	1.328	1.324	1.319	1.318	1.316	.25	***18***
1.837	1.805	1.783	1.754	1.736	1.698	1.691	1.678	1.665	1.661	1.657	.10	
2.191	2.141	2.107	2.063	2.035	1.978	1.968	1.948	1.929	1.923	1.917	.05	
2.559	2.491	2.445	2.384	2.347	2.269	2.256	2.229	2.204	2.195	2.187	.025	
3.077	2.983	2.919	2.835	2.784	2.678	2.660	2.623	2.589	2.577	2.566	.010	
4.590	4.418	4.301	4.151	4.059	3.869	3.836	3.770	3.710	3.690	3.670	.001	
1.367	1.356	1.349	1.339	1.333	1.320	1.317	1.312	1.308	1.306	1.305	.25	***19***
1.814	1.782	1.759	1.730	1.711	1.673	1.666	1.652	1.639	1.635	1.631	.10	
2.155	2.106	2.071	2.026	1.999	1.940	1.930	1.910	1.891	1.884	1.878	.05	
2.509	2.441	2.394	2.333	2.295	2.217	2.203	2.176	2.150	2.142	2.133	.025	
3.003	2.909	2.844	2.761	2.709	2.602	2.584	2.547	2.512	2.501	2.489	.010	
4.430	4.259	4.143	3.994	3.902	3.713	3.680	3.615	3.555	3.534	3.514	.001	
1.358	1.348	1.340	1.330	1.324	1.310	1.307	1.302	1.298	1.296	1.294	.25	***20***
1.794	1.761	1.738	1.708	1.690	1.650	1.643	1.629	1.616	1.612	1.607	.10	
2.124	2.074	2.039	1.994	1.966	1.907	1.896	1.875	1.856	1.850	1.843	.05	
2.464	2.396	2.349	2.287	2.249	2.170	2.156	2.128	2.103	2.094	2.085	.025	
2.938	2.843	2.778	2.695	2.643	2.535	2.517	2.479	2.445	2.433	2.421	.010	
4.290	4.121	4.005	3.856	3.765	3.576	3.544	3.478	3.418	3.398	3.378	.001	
1.343	1.332	1.324	1.314	1.307	1.293	1.290	1.285	1.280	1.278	1.276	.25	***22***
1.759	1.726	1.702	1.671	1.652	1.611	1.604	1.590	1.576	1.571	1.567	.10	
2.071	2.020	1.984	1.938	1.909	1.849	1.838	1.817	1.797	1.790	1.783	.05	
2.389	2.320	2.272	2.210	2.171	2.090	2.076	2.047	2.021	2.012	2.003	.025	
2.827	2.733	2.667	2.583	2.531	2.422	2.403	2.365	2.329	2.317	2.305	.010	
4.058	3.891	3.776	3.628	3.537	3.349	3.317	3.251	3.191	3.171	3.150	.001	

APPENDIX A.4 The *F*-distribution, continued.

Outside column is *df* for the denominator; adjacent column is alpha; remaining columns have the *df* for the numerator in the first row.

df	α	1	2	3	4	5	6	7	8	9	10	12	15
24	.25	1.390	1.470	1.462	1.445	1.428	1.414	1.402	1.392	1.383	1.375	1.362	1.347
	.10	2.927	2.538	2.327	2.195	2.103	2.035	1.983	1.941	1.906	1.877	1.832	1.783
	.05	4.260	3.403	3.009	2.776	2.621	2.508	2.423	2.355	2.300	2.255	2.183	2.108
	.025	5.717	4.319	3.721	3.379	3.155	2.995	2.874	2.779	2.703	2.640	2.541	2.437
	.010	7.823	5.614	4.718	4.218	3.895	3.667	3.496	3.363	3.256	3.168	3.032	2.889
	.001	14.03	9.340	7.554	6.589	5.977	5.551	5.235	4.991	4.797	4.638	4.393	4.139
26	.25	1.384	1.463	1.454	1.437	1.420	1.406	1.393	1.383	1.374	1.366	1.352	1.337
	.10	2.909	2.519	2.307	2.174	2.082	2.014	1.961	1.919	1.884	1.855	1.809	1.760
	.05	4.225	3.369	2.975	2.743	2.587	2.474	2.388	2.321	2.265	2.220	2.148	2.072
	.025	5.659	4.265	3.670	3.329	3.105	2.945	2.824	2.729	2.653	2.590	2.491	2.387
	.010	7.721	5.526	4.637	4.140	3.818	3.591	3.421	3.288	3.182	3.094	2.958	2.815
	.001	13.74	9.117	7.357	6.406	5.802	5.381	5.070	4.829	4.637	4.480	4.238	3.986
28	.25	1.380	1.457	1.448	1.430	1.413	1.399	1.386	1.375	1.366	1.358	1.344	1.329
	.10	2.894	2.503	2.291	2.157	2.064	1.996	1.943	1.900	1.865	1.836	1.790	1.740
	.05	4.196	3.340	2.947	2.714	2.558	2.445	2.359	2.291	2.236	2.190	2.118	2.041
	.025	5.610	4.221	3.626	3.286	3.063	2.903	2.782	2.687	2.611	2.547	2.448	2.344
	.010	7.636	5.453	4.568	4.074	3.754	3.528	3.358	3.226	3.120	3.032	2.896	2.753
	.001	13.50	8.930	7.193	6.253	5.657	5.241	4.933	4.695	4.505	4.349	4.109	3.859
30	.25	1.376	1.452	1.443	1.424	1.407	1.392	1.380	1.369	1.359	1.351	1.337	1.321
	.10	2.881	2.489	2.276	2.142	2.049	1.980	1.927	1.884	1.849	1.819	1.773	1.722
	.05	4.171	3.316	2.922	2.690	2.534	2.421	2.334	2.266	2.211	2.165	2.092	2.015
	.025	5.568	4.182	3.589	3.250	3.026	2.867	2.746	2.651	2.575	2.511	2.412	2.307
	.010	7.562	5.390	4.510	4.018	3.699	3.473	3.305	3.173	3.067	2.979	2.843	2.700
	.001	13.29	8.773	7.054	6.125	5.534	5.122	4.817	4.582	4.393	4.239	4.001	3.753
40	.25	1.362	1.435	1.424	1.404	1.386	1.371	1.357	1.345	1.335	1.327	1.312	1.295
	.10	2.835	2.440	2.226	2.091	1.997	1.927	1.873	1.829	1.793	1.763	1.715	1.662
	.05	4.085	3.232	2.839	2.606	2.449	2.336	2.249	2.180	2.124	2.077	2.003	1.924
	.025	5.424	4.051	3.463	3.126	2.904	2.744	2.624	2.529	2.452	2.388	2.288	2.182
	.010	7.314	5.178	4.313	3.828	3.514	3.291	3.124	2.993	2.888	2.801	2.665	2.522
	.001	12.61	8.251	6.595	5.698	5.128	4.731	4.436	4.207	4.024	3.874	3.643	3.400
60	.25	1.349	1.419	1.405	1.385	1.366	1.349	1.335	1.323	1.312	1.303	1.287	1.269
	.10	2.791	2.393	2.177	2.041	1.946	1.875	1.819	1.775	1.738	1.707	1.657	1.603
	.05	4.001	3.152	2.758	2.525	2.368	2.254	2.167	2.097	2.040	1.993	1.917	1.836
	.025	5.286	3.925	3.43	3.008	2.786	2.627	2.507	2.412	2.334	2.270	2.169	2.061
	.010	7.077	4.977	4.126	3.649	3.339	3.119	2.953	2.823	2.718	2.632	2.496	2.352
	.001	11.97	7.768	6.171	5.307	4.757	4.372	4.086	3.865	3.687	3.542	3.315	3.078
100	.25	1.339	1.406	1.391	1.369	1.349	1.332	1.317	1.304	1.293	1.283	1.267	1.248
	.10	2.756	2.356	2.139	2.002	1.906	1.834	1.778	1.732	1.695	1.663	1.612	1.557
	.05	3.936	3.087	2.696	2.463	2.305	2.191	2.103	2.032	1.975	1.927	1.850	1.768
	.025	5.179	3.828	3.250	2.917	2.696	2.537	2.417	2.321	2.244	2.179	2.077	1.968
	.010	6.895	4.824	3.984	3.513	3.206	2.988	2.823	2.694	2.590	2.503	2.368	2.223
	.001	11.496	7.408	5.857	5.017	4.482	4.107	3.829	3.612	3.439	3.296	3.074	2.840

The F-distribution, continued. APPENDIX A.4

Outside column is *df* for the denominator; adjacent column is alpha; remaining columns have the *df* for the numerator in the first row.

20	25	30	40	50	100	120	200	500	1000	∞	α	*df*
1.331	1.319	1.311	1.300	1.293	1.278	1.275	1.270	1.264	1.263	1.261	.25	***24***
1.730	1.696	1.672	1.641	1.621	1.579	1.571	1.556	1.542	1.538	1.533	.10	
2.027	1.975	1.939	1.892	1.863	1.800	1.790	1.768	1.747	1.740	1.733	.05	
2.327	2.257	2.209	2.146	2.107	2.024	2.010	1.981	1.954	1.945	1.935	.025	
2.738	2.643	2.577	2.492	2.440	2.329	2.310	2.271	2.235	2.223	2.211	.010	
3.873	3.707	3.593	3.447	3.356	3.168	3.136	3.070	3.010	2.989	2.969	.001	
1.320	1.309	1.300	1.289	1.282	1.266	1.263	1.257	1.251	1.249	1.247	.25	***26***
1.706	1.671	1.647	1.615	1.594	1.551	1.544	1.528	1.514	1.509	1.504	.10	
1.990	1.938	1.901	1.853	1.823	1.760	1.749	1.726	1.705	1.698	1.691	.05	
2.276	2.205	2.157	2.093	2.053	1.969	1.954	1.925	1.897	1.888	1.878	.025	
2.664	2.569	2.503	2.417	2.364	2.252	2.333	2.193	2.156	2.144	2.131	.010	
3.723	3.558	3.445	3.299	3.208	3.020	2.987	2.922	2.861	2.840	2.819	.001	
1.311	1.299	1.291	1.279	1.271	1.255	1.252	1.246	1.240	1.238	1.236	.25	***28***
1.685	1.650	1.625	1.592	1.572	1.528	1.520	1.504	1.489	1.484	1.478	.10	
1.959	1.906	1.869	1.820	1.790	1.725	1.714	1.691	1.669	1.662	1.654	.05	
2.232	2.161	2.112	2.048	2.007	1.922	1.907	1.877	1.848	1.839	1.829	.025	
2.602	2.506	2.440	2.354	2.300	2.187	2.167	2.127	2.090	2.077	2.064	.010	
3.598	3.434	3.321	3.176	3.085	2.897	2.864	2.798	2.736	2.716	2.695	.001	
1.303	1.291	1.282	1.270	1.263	1.245	1.242	1.236	1.230	1.228	1.226	.25	***30***
1.667	1.632	1.606	1.573	1.552	1.507	1.499	1.482	1.467	1.462	1.456	.10	
1.932	1.878	1.841	1.792	1.761	1.695	1.683	1.660	1.637	1.630	1.622	.05	
2.195	2.124	2.074	2.009	1.968	1.882	1.866	1.835	1.806	1.797	1.787	.025	
2.549	2.453	2.386	2.299	2.245	2.131	2.111	2.070	2.032	2.019	2.006	.010	
3.493	3.330	3.217	3.072	2.981	2.792	2.760	2.693	2.631	2.610	2.589	.001	
1.276	1.263	1.253	1.240	1.231	1.212	1.208	1.201	1.193	1.191	1.188	.25	***40***
1.605	1.568	1.541	1.506	1.483	1.434	1.425	1.406	1.389	1.383	1.377	.10	
1.839	1.783	1.744	1.693	1.660	1.589	1.577	1.551	1.526	1.517	1.509	.05	
2.068	1.994	1.943	1.875	1.832	1.741	1.724	1.691	1.659	1.648	1.637	.025	
2.369	2.271	2.203	2.114	2.058	1.938	1.917	1.874	1.833	1.819	1.805	.010	
3.145	2.984	2.872	2.727	2.636	2.444	2.410	2.341	2.277	2.255	2.233	.001	
1.248	1.234	1.223	1.208	1.198	1.176	1.172	1.163	1.154	1.151	1.147	.25	***60***
1.543	1.504	1.476	1.437	1.413	1.358	1.348	1.326	1.306	1.299	1.291	.10	
1.748	1.690	1.649	1.594	1.559	1.481	1.467	1.438	1.409	1.399	1.389	.05	
1.944	1.869	1.815	1.744	1.699	1.599	1.581	1.543	1.507	1.495	1.482	.025	
2.198	2.098	2.028	1.936	1.877	1.749	1.726	1.678	1.633	1.617	1.601	.010	
2.826	2.667	2.555	2.409	2.316	2.118	2.082	2.009	1.939	1.915	1.890	.001	
1.226	1.210	1.198	1.182	1.171	1.145	1.140	1.129	1.118	1.114	1.109	.25	***100***
1.494	1.453	1.423	1.382	1.355	1.293	1.282	1.257	1.232	1.223	1.214	.10	
1.676	1.616	1.573	1.515	1.477	1.392	1.376	1.342	1.308	1.296	1.283	.05	
1.849	1.770	1.715	1.640	1.592	1.483	1.463	1.420	1.378	1.363	1.347	.025	
2.067	1.965	1.893	1.797	1.735	1.598	1.572	1.518	1.466	1.447	1.427	.010	
2.591	2.431	2.319	2.170	2.076	1.867	1.892	1.749	1.671	1.644	1.615	.001	

APPENDIX A.4 The *F*-distribution, continued.

Outside column is *df* for the denominator; adjacent column is alpha; remaining columns have the *df* for the numerator in the first row.

df	α	1	2	3	4	5	6	7	8	9	10	12	15
120	.25	1.336	1.402	1.387	1.365	1.345	1.328	1.313	1.300	1.289	1.279	1.262	1.243
	.10	2.748	2.347	2.130	1.992	1.896	1.824	1.767	1.722	1.684	1.652	1.601	1.545
	.05	3.920	3.072	2.680	2.447	2.290	2.175	2.087	2.016	1.959	1.910	1.834	1.750
	.025	5.152	3.805	3.227	2.894	2.674	2.515	2.395	2.299	2.222	2.157	2.055	1.945
	.010	6.851	4.787	3.949	3.480	3.174	2.956	2.792	2.663	2.559	2.472	2.336	2.191
	.001	11.38	7.321	5.781	4.947	4.416	4.044	3.767	3.552	3.379	3.237	3.016	2.783
200	.25	1.331	1.396	1.380	1.358	1.337	1.319	1.304	1.291	1.279	1.269	1.252	1.232
	.10	2.731	2.329	2.111	1.973	1.876	1.804	1.747	1.701	1.663	1.631	1.579	1.522
	.05	3.888	3.041	2.650	2.417	2.259	2.144	2.056	1.985	1.927	1.878	1.801	1.717
	.025	5.100	3.758	3.182	2.850	2.630	2.472	2.351	2.256	2.178	2.113	2.010	1.900
	.010	6.763	4.713	3.881	3.414	3.110	2.893	2.730	2.601	2.497	2.411	2.275	2.129
	.001	11.15	7.152	5.634	4.812	4.287	3.920	3.647	3.434	3.263	3.123	2.904	2.672
500	.25	1.326	1.390	1.374	1.351	1.330	1.312	1.296	1.283	1.271	1.261	1.243	1.223
	.10	2.716	2.313	2.095	1.956	1.859	1.786	1.729	1.683	1.644	1.612	1.559	1.501
	.05	3.860	3.014	2.623	2.390	2.232	2.117	2.028	1.957	1.899	1.850	1.772	1.686
	.025	5.054	3.716	3.142	2.811	2.592	2.434	2.313	2.217	2.139	2.074	1.971	1.859
	.010	6.686	4.648	3.821	3.357	3.054	2.838	2.675	2.547	2.443	2.356	2.220	2.075
	.001	10.96	7.004	5.506	4.693	4.175	3.813	3.542	3.332	3.163	3.023	2.806	2.576
1000	.25	1.325	1.388	1.372	1.349	1.328	1.309	1.294	1.280	1.268	1.258	1.240	1.220
	.10	2.711	2.308	2.089	1.950	1.853	1.780	1.723	1.676	1.638	1.605	1.552	1.494
	.05	3.851	3.005	2.614	2.381	2.223	2.108	2.019	1.948	1.889	1.840	1.762	1.676
	.025	5.039	3.703	3.129	2.799	2.579	2.421	2.300	2.204	2.126	2.061	1.958	1.846
	.010	6.660	4.626	3.801	3.338	3.036	2.820	2.657	2.529	2.425	2.339	2.203	2.056
	.001	10.89	6.956	5.464	4.655	4.139	3.778	3.508	3.299	3.130	2.991	2.774	2.544
∞	.25	1.323	1.386	1.369	1.346	1.325	1.307	1.291	1.277	1.265	1.255	1.237	1.216
	.10	2.706	2.303	2.084	1.945	1.847	1.774	1.717	1.670	2.632	1.599	1.546	1.487
	.05	3.841	2.996	2.605	2.372	2.214	2.099	2.010	1.938	1.880	1.831	1.752	1.666
	.025	5.024	3.689	3.116	2.786	2.566	2.408	2.288	2.192	2.114	2.048	1.945	1.833
	.010	6.635	4.605	3.782	3.319	3.017	2.802	2.639	2.511	2.407	2.321	2.185	2.039
	.001	10.83	6.908	5.422	4.617	4.103	3.743	3.474	3.266	3.098	2.959	2.742	2.513

The F-distribution, continued. APPENDIX A.4

Outside column is *df* for the denominator; adjacent column is alpha; remaining columns have the *df* for the numerator in the first row.

20	25	30	40	50	100	120	200	500	1000	∞	α	*df*
1.220	1.204	1.912	1.175	1.164	1.137	1.131	1.120	1.108	1.103	1.099	.25	**120**
1.482	1.440	1.409	1.368	1.340	1.277	1.265	1.239	1.212	1.203	1.193	.10	
1.659	1.598	1.554	1.495	1.457	1.369	1.352	1.316	1.280	1.267	1.254	.05	
1.825	1.746	1.690	1.614	1.565	1.454	1.433	1.388	1.343	1.327	1.310	.025	
2.035	1.932	1.860	1.763	1.700	1.559	1.533	1.477	1.421	1.401	1.381	.010	
2.534	2.375	2.262	2.113	2.017	1.806	1.767	1.684	1.603	1.574	1.543	.001	
1.209	1.192	1.179	1.162	1.149	1.120	1.114	1.100	1.086	1.080	1.074	.25	**200**
1.458	1.414	1.383	1.339	1.310	1.242	1.228	1.199	1.168	1.157	1.144	.10	
1.623	1.561	1.516	1.455	1.415	1.321	1.302	1.263	1.221	1.205	1.189	.05	
1.778	1.698	1.640	1.562	1.511	1.393	1.370	1.320	1.269	1.250	1.229	.025	
1.971	1.868	1.794	1.694	1.629	1.481	1.453	1.391	1.328	1.304	1.279	.010	
2.424	2.264	2.151	2.000	1.902	1.682	1.641	1.552	1.460	1.427	1.390	.001	
1.198	1.181	1.168	1.149	1.136	1.103	1.096	1.081	1.062	1.055	1.045	.25	**500**
1.435	1.391	1.358	1.313	1.282	1.209	1.194	1.160	1.122	1.106	1.087	.10	
1.592	1.528	1.482	1.419	1.376	1.275	1.255	1.210	1.159	1.138	1.113	.05	
1.736	1.655	1.596	1.515	1.462	1.336	1.311	1.254	1.192	1.166	1.137	.025	
1.915	1.810	1.735	1.633	1.566	1.408	1.377	1.308	1.232	1.201	1.164	.010	
2.328	2.168	2.054	1.900	1.800	1.571	1.526	1.427	1.319	1.276	1.226	.001	
1.195	1.177	1.164	1.145	1.131	1.097	1.090	1.073	1.053	1.044	1.031	.25	**1000**
1.428	1.383	1.350	1.304	1.273	1.197	1.181	1.145	1.103	1.084	1.060	.10	
1.581	1.517	1.471	1.406	1.363	1.260	1.239	1.190	1.134	1.110	1.078	.05	
1.722	1.640	1.581	1.499	1.445	1.316	1.290	1.230	1.162	1.132	1.094	.025	
1.897	1.791	1.716	1.613	1.544	1.383	1.351	1.278	1.195	1.159	1.112	.010	
2.297	2.136	2.022	1.686	1.767	1.533	1.487	1.383	1.266	1.216	1.153	.001	
1.191	1.174	1.160	1.140	1.127	1.091	1.066	1.066	1.042	1.030	1.0097	.25	∞
1.421	1.375	1.342	1.295	1.263	1.185	1.130	1.130	1.082	1.058	1.0006	.10	
1.571	1.506	1.459	1.394	1.350	1.243	1.170	1.170	1.106	1.075	1.0007	.05	
1.708	1.626	1.566	1.484	1.428	1.296	1.205	1.205	1.128	1.090	1.0009	.025	
1.878	1.773	1.696	1.592	1.523	1.358	1.247	1.247	1.53	1.107	1.0010	.010	
2.266	2.105	1.990	1.835	1.733	1.494	1.338	1.338	1.207	1.144	1.0014	.001	

Source: From *Biometrika Tables for Statisticians* (Vol. 1) (3rd ed.) by E.S. Pearson and H.O. Hartley (Eds.), 1966, London: Biometrika Trustees.

APPENDIX A.5 Values of the studentized range (q), ($\alpha = .10$).

	Number of groups (k)								
df_E	2	3	4	5	6	7	8	9	10
1	8.93	13.4	16.4	18.5	20.2	21.5	22.6	23.6	24.5
2	4.13	5.73	6.77	7.54	8.14	8.63	9.05	9.41	9.72
3	3.33	4.47	5.20	5.74	6.06	6.51	6.81	7.06	7.29
4	3.01	3.98	4.59	5.03	5.39	5.68	5.93	6.14	6.33
5	2.85	3.72	4.26	4.66	4.98	5.24	5.46	5.65	5.82
6	2.75	3.56	4.07	4.44	4.73	4.97	5.17	5.34	5.50
7	2.68	3.45	3.93	4.28	4.55	4.78	4.97	5.14	5.28
8	2.63	3.37	3.83	4.17	4.43	4.65	4.83	4.99	5.13
9	2.59	3.32	3.76	4.08	4.34	4.54	4.72	4.87	5.01
10	2.56	3.27	3.70	4.02	4.26	4.47	4.64	4.78	4.91
11	2.54	3.23	3.66	3.96	4.20	4.40	4.57	4.71	4.84
12	2.52	3.20	3.62	3.92	4.16	4.35	4.51	4.65	4.78
13	2.50	3.18	3.59	3.88	4.12	4.30	4.46	4.60	4.72
14	2.49	3.16	3.56	3.85	4.08	4.27	4.42	4.56	4.68
15	2.48	3.14	3.54	3.83	4.05	4.23	4.39	4.52	4.64
16	2.47	3.12	3.52	3.80	4.03	4.21	4.36	4.49	4.61
17	2.46	3.11	3.50	3.78	4.00	4.18	4.33	4.46	4.58
18	2.45	3.10	3.49	3.77	3.98	4.16	4.31	4.44	4.55
19	2.45	3.09	3.47	3.75	3.97	4.14	4.29	4.42	4.53
20	2.44	3.08	3.46	3.74	3.95	4.12	4.27	4.40	4.51
24	2.42	3.05	3.42	3.69	3.90	4.07	4.21	4.34	4.44
30	2.40	3.02	3.39	3.65	3.85	4.02	4.16	4.28	4.38
40	2.38	2.99	3.35	3.60	3.80	3.96	4.10	4.21	4.32
60	2.36	2.96	3.31	3.56	3.75	3.91	4.04	4.16	4.25
120	2.34	2.93	3.28	3.52	3.71	3.86	3.99	4.10	4.19
∞	2.33	2.90	3.24	3.48	3.66	3.81	3.93	4.04	4.13

APPENDIX A.5

Values of the studentized range (q), continued, ($\alpha = .05$).

df_E	Number of groups (k)								
	2	3	4	5	6	7	8	9	10
1	18.0	27.0	32.8	37.1	40.4	43.1	45.4	47.4	49.1
2	6.09	8.3	9.8	10.9	11.7	12.4	13.0	13.5	14.0
3	4.50	5.91	6.82	7.50	8.04	8.48	8.85	9.18	9.46
4	3.93	5.04	5.76	6.29	6.71	7.05	7.35	7.60	7.83
5	3.64	4.60	5.22	5.67	6.03	6.33	6.58	6.80	6.99
6	3.46	4.34	4.90	5.31	5.63	5.89	6.12	6.32	6.49
7	3.34	4.16	4.68	5.06	5.36	5.61	5.82	6.00	6.16
8	3.26	4.04	4.53	4.89	5.17	5.40	5.60	5.77	5.92
9	3.20	3.95	4.42	4.76	5.02	5.24	5.43	5.60	5.74
10	3.15	3.88	4.33	4.65	4.91	5.12	5.30	5.46	5.60
11	3.11	3.82	4.26	4.57	4.82	5.03	5.20	5.35	5.49
12	3.08	3.77	4.20	4.51	4.75	4.95	5.12	5.27	5.40
13	3.06	3.73	4.15	4.45	4.69	4.88	5.05	5.19	5.32
14	3.03	3.70	4.11	4.41	4.64	4.83	4.99	5.13	5.25
15	3.01	3.67	4.08	4.37	4.60	4.78	4.94	5.08	5.20
16	3.00	3.65	4.05	4.33	4.56	4.74	4.90	5.03	5.15
17	2.98	3.63	4.02	4.30	4.52	4.71	4.86	4.99	5.11
18	2.97	3.61	4.00	4.29	4.49	4.67	4.82	4.96	5.07
19	2.96	3.59	3.98	4.25	4.47	4.65	4.79	4.92	5.04
20	2.95	3.58	3.96	4.23	4.45	4.62	4.77	4.90	5.01
24	2.92	3.53	3.90	4.17	4.37	4.54	4.68	4.81	4.92
30	2.89	3.49	3.84	4.10	4.30	4.46	4.60	4.72	4.83
40	2.86	3.44	3.79	4.04	4.23	4.39	4.54	4.63	4.74
60	2.83	3.40	3.74	3.98	4.16	4.31	4.44	4.55	4.65
120	2.80	3.36	3.69	3.92	4.10	4.24	4.36	4.48	4.56
∞	2.77	3.31	3.63	3.86	4.03	4.17	4.29	4.39	4.47

APPENDIX A.5 Values of the studentized range (q), continued, ($\alpha = .01$).

	Number of groups (k)								
df_E	2	3	4	5	6	7	8	9	10
1	90.02	135.04	164.26	185.58	202.21	215.77	227.17	236.97	245.54
2	14.04	19.02	22.29	24.72	26.63	28.20	29.53	30.68	31.69
3	8.26	10.62	12.17	13.32	14.24	15.00	15.64	16.20	16.69
4	6.51	8.12	9.17	9.96	10.58	11.10	11.54	11.93	12.26
5	5.70	6.98	7.80	8.42	8.91	9.32	9.67	9.97	10.24
6	5.24	6.33	7.03	7.56	7.97	8.32	8.61	8.87	9.10
7	4.95	5.92	6.54	7.01	7.37	7.68	7.94	8.17	8.37
8	4.75	5.64	6.20	6.63	6.96	7.24	7.47	7.68	7.86
9	4.60	5.43	5.96	6.35	6.66	6.92	7.13	7.33	7.49
10	4.48	5.27	5.77	6.14	6.43	6.67	6.88	7.05	7.21
11	4.39	5.15	5.62	5.97	6.25	6.48	6.67	6.84	6.99
12	4.32	5.05	5.50	5.84	6.10	6.32	6.51	6.67	6.81
13	4.26	4.96	5.40	5.73	5.98	6.19	6.37	6.53	6.67
14	4.21	4.90	5.32	5.63	5.88	6.09	6.26	6.41	6.54
15	4.17	4.84	5.25	5.56	5.80	5.99	6.16	6.31	6.44
16	4.13	4.79	5.19	5.49	5.72	5.92	6.08	6.22	6.35
17	4.10	4.74	5.14	5.43	5.66	5.85	6.01	6.15	6.27
18	4.07	4.70	5.09	5.38	5.60	5.79	5.94	6.08	6.20
19	4.05	4.67	5.05	5.33	5.55	5.74	5.89	6.02	6.14
20	4.02	4.64	5.02	5.29	5.51	5.69	5.84	5.97	6.09
24	3.96	4.55	4.91	5.17	5.37	5.54	5.69	5.81	5.92
30	3.89	4.46	4.80	5.05	5.24	5.40	5.54	5.65	5.76
60	3.76	4.28	4.59	4.82	4.99	5.13	5.25	5.36	5.45
120	3.70	4.20	4.50	4.71	4.87	5.01	5.12	5.21	5.30
∞	3.64	4.12	4.40	4.60	4.76	4.88	4.99	5.08	5.16

Source: From *Biometrika Tables for Statisticians* (Vol. 1) (3rd ed.) by E.S. Pearson and H.O.Hartley (Eds.), 1966, London: Biometrika Trustees.

Values of the chi-square distribution (χ^2). APPENDIX A.6

df	0.10	0.05	0.01
1	2.71	3.84	6.63
2	4.61	5.99	9.21
3	6.25	7.81	11.34
4	7.78	9.49	13.28
5	9.24	11.07	15.09
6	10.64	12.59	16.81
7	12.02	14.07	18.48
8	14.36	15.51	20.09
9	14.68	16.92	21.67
10	15.99	18.31	23.21
11	17.28	19.68	24.73
12	18.55	21.03	26.22
13	19.81	22.36	27.69
14	21.06	23.68	29.14
15	22.31	25.00	30.58
16	23.54	26.30	32.00
17	24.77	27.59	33.41
18	25.99	28.87	34.81
19	27.20	30.14	36.19
20	28.41	31.41	37.57
25	34.38	37.65	44.31
30	40.26	43.77	50.89
40	51.81	55.76	63.69
50	63.17	67.50	76.15
60	74.40	79.08	88.80
70	85.53	90.53	100.43
80	96.58	101.88	112.33
90	107.57	113.15	124.12
∞	118.50	124.34	135.81

Source: From *Biometrika Tables for Statisticians* (Vol. 1) (3rd ed.) by E.S. Pearson and H.O. Hartley (Eds.), 1966, London: Biometrika Trustees.

APPENDIX A.7 Critical values of ρ (ρ_s).

	α	.25	.10	.05	.025	.01	.005	.0025	.001	.005 (one-tailed)
N	α	.50	.20	.10	.05	.02	.01	.005	.002	.001 (two-tailed)
4		.600	1.000	1.000						
5		.500	.800	.900	1.000	1.000				
6		.371	.657	.829	.886	.943	1.000	1.000		
7		.321	.571	.714	.786	.893	.929	.964	1.000	1.000
8		.310	.524	.643	.738	.833	.881	.905	.952	.976
9		.267	.483	.600	.700	.783	.833	.867	.917	.933
10		.248	.455	.564	.648	.745	.794	.830	.879	.903
11		.236	.427	.536	.618	.709	.755	.800	.845	.873
12		.224	.406	.503	.587	.671	.727	.776	.825	.860
13		.209	.385	.484	.560	.648	.703	.747	.802	.835
14		.200	.367	.464	.538	.622	.675	.723	.776	.811
15		.189	.354	.443	.521	.604	.654	.700	.754	.786
16		.182	.341	.429	.503	.582	.635	.679	.732	.765
17		.176	.328	.414	.485	.566	.615	.662	.713	.748
18		.170	.317	.401	.472	.550	.600	.643	.695	.728
19		.165	.309	.391	.460	.535	.584	.628	.677	.712
20		.161	.299	.380	.447	.520	.570	.612	.662	.696
21		.156	.292	.370	.435	.508	.556	.599	.648	.681
22		.152	.284	.361	.425	.496	.544	.586	.634	.667
23		.148	.278	.353	.415	.486	.532	.573	.622	.654
24		.144	.271	.344	.406	.476	.521	.562	.610	.642
25		.142	.265	.337	.398	.466	.511	.551	.598	.630

Critical values of $\rho(\rho_s)$, continued. APPENDIX A.7

N	α	.25	.10	.05	.025	.01	.005	.0025	.001	.005 (one-tailed)
	α	.50	.20	.10	.05	.02	.01	.005	.002	.001 (two-tailed)
26		.138	.259	.331	.390	.457	.501	.541	.587	.619
27		.136	.255	.324	.382	.448	.491	.531	.577	.608
28		.133	.250	.317	.375	.440	.483	.522	.567	.598
29		.130	.245	.312	.368	.433	.475	.513	.588	.589
30		.128	.240	.306	.362	.425	.467	.504	.549	.580
31		.126	.236	.301	.356	.418	.459	.496	.541	.571
32		.124	.237	.296	.350	.411	.452	.489	.533	.563
33		.121	.229	.291	.345	.405	.446	.482	.525	.554
34		.120	.225	.287	.340	.399	.439	.475	.517	.547
35		.118	.222	.283	.335	.394	.433	.468	.510	.539
36		.116	.219	.279	.330	.388	.427	.462	.504	.533
37		.114	.216	.275	.325	.383	.421	.456	.497	.526
38		.113	.212	.271	.321	.378	.415	.450	.491	.519
39		.111	.210	.267	.317	.373	.410	.444	.485	.513
40		.110	.207	.264	.313	.368	.405	.439	.479	.507
41		.108	.204	.261	.309	.365	.400	.433	.473	.501
42		.107	.202	.257	.305	.359	.395	.428	.468	.495
43		.105	.199	.254	.301	.355	.391	.423	.463	.490
44		.104	.197	.251	.298	.351	.386	.419	.458	.484
45		.103	.194	.248	.294	.347	.382	.414	.453	.479
46		.102	.192	.246	.291	.343	.378	.410	.448	.474
47		.101	.190	.243	???	.341	.374	.405	.443	.469

Source: Zar, J. H. (1972). Significance testing of the Spearman rank correlation coefficient. *Journal of the American Statistical Association, 67*, 578–580. Reprinted with permission.

APPENDIX A.8 Table of probabilities for the sign test (one-tailed test).

k **= the number of plus or minus signs, whichever is smaller.**

N	0	1	2	3	4	5	6	7	8	9	10	11	12	13	14	15	16	17
4	.062	.312	.688	.938	1.0													
5	.031	.188	.500	.812	.969	1.0												
6	.016	.109	.344	.656	.891	.984	1.0											
7	.008	.062	.227	.500	.773	.938	.992	1.0										
8	.004	.035	.145	.363	.637	.855	.965	.996	1.0									
9	.002	.020	.090	.254	.500	.746	.910	.980	.998	1.0								
10	.001	.011	.055	.172	.377	.623	.828	.945	.989	.999	1.0							
11		.006	.033	.113	.274	.500	.726	.887	.967	.994	.999+	1.0						
12		.003	.019	.073	.194	.387	.613	.806	.927	.981	.997	.999+	1.0					
13		.002	.011	.046	.133	.291	.500	.709	.867	.954	.989	.998	.999+	1.0				
14		.001	.006	.029	.090	.212	.395	.605	.788	.910	.971	.994	.999	.999+	1.0			
15			.004	.018	.059	.151	.304	.500	.696	.849	.941	.982	.996	.999+	.999+	1.0		
16			.002	.011	.038	.105	.227	.402	.598	.773	.895	.962	.989	.998	.999+	.999+	1.0	
17			.001	.006	.025	.072	.166	.315	.500	.685	.834	.928	.975	.994	.999	.999+	.999+	1.0
18			.001	.004	.015	.048	.119	.240	.407	.593	.760	.881	.952	.985	.996	.999	.999+	.999+
19				.002	.010	.032	.084	.180	.324	.500	.676	.820	.916	.968	.990	.998	.999+	.999+
20				.001	.006	.021	.058	.132	.252	.412	.588	.748	.868	.942	.979	.994	.999+	.999+

Table of probabilities for the sign test (one-tailed test), continued. APPENDIX A.8

N	0	1	2	3	4	5	6	7	8	9	10	11	12	13	14	15	16	17
21				.001	.004	.013	.039	.095	.192	.332	.500	.668	.808	.905	.961	.987	.996	.999
22					.002	.008	.026	.067	.143	.262	.416	.584	.738	.857	.933	.974	.992	.998
23					.001	.005	.017	.047	.105	.202	.339	.500	.661	.798	.895	.953	.983	.995
24					.001	.003	.011	.032	.076	.154	.271	.419	.581	.729	.846	.924	.968	.989
25						.002	.007	.022	.054	.115	.212	.345	.500	.655	.788	.885	.946	.978
26						.001	.005	.014	.038	.084	.163	.279	.423	.577	.721	.837	.916	.962
27						.001	.003	.010	.026	.061	.124	.221	.351	.500	.649	.779	.876	.939
28							.002	.006	.018	.044	.092	.172	.286	.425	.575	.714	.828	.908
29							.001	.004	.012	.031	.068	.132	.229	.356	.500	.644	.771	.868
30							.001	.003	.008	.021	.049	.100	.181	.292	.428	.572	.708	.819
31								.002	.005	.015	.035	.075	.141	.237	.360	.500	.640	.763
32								.001	.004	.010	.025	.055	.108	.189	.298	.430	.570	.702
33								.001	.002	.007	.018	.040	.081	.148	.243	.364	.500	.636
34									.001	.005	.012	.029	.061	.115	.196	.304	.432	.568
35									.001	.003	.008	.020	.045	.088	.155	.250	.368	.500
34									.001	.005	.012	.029	.061	.115	.196	.304	.432	.568
35									.001	.003	.008	.020	.045	.088	.155	.250	.368	.500

Source: Adapted from Table D, pp. 324–325, of *Nonparametric Statistics for the Behavioral Sciences*, 2nd ed. (1988) by N.J. Castellan and S. Siegal. Copyright McGraw-Hill, pp. 324–325. Used with permission.

APPENDIX A.9 Wilcoxon signed-rank test table (one-tailed test).

	N								
C*	3	4	5	6	7	8	9	10	11
3	.6250								
4	.3750								
5	.2500	5.625							
6	.1250	.4375							
7		.3125							
8		.1875	.5000						
9		.1250	.4063						
10		.0625	.3125						
11			.2188	.5000					
12			.1563	.4219					
13			.0938	.3438					
14			.0625	.2813	.5313				
15			.0313	.2188	.4688				
16				.1563	.4063				
17				.1094	.3438				
18				.0781	.2891	.5273			
19				.0469	.2344	.4727			
20				.0313	.1875	.4219			
21				.0156	.1484	.3711			
22					.1094	.3203			
23					.0781	.2734	.5000		
24					.0547	.2305	.4551		
25					.0391	.1914	.4102		
26					.0234	.1563	.3672		
27					.0156	.1250	.3262		
28					.0078	.0977	.2852	.5000	
29						.0742	.2480	.4609	
30						.0547	.2129	.4229	
31						.0391	.1797	.3848	
32						.0273	.1504	.3477	
33						.0195	.1250	.3125	.5171
34						.0117	.1016	.2783	.4829
35						.0078	.0820	.2461	.4492
36						.0039	.064	.2158	.4155
37							.0488	1.875	.3823
38							.0371	.1611	.3501

*C = the sum of the positive ranks.

Wilcoxon signed-rank test table (one-tailed test), continued. APPENDIX A.9

	N						
C*	9	10	11	12	13	14	15
39	.0273	1.377	.3188	.5151			
40	.0195	.1162	.2886	.4849			
41	.0137	.0967	.2598	.4548			
42	.0098	.0801	.2324	.4250			
43	.0059	.0654	.2065	.3955			
44	.0039	.0527	.1826	.3667			
45	.0020	.0420	.1602	.3386			
46		.0322	1.392	.3110	.5000		
47		.0244	1.201	.2847	.4730		
48		.0186	1.030	.2593	.4463		
49		.0137	.0874	.2349	.4197		
50		.0098	.0737	.2119	.3934		
51		.0068	.0615	.1902	.3677		
52		.0049	.0508	.1697	.3424		
53		.0029	.0415	.1506	.3177	.5000	
54		.0020	.0337	.1331	.2939	.4758	
55		.0010	.0269	.1167	.2709	.4516	
56			.0210	.1018	.2487	.4276	
57			.0161	.0881	.2274	.4039	
58			.0122	.0757	.2072	.3804	
59			.0093	.0647	.1879	.3574	
60			.0068	.0549	.1698	.3349	.5110
61			.0049	.0461	.1527	.3129	.4890
62			.0034	.0386	.1367	.2915	.4670
63			.0024	.0320	.1219	.2708	.4452
64			.0015	.0261	.1082	.2508	.4235
65			.0010	.0212	.0955	.2316	.4020
66			.0005	.0171	.0839	.2131	.3808
67				.0134	.0732	.1955	.3599
68				.0105	.0636	.1788	.3394
69				.0081	.0549	.1629	.3193
70				.0061	.0471	.1479	.2997
71				.0046	.0402	.1338	.2807
72				.0034	.0341	.1206	.2622
73				.0024	.0287	.1083	.2444
74				.0017	.0239	.0969	.2271
75				.0012	.0199	.0863	.2106

*C = the sum of the positive ranks.

Source: Table H, pp. 332–333, of *Nonparametric Statistics for the Behavioral Sciences*, 2nd ed. (1988) by N.J. Castellan and S. Siegal. Copyright McGraw-Hill, pp. 332–333. Used with permission.

APPENDIX A.10 Critical values for Kruskal-Wallis one-way analysis of variance by rank.

SAMPLE SIZES			α			
n_1	n_2	n_3	.10	.05	.01	.005
2	2	2	4.25			
3	2	1	4.29			
3	2	2	4.71	4.71		
3	3	1	4.57	5.14		
3	3	2	4.56	5.36		
3	3	3	4.62	5.60	7.20	7.20
4	2	1	4.50			
4	2	2	4.46	5.33		
4	3	1	4.06	5.21		
4	3	2	4.51	5.44	6.44	7.00
4	3	3	4.71	5.73	6.75	7.32
4	4	1	4.17	4.97	6.67	
4	4	2	4.55	5.45	7.04	7.28
4	4	3	4.55	5.60	7.14	7.59
4	4	4	4.65	5.69	7.66	8.00
5	2	1	4.20	5.00		
5	2	2	4.36	5.16	6.53	
5	3	1	4.02	4.96		
5	3	2	4.65	5.25	6.82	7.18
5	3	3	4.53	5.65	7.08	7.51
5	4	1	3.99	4.99	6.95	7.36
5	4	2	4.54	5.27	7.12	7.57
5	4	3	4.55	5.63	7.44	7.91
5	4	4	4.62	5.62	7.76	8.14
5	5	1	4.11	5.13	7.31	7.75
5	5	2	4.62	5.34	7.27	8.13
5	5	3	4.54	5.71	7.54	8.24
5	5	4	4.53	5.64	7.77	8.37
5	5	5	4.56	5.78	7.98	8.72
Large samples			4.61	5.99	9.21	10.60

Sources: C.H. Kraft and C. van Eeden (1968). *A nonparametric introduction to statistics*. Macmillan./J.H. Zar (1984). *Biostatistical analysis*. Pearson./N.J. Castellan and S. Siegal (1988), *Nonparametric Statistics for the Behavioral Sciences*, 2nd ed. McGraw-Hill. Used with permission.

Critical values for Friedman's two-way analysis of variance by ranks. APPENDIX A.11

k*	N*	$\alpha \leq .10$	$\alpha \leq .05$	$\alpha \leq .01$
3	3	6.00	6.00	—
	4	6.00	6.50	8.00
	5	5.20	6.40	8.40
	6	5.33	7.00	9.00
	7	5.43	7.14	8.86
	8	5.25	6.25	9.00
	9	5.56	6.22	8.67
	10	5.00	6.20	9.60
	11	4.91	6.54	8.91
	12	5.17	6.17	8.67
	13	4.77	6.00	9.39
	∞	4.61	5.99	9.21
4	2	6.00	6.00	—
	3	6.60	7.40	8.60
	4	6.30	7.80	9.60
	5	6.36	7.80	9.96
	6	6.40	7.60	10.00
	7	6.26	7.80	10.37
	8	6.30	7.50	10.35
	∞	6.25	7.82	11.34
5	3	7.47	8.53	10.13
	4	7.60	8.80	11.00
	5	7.68	8.96	11.52
	∞	7.78	9.49	13.28

*k = number of columns; N = number of rows

Sources: Hollander, M., & Wolfe, D.A. (1973). *Nonparametric statistics.* New York: John Wiley./Siegel, S., & Castellan, N. J. (1988). *Nonparametric statistics for the behavioral sciences* (2nd ed.). New York: McGraw-Hill./Kendall, M.G. (1970). *Rank correlation methods* (4th ed.). London: Charles Griffin & Co.

B Choosing Statistical Tests

APPENDIX B.1 How to choose a parametric statistical test.

DEPENDENT VARIABLE(S)		INDEPENDENT VARIABLES						
		Nominal Level of Measurement					Continuous Measurement	
			One IV		Two or	more IV	One IV	Two or more IV
		2 Levels	2 or more levels	Covariate (nominal or continuous)	No covariate	Covariate (nominal or continuous)		
Nominal measurement level	2 levels				Logistic regression	Logistic regression		Logistic regression or Discriminant analysis
	2 or more levels							Discriminant analysis
Continuous measurement level	One DV	*t*-test	One-way ANOVA	One-way ANCOVA	Factorial ANOVA	Factorial ANCOVA	Pearson product-moment correlation	Multiple regression or Path analysis
	2 or more DVs	One-way MANOVA		One-way MANCOVA	Factorial MANOVA	Factorial MANCOVA		Path analysis

How to choose a nonparametric statistical test. APPENDIX B.2

Level of Measurement	Measures of Association	One-Sample Case	Two-Sample Case		N Sample Case	
			Independent groups	Repeated measures or matched	Independent groups	Repeated measures or matched
Nominal	Phi coefficient Kappa coefficient	Binomial test Chi-square goodness of fit	Chi-square test		Chi-square test	Cochran *Q* test
Ordinal	Spearman rank order correlation coefficient Kendall coefficient of concordance		Median test Mann–Whitney U test	Sign test Wilcoxon signed-rank test	Kruskal–Wallis one-way ANOVA	Friedman two-way ANOVA by ranks test

C Answer Key

This appendix contains answers to Part A questions. Your instructor can furnish answers to Part B questions.

Chapter 1, Part A

1. A
2. D
3. B
4. D
5. A

Chapter 2, Part A

1. A
2. D
3. B
4. B
5. D
6. A

Chapter 3, Part A

1. A
2. A
3. B
4. A or B
5. C
6. D
7. C

Chapter 4, Part A

1. A
2. B
3. B
4. B

Chapter 5, Part A

1. D
2. B
3. C; both are linear transformations.
4. C; $r = .80$; if 68% level of confidence is 44–56, the SEE must be 6.
5. Solve for b and c. $b = .5$ and $c = 25$, thus $\hat{Y} = .5X + 25$.
6. B

Chapter 6, Part A

1. There is no difference between the national mean and your university mean.
2. There is a difference between the national mean and your university mean.
3. $\Sigma_{\bar{X}} = \frac{15}{\sqrt{25}} = \frac{15}{5} = 3$
4. $z = \frac{83 - 75}{3} = \frac{8}{3} = 2.67$

   Since $2.67 > 1.96$, you reject the null hypothesis that the means are equivalent. Your university students scored significantly different from 75.
5. B
6. A
7. B
8. A

Chapter 7, Part A

1. A
2. B
3. D
4. C
5. B

Chapter 8, Part A

1. $S_{\bar{X}} = \frac{14}{\sqrt{25}} = \frac{14}{5} = 2.8$

   $t = \frac{81 - 75}{2.8} = \frac{6}{2.8} = 2.14$

   $t_{(.05)}$ with 24 $df = 2.064$.
   Thus, reject the null hypothesis. Human performance majors have a more positive attitude toward physical activity than the general population (of course!).
2. Type I error
3. 5%
4. D
5. A. Change 10 to 12. We now need 72 subjects per group.
   B. Change 1.96 to 1.65. We now need approximately 40 subjects per group.
   C. Change 1.04 to 1.28. We now need approximately 58 subjects per group.
   D. Change 6 to 4. We now need approximately 113 subjects per group.

Chapter 9, Part A

1. 2, 4, 6
2. 4
3. There is no significant difference among the three means.
4. $SS_T = 52$; $SS_B = 24$; $SS_W = 28$
5. $df_T = 8$; $df_B = 2$; $df_W = 6$
6. $MS_B = 24/2 = 12$; $MS_W = 28/6 = 4.67$
7. F-ratio = 12/4.67 = 2.57
8. Critical F (2, 6) at $\alpha = .05$ is 5.14
9. You do not reject the null hypothesis. You conclude that the difference between the means is attributable to sampling error.
10. Type II, because you did not reject the null hypothesis

Chapter 10, Part A

1. B
2. D
3. C

Chapter 11, Part A

1. A
2. B
3. D
4. A: 3 x 5 between-between design
5.

Source	*SS*	*df*	*MS*	*F-ratio*
Factor A	21 (A)	2	10.5	3.5 (E)
Factor B	24 (I)	4	6 (J)	2 (K)
Interaction	60 (H)	8 (B)	7.5 (G)	2.5
Error	225	75 (C)	3 (D)	
Total	330	89		

   A = 2 * 10.5 = 21
   B = 2 * 4 = 8
   C = 89 − (2 + 4 + 8) = 75
   D = 225/75 = 3
   E = 10.5/3 = 3.5
   G = 3 * 2.5 = 7.5
   H = 7.5 * 8 = 60
   I = 330 − (225 + 60 + 21) = 24
   J = 24/4 = 6
   K = 6/3 = 2

6. Factor A (critical F = .05, df = 2, 75 = 3.15): significant

   Factor B (critical F = .05, df = 4, 75 = 2.53): not significant

   Interaction (critical F = .05, df = 8, 75 = 2.10): significant

7. That the pattern of learning the tennis serve with the different methods of instruction was not the same for the subjects of various ages.

Chapter 12, Part A

1. D
2. A
3. C
4. C

Chapter 13, Part A

1. A
2. D
3. D, loss of power
4. D
5. C
6. B

Chapter 14, Part A

1. D
2. Use the chi-square statistic in this one-sample case. Chance on the examination is 25%. The expected value if the student did not do better than chance is 25 out of 100 items. The student got 33 correct out of 100 items (33%). The calculated chi-square is 3.41. The critical value with 1 *df* is 3.84. Therefore, the student did not do better than chance. His score of 33 correct is likely for someone who did not study.
3. Use the chi-square statistic in this 3 x 3 case. The calculated chi-square value is 18.51. The critical value with 4 *df* is 9.41. Therefore, you reject the null hypothesis of no relationship and conclude that there is a relationship between SES and smoking.

Glossary

***a priori* contrasts** See *planned comparisons*.

alternative hypothesis Generally, the antithesis of the null hypothesis. The alternative hypothesis is assumed to reflect the true status of the relationship between variables when the null hypothesis is rejected. Typically, the alternative hypothesis is the research hypothesis.

animal research conducted with animal models.

ANOVA Analysis of variance. An inferential statistical procedure involving the examination of mean differences when the independent variable has three or more levels to be compared.

applied research Seeks to determine an immediate solution to be applied to a present problem; generally conducted to answer a precisely stated question.

asymptotic Describes a curve that approaches but never touches the *X*-axis.

basic research Seeks to increase the depth of understanding of a process by exploring fundamental principles; generally theoretical in nature.

between-groups design Research design in which each subject appears in one group only.

biased statistic A statistic is biased if the mean of its sampling distribution is *not* equal to the population parameter it is estimating.

Bonferroni adjustment A procedure to reduce the probability of a Type I error when performing multiple tests of significance.

canonical model The general model of relating any number of predictor *(X)* variables with any number of predicted variables *(Y)*.

central limit theorem A theory stating that with repeated samples, the sampling distribution of a statistic will become normally distributed regardless of the shape of the original population's distribution.

cluster sampling Method of deriving a sample from a population when natural groups of subjects are used as the sampling unit.

coefficient of determination (r^2) Statistic representing the percentage of variation in the predicted variable *(Y)* that can be accounted for by the predictor variable *(X)*; the square of the correlation coefficient.

complex comparison A multiple comparison procedure involving more than two means.

conceptual replication A method of increasing external validity by modifying the independent variables or by measuring the dependent variable differently.

concurrent validity A measurement procedure has concurrent validity when a correlation exists between the measure and a criterion measure obtained at the same point in time.

confidence level A researcher-selected percentage (typically, 95% or 99%) indicating the relative confidence of a decision.

consistency An estimate is consistent if its accuracy increases as the sample size increases.

construct validity A measurement procedure has construct validity when there is sufficient evidence suggesting that an unobservable construct exists.

content validity A measurement procedure has content validity when it adequately assesses the content of the attribute being measured.

continuous measurement Interval and ratio measurement scales are said to be continuous because theoretically there is no limit to how far they can be subdivided. This is in contrast to discrete data, which are reported in whole numbers. (See *interval* and *ratio*.)

contrast-based error rate A multiple comparison procedure where the alpha level is applied to each contrast conducted.

convenience sampling Selecting subjects who are conveniently available. Sometimes called *accidental sampling*.

convergent validity A measurement has convergent validity when there is correlational evidence suggesting that two measures are assessing the same construct.

correlation coefficient A statistic indicating the magnitude and direction of the relationship between two or more variables.

correlational statistics Measurements obtained from a sample that seek to describe the relationship between two or more variables.

Cronbach's alpha An intraclass reliability coefficient having values from 0.0 to +1.0.

cross-sectional research A procedure in which different groups of subjects are measured at the same point in time and then group differences are extrapolated through comparison of the results. It provides a snapshot in time.

deductive reasoning that moves from observations of general information to the explanation of specific events; opposite of inductive reasoning.

degrees of freedom The number of values in the calculation of a statistic that are free to vary when restrictions are placed on the data.

dependent *t*-test A statistical procedure used to test mean differences between two matched, paired, or correlated groups of subjects.

dependent variable The variable that is expected to change and is measured (the outcome). The symbol for the dependent variable is usually Y.

descriptive research Describes the current status of behavior and events for variables of interest.

descriptive statistics Measurements obtained from a sample that seek to describe the data set.

differential selection of subjects A threat to internal validity when the characteristics of the subjects are initially different in the treatment conditions because of improper or unavoidable selection and/or assignment procedures.

discriminant validity A measurement has discriminant validity when there is correlational evidence suggesting that two measures do not assess the same construct.

effect size (ES) A statistic that reflects the amount of association between the independent and dependent variables in terms of mean differences (Cohen's *d*) or variance accounted for (R^2, eta squared, omega squared).

efficiency The relative precision with which an estimator estimates a parameter.

estimate See *statistic*.

evidence-based Grounded on sound theory and data obtained from research conducted according to the scientific method

expectancy A threat to external validity where the researcher's bias or prior beliefs about the study outcome cause unintentional effects on the study; also referred to as "the self-fulfilling prophecy".

expected mean square The mathematical expected value of the variance resulting from repeated sampling from a distribution.

expected value The mean of a statistic's sampling distribution.

experimental mortality A threat to internal validity where uneven loss of subjects occurs from the various treatment conditions during the course of the study.

experimental research Seeks to determine cause-and effect relationships between and among variables.

external validity Evidence that the findings of a study can be generalized to other situations, subjects, or environments.

face validity A measurement procedure has face validity when it appears to measure what it is supposed to measure.

family-based error rate A multiple comparison procedure where the alpha level is applied to the entire set of contrasts conducted.

fixed independent variables Independent variable levels that are selected by the researcher and thus are not selected at random.

generalizability The certainty with which the findings of a study can be applied to other situations, subjects, or environments. *See also* external validity.

Hawthorne effect A threat to external validity where subjects may change their behaviors or outcome because they are aware they are subjects in a study.

histogram A graphical representation of the frequency of scores in a distribution; sometimes called a *bar graph*.

homogeneity of variance A parametric statistic assumption for the *t*-test and ANOVA requiring that the variances of all groups are equivalent.

human research conducted with human subjects.

independent *t*-test A statistical procedure used to test mean differences between two independent groups of subjects.

independent variable The variable that is manipulated by the researcher. In some situations, this is called the treatment, exposure, or predictor variable. The symbol for the independent variable is usually *X*.

inductive reasoning that moves from observations of specific events to predictions about general principles; opposite of deductive reasoning.

inferential statistics Process of obtaining data on a sample and generalizing (i.e., inferring) its characteristics to the population from which the sample was drawn.

instrumentation A threat to internal validity where measurement errors due to faulty equipment or a change in the definition of the measured variable result in incorrect data.

interaction The combined effects of two or more independent variables on a dependent variable.

interclass correlation A measure of the relationship between two variables having values from −1.0 to +1.0. Often used to estimate reliability with the Pearson product-moment correlation coefficient.

internal validity Evidence that the results of an experiment can be attributed to the effect of the independent variable rather than some confounding variable(s).

interval estimate A range of values around a sample statistic developed with a selected level of confidence.

interval measurement A measurement made by using a scale to quantify the amount of some characteristic possessed by the subjects. The zero value on the scale is arbitrarily chosen.

intraclass correlation A measure of the relationship between two or more repeatedly measured variables having values from 0.0 to +1.0. Often used to estimate reliability. (*See also* Cronbach's alpha.)

kurtosis A shape characteristic related to the variability and peakedness of a distribution; types of kurtosis are mesokurtic, leptokurtic, and platykurtic.

leptokurtic The shape of a distribution that is peaked, with scores being more homogeneous (less scattered) than in a normal distribution.

local history A threat to internal validity in which unanticipated events occurring during a study may alter the subjects' behaviors in an uncontrolled and unaccountable way.

longitudinal research A procedure in which changes are measured repeatedly over a period of time (e.g., weeks, months, even years) on the same subjects.

MANOVA Multivariate ANOVA; an extension of ANOVA that permits the testing of multiple dependent variables.

maturation of subjects A threat to internal validity when change affects the subjects' characteristics due to growth or development.

mean The arithmetic average of the scores in a distribution.

mean square An estimate of the population variance calculated by dividing the sum of squares by the associated degrees of freedom.

measures of central tendency Statistics that serve to quantify the center of a distribution.

measures of variability Statistics that serve to quantify the dispersion or spread in a distribution of scores (e.g., range, variance, standard deviation).

median The point below and above which 50 percent of the scores in a distribution fall.

The median is the 50th percentile and most typical score.

mesokurtic The shape of a distribution resembling the normal distribution.

mode The most frequently occurring score in a distribution.

multiple comparisons (MC) Statistical methods used to investigate which means are significantly different following a significant omnibus test for ANOVA or related procedures.

multiple correlation A statistic representing the correlation between a predicted variable *(Y)* and more than one predictor variable *(X)*. The multiple correlation ranges from 0.0 to +1.0.

multiple regression See *multiple correlation.*

nominal measurement A measurement made by assessment of equality or difference; often uses descriptors to classify subjects into categories.

nonparametric statistics A branch of statistics in which the data do not have to meet certain assumptions required for parametric statistics (e.g., continuous measurement, normality).

null hypothesis A statement that the independent variable and the dependent variable are not related. It is the null hypothesis that researchers actually test and make decisions about, based on probability.

objectivity The reliability of raters to record the same value for the same observation. Objectivity ranges from 0.0 to +1.0.

omega squared (ω^2) An estimate of the percentage of variability in the data that can be attributed to the influence of the independent variable.

one-tailed test A statistical test where all the probability of making a Type I error is located in one end of the distribution.

ordinal measurement A measurement made by ranking based on whether one subject has more or less of some characteristic than another.

overgeneralizing A threat to external validity resulting from a tendency to generalize beyond the levels of the independent and/or dependent variables in the study.

parameter A fact about a population, typically represented by a Greek symbol.

parametric statistics A branch of statistics whose accuracy is predicated on the validity of certain assumptions about the population from which the samples have been drawn and about the samples themselves.

Pearson product-moment (PPM) correlation coefficient Correlation coefficient widely used in all sciences to quantify the magnitude and direction of the linear relationship between any two variables; also referred to as Pearson's *r*. Ranges from –1.0 (perfect negative correlation) to +1.0 (perfect positive correlation). Zero indicates no correlation.

percent improvement Reflects the percentage of change that occurred between measures from one occasion to another or the percentage of difference from baseline or control to subsequent measurements.

percentile The percentage of scores falling at or below a given value.

planned comparisons A multiple comparison procedure where the contrasts to be tested are identified before the study is conducted. Also referred to as *a priori* contrasts.

platykurtic The shape of a distribution that is flatter and more heterogeneous than the normal curve, with more values in the tails of the distribution than in the normal distribution.

point estimate The value of a sample statistic used to estimate a parameter.

population Any set of subjects that have at least one attribute in common.

***post hoc* test** Multiple comparison procedure designed to test the significance of group differences following the finding of a significant *F*-ratio.

power The probability of rejecting the null hypothesis when it is false.

predictive validity A measurement procedure has predictive validity when a correlation exists between the measure and a criterion measure obtained at some future point in time.

pretest sensitization A threat to external validity that occurs when subjects' subsequent behavior is influenced by having completed a pretest.

pretesting A threat to internal validity where subjects' characteristics may change due to administration of a pretest. Also, a threat to external validity because subjects who have been pretested may no longer represent the population.

qualitative research Seeks to describe and to qualify what occurs; relies on subjective and observational information. Sometimes called *ethnographic research*.

quantitative research Depends on virtually complete control over all events except those that are being examined; relies on numerical information.

random independent variable Independent variable in which levels are selected by the researcher with random procedures.

random sampling Method of deriving a sample from a population where every subject in the population has an equally likely chance of being selected to be in the sample and every subject in the population has an independent chance of being selected.

range The difference between the high score and the low score. The exclusive range is the difference between the high and low score; the inclusive range is the exclusive range plus one.

ratio measurement A measurement scale containing an absolute zero. Ratio measurements permit statements of comparison, such as "twice as much".

reliable (reliability) A characteristic indicating the consistency of measurement. Reliability ranges from 0.0 to +1.0.

repeated-measures design An experimental design used to discover mean differences between a set of subjects measured on more than one occasion. Also called a *within-groups* design.

research hypothesis A prediction derived from theory or a researcher's speculation regarding the likely outcome of an experiment.

robust The ability of a statistical procedure to be valid even when the assumptions required for that procedure are not met.

sample A subset of subjects selected from a population.

sampling distribution A frequency distribution of a statistic developed from repeated samples of the same size taken from a defined population.

sampling error The difference between a statistic and a parameter.

scattergram Graph representing the relationship between two variables (*X* and *Y*). Also called *scatterplot*.

scatterplot See *scattergram*.

scientific method A method of knowing based on hypothesis development, data collection, and decision-making. It is verifiable and reproducible.

semi-interquartile range A measure of variability calculated as half the distance between the scores representing the 75th and the 25th percentiles.

significant A statistical result concluding that the observed differences are not attributable to chance. When the null hypothesis is rejected, the results are said to be statistically significant. Commonly used levels of significance are .05 and .01.

simple comparison A multiple comparison procedure involving only two means.

skewness A shape characteristic related to the degree to which a distribution departs from symmetry around its mean.

sphericity An assumption required for repeated-measures (within-groups) designs that relates to the pattern of the variances, the correlations, and score differences from all of the levels of the independent variable.

standard deviation A linear measure of variability. The standard deviation is equal to the square root of the variance.

standard error (SE) See *standard error of estimate*.

standard error of estimate (SEE) A statistic indicating the amount of error present when predicting *Y* from one or more predictor variables *(X)*. Also called the *standard error* (SE) or *standard error of prediction* (SEP). The SEE is a standard deviation.

standard error of measurement (SEM) A value reflecting the amount of variation in an observed score that is attributable to errors of measurement. The SEM is a standard deviation.

standard error of prediction (SEP) See *standard error of estimate*.

standard error of the mean A statistic reflecting the variability around a sample estimate. It is the standard deviation of a sampling distribution of means.

standard scores Scores derived from an original distribution that results in a given mean and standard deviation. See *z-score* and *T-score*.

statistic A fact about a sample, typically represented by a Roman letter.

statistical regression A threat to internal validity resulting from the tendency for extreme scores to move toward the group mean when measured a second time; also referred to as *regression to the mean*.

stratified sampling Method of deriving a sample from a population that ensures the sample is representative of the population for selected characteristics.

sums of squares The sum of the squared deviations of each score from the mean. The numerator in a variance estimate.

systematic sample Method of deriving a sample from a population by selecting every *n*th subject from a list of every subject in the population.

theory An educated supposition about the relationship among some natural phenomena, generally derived through observation, experimentation, and reflective thinking.

transferability How well the results of a study can describe, explain, or predict the behaviors of individuals different from those in the study or in dissimilar situations and environments.

***T*-score** A standardized score with a mean of 50 and standard deviation of 10.

two-tailed test A statistical test where the probability of making a Type I error is divided equally on both ends of the distribution.

Type I error Rejecting the null hypothesis when it is actually true. An alpha (α) error.

Type II error Failing to reject the null hypothesis when it is actually false. A beta error (B).

valid A measurement that assesses what it is intended to assess (i.e., it is truthful).

variable A measured characteristic that can take on different values (e.g., height, weight, BMI).

variance *General definition:* The fact that that not all values are the same for measured characteristics; there is variability in things, persons, and observations. *Statistical definition:* A statistic quantifying the amount of variability in a set of scores. It is the average of the squared deviation from the mean.

within-groups design Research design in which subjects appear in more than one group and are measured more than once. See also *repeated-measures design*.

Index

Printed in the United States
by Baker & Taylor Publisher Services